The 42nd International Workshop on Bayesian Inference and Maximum Entropy Methods in Science and Engineering

Volume Editors

Roland Preuss
Udo von Toussaint

MDPI

Basel • Beijing • Wuhan • Barcelona • Belgrade • Novi Sad • Cluj • Manchester

Volume Editors

Roland Preuss
NMPP
Max-Planck-Institute for
Plasma Physics
Garching
Germany

Udo von Toussaint
NMPP
Max-Planck-Institute for
Plasma Physics
Garching
Germany

Editorial Office
MDPI AG
Grosspeteranlage 5
4052 Basel, Switzerland

This is a reprint of the Proceedings, published open access by the journal *Physical Sciences Forum* (ISSN 2673-9984), freely accessible at: https://www.mdpi.com/2673-9984/9/1.

For citation purposes, cite each article independently as indicated on the article page online and as indicated below:

Lastname, A.A.; Lastname, B.B. Article Title. *Journal Name* **Year**, *Volume Number*, Page Range.

ISBN 978-3-7258-4179-0 (Hbk)
ISBN 978-3-7258-4180-6 (PDF)
https://doi.org/10.3390/books978-3-7258-4180-6

Cover image courtesy of Roland Preuss, MPI f. Plasma Physics

Contents

physical sciences forum

MDPI

Editorial

Preface of the 42nd International Workshop on Bayesian Inference and Maximum Entropy Methods in Science and Engineering—MaxEnt 2023

Udo von Toussaint and Roland Preuss *

Max-Planck-Institut für Plasmaphysik, 85748 Garching, Germany; udt@ipp.mpg.de
* Correspondence: preuss@ipp.mpg.de

1. Introduction

The forty-second International Conference on Bayesian Inference and Maximum Entropy Methods in Science and Engineering (42nd MaxEnt'23) was held at the Max Planck Institute for Plasmaphysics (IPP) in Garching, Germany, from 3rd to 7th of July 2023 (https://www.ipp.mpg.de/maxent2023) (Figure 1). This conference continued a long series of MaxEnt workshops which started in the late 1970s centered on ill-conditioned data analysis tasks, thus making this workshop series one of the oldest (if not the oldest) conferences focusing on areas which are now commonly (but not always correctly) denoted as ML/AI.

Figure 1. Conference photo.

The MaxEnt event at the IPP provided an opportunity for scientists, researchers, academics, and professionals to meet and to exchange and discuss the latest ideas and approaches in the areas of Bayesian inference and maximum entropy methods in data analysis, information processing, and inverse problems from a broad range of diverse disciplines, including astronomy and astrophysics, geophysics, medical imaging, acoustics, molecular imaging and genomics, non-destructive evaluation, particle and quantum physics, physical and chemical measurement techniques, economics, econometrics, and robust estimation.

In particular, topics of interest also included new approaches to strong AI, physics-inspired neural networks (PINNs), and the development of efficient closed-loop Bayesian experimental design approaches.

The following list of talks given at the workshop illustrates the wide range of topics addressed.

2. 32 Talks

- Danielle Hodgkinson: A Bayesian Data Analysis Method for an Experiment to Measure the Gravitational Acceleration of Antihydrogen.
- Fabian Sigler: Second Order Theory of Mind a Sociophysical Model.
- David Iagaru: Uncertainty Quantification for Super-Resolution.
- Zahra Amini Farsani: Advanced Maximum Entropy Algorithms for Bioimaging Issues.
- Nhan Le: An Iterative Bayesian Algorithm for 3D Image Reconstruction Using Multi-View Compton Data.
- Peter Dieterich: Bayesian inference of anomalous dynamics and tempering memory of migrating cells.
- Yannis Kalaidzidis: Off-target level estimation for siRNA mediated downregulation of gene expression.
- Chang Sub Kim: Bayesian Inversion in the Brain.
- Robert Köberl: Magnetohydrodynamic Equilibrium Reconstruction with Consistent Uncertainties.
- Robert Niven: Construction of Dimensionless Groups by Entropic Similarity.
- Gerd Verdoolaege: Geodesic Least Squares Robust Regression Using Information Geometry.
- Theo Grente: Analysis of Ecological Networks and Information Theory Tools.
- Richard Grumitt: Flow Annealed Kalman Filters.
- Nick Candau: Machine Learning Using Maximum Entropy and Bayesian Inference.
- Sascha Ranftl: Physics-consistent infinite neural networks.
- Ali Mohammad-Djafari: Deep Learning with Bayesian Principles and Inverse Problems.
- Tiago Silva: Data Fusion and Information Content in Self-consistent Processing of Ion Beam Analysis Data.
- Louis Duval: Study of B-like ions X-ray Emission Spectra in an Electron-Cyclotron Resonance Ion Source plasma.
- Ning Chu: A fast-rotating sound localization method based on modal composition beamforming under Bayesian framework.
- Minas Karamanis: Preconditioned Monte Carlo.
- Cesar Godinho: Bayesian Analysis of Strong Field QED Tests for He-Like Ions Using the Nested Sampling Algorithm.
- Nico Unglert: Exploration of Materials Configuration Spaces using Neural Network Force Fields and Nested Sampling.
- Anthony Garrett: Improved Derivation of the Laws of Probability.
- Romke Bontekoe: Kangaroos in Cambridge.
- Stefan Behringer: Value of Information in Zero-Sum Games.
- Kevin Knuth: Using Nested Sampling to Simulate the Characteristics of Extra-Terrestrial Civilizations that Encounter Earth.
- Andrew Fowlie: Origins of Parameters in Adimensional Models.
- Vishal Johnson: Quantum Measurement and Objective Classical Reality.
- Alexander Vosseler: BHAD: Explainable anomaly detection using Bayesian histograms.
- Margret Westerkamp: Inferring Evidence from Nested Sampling Data via Information Field Theory.
- Oliver Schulz: BAT.jl, the Bayesian Analysis Toolkit in Julia.
- John Skilling: Arithmetic and Quantum Theory.

3. Nested Sampling Workshop

As the innovative idea of the Nested Sampling algorithm was initially presented by John Skilling at the MaxEnt workshop back in 2004 at the IPP in Garching (i.e., at the same site), it appeared appropriate to include a special session specifically dedicated to Nested Sampling. Though initially considered with some reservation, the ideas underlying Nested Sampling have been proven to be very fruitful and have meanwhile become widely adopted. This can be seen from the range of topics covered in this 1.5-day special session:

- Johannes Buchner, Will Handley: Welcome, goals of the workshop, overview.
- John Skilling: Nested Sampling—The idea.
- Johannes Buchner: Sampling strategies: ideas from the past and current frontiers.
- Josh Speagle: Musings on dynesty, Dynamic Nested Sampling, and Challenges with Practical Implementations of Nested Sampling Algorithms.
- Leah South (remote, 18:30 local time): Unbiased and Consistent Nested Sampling via Sequential Monte Carlo.
- Doris Schneider: Plateaus and rare event estimation.
- Nicolas Chopin: Waste-free Sequential Monte Carlo, and its connection to NS.
- Michael Habeck: Adaptive Ensemble Annealing.
- Discussion: Open issues on statistical foundations, synergies with work on related algorithms (bridge/path sampling, SMC, rare event simulation).
- Jack O'Brien: Probabilistic Inference of Type Ia Supernova Abundances.
- Pablo Lemos: Gradient-based Nested Sampling.
- Xi Chen: Bayesian posterior repartitioning for nested sampling.
- Aleksandr Petrosyan: Stress testing SuperNest.
- Michael J. Williams: Importance NS with normalizing flows.
- Johannes Lange: Boosting importance nested sampling with deep learning.
- Lukas Hergt: Post-Processing Tools for Nested Sampling.
- Discussion: The relation between machine learning and NS.
- Livia Bartok-Partay: Nested sampling of atomic potential energy surfaces.
- George Marchant: Nested sampling of magnetic materials.
- Omar Adesida: NS as a tool for exploring the thermodynamic behaviour of hard sphere dimers.
- Ning Xiang: Nested sampling for detection, enumeration, and localization of multiple sound sources using a spherical microphone array.
- Mingrui Yang: Surface Phase Diagrams from Nested Sampling.
- Jason McEwen: Proximal nested sampling for high-dimensional Bayesian model selection.
- Andrew Fowlie: Opening up nested sampling.
- Brendon Brewer: Computing entropies with NS.
- Will Handley: The scaling frontier of nested sampling.
- Will Handley: workshop summary talk: Frontiers of nested sampling.

As this Editorial exemplifies, the intersections between Bayesian inference, ML/AI, and data science will become increasingly interconnected in the future. As such, it is our hope that these conference proceedings will serve as a valuable source of inspiration for future endeavors in this field.

Funding: The MaxEnt 2023 conference was supported by the Max-Planck-Institut für Plasmaphysik in Garching, Germany.

Conflicts of Interest: The authors declare no conflict of interest.

physical sciences forum

MDPI

Proceeding Paper

An Iterative Bayesian Algorithm for 3D Image Reconstruction Using Multi-View Compton Data †

Nhan Le [1,*], Hichem Snoussi [1] and Alain Iltis [2]

[1] LIST3N, Troyes University of Technology, 12 Rue Marie Curie, 10004 Troyes, France; hichem.snoussi@utt.fr
[2] Damavan Imaging, 2 Rue Gustave Eiffel, 10430 Rosières-Près-Troyes, France; alain.iltis@damavan-imaging.com
* Correspondence: thi-ai-nhan.le@utt.fr; Tel.: +33-3-25-71-84-42

† Presented at the 42nd International Workshop on Bayesian Inference and Maximum Entropy Methods in Science and Engineering, Garching, Germany, 3–7 July 2023.

Abstract: Conventional maximum likelihood-based algorithms for 3D Compton image reconstruction are often stuck with slow convergence and large data volume, which could be unsuitable for some practical applications, such as nuclear engineering. Taking advantage of the Bayesian framework, we propose a fast-converging iterative maximum a posteriori reconstruction algorithm under the assumption of the Poisson data model and Markov random field-based convex prior in this paper. The main originality resides in developing a new iterative maximization scheme with simultaneous updates following the line search strategy to bypass the spatial dependencies among neighboring voxels. Numerical experiments on real datasets conducted with hand-held Temporal Compton cameras developed by Damavan Imaging company and punctual 0.2 MBq ^{22}Na sources with zero-mean Gaussian Markov random field confirm the outperformance of the proposed maximum a posteriori algorithm over various existing expectation–maximization type solutions.

Keywords: 3D image reconstruction; line search maximization; maximum a posteriori; Markov random field; real Compton data; simultaneous update

1. Introduction

Compton reconstruction is a method to locate gamma ray sources by measuring a Compton scatter and a photoelectric absorption for each radioactive decay, in terms of positions and deposited energies. The literature of 3D Compton reconstruction methods has been extensively developed following two main streams: (*i*) *analytical inversion* [1], and (*ii*) *iterative model-based* [2]. The first stream relies mostly on the relationship between a function and its line integrals. Although it typically leads to fast reconstruction algorithms such as filtered back-projection [3], the accuracy of the resulting reconstructed image is usually limited because of approximations needed in the line-integral model and the inability to take account of uncertainties inherent in photon-limited detection. On the contrary, iterative reconstruction algorithms in the second stream exhibit significant accuracy thanks to the explicit modeling of statistical variations in the photon detection process and the progressive refining of reconstructed images through repetitive calculations. However, the price of this added refinement is the substantially higher computation cost.

In the current state of the art, great interest has put in 3D Compton reconstruction using statistical iterative algorithms [4]. The maximum-likelihood expectation–maximization (MLEM) algorithm and its variants have been the leading statistical iterative reconstruction algorithms for several years. Being applicable for Poisson Compton data in both the bin-mode [5] and list-mode [6], the conventional MLEM algorithm exhibits attractive properties, including consistent and predictable convergence behavior, guaranteed non-negativity, and count preservation at each iteration. Nevertheless, it converges slowly, and requires a large number of Compton events for noise reduction. These two issues are, however, of great concern for various practical applications, especially in nuclear engineering [7]. A potential solution to these issues is to formulate the 3D reconstruction in the Bayesian framework with a Markov random field (MRF)-based prior, reflecting the smoothness in the expected image. Accordingly, the reconstructed image can be seen as a maximum a posteriori (MaP) estimate from Compton data [2]. Equivalent to the Gibbs distribution [8], the MRF framework provides a simple, yet consistent and highly flexible, formalism for describing the image local structure, where neighboring voxels tend to share similar values [9]. The main difficulty for the MaP estimation with MRF priors is the inseparable neighboring voxels in a so-called energy function that makes a direct application of the conventional EM algorithm generally impossible. To remedy this, the one-step-late method [10] could be used at the risk that the EM algorithm converges to the MAP solution for only certain MRF priors [11]. Another alternative is to replace the energy function by a surrogate enabling separable neighboring voxels [12]. If properly chosen, it can lead to a closed-form update of the EM algorithm even for non-quadratic energies [13]. Additionally, we can also adopt a greedy pixel-wise maximization scheme such as iterative coordinate ascent (ICA) [14] to break interactions among neighboring voxels. Nevertheless, such a sequential greedy maximization is often time consuming due to the large number of voxels considered in 3D reconstruction.

Motivated by the Bayesian framework advantages, our aim is to develop a fast-converging iterative MaP reconstruction algorithm called LMMRFMaP under the assumption of the list-mode Poisson data model and MRF-based convex prior. This LMMRFMaP algorithm separates interactive voxels in a similar way as the ICA [14]; however, in the maximization process, a new simultaneous update with line search is proposed for all voxels, rather than sequentially updating voxels one by one. This improvement allows us to very quickly locate radioactive sources from the list-mode Compton data space, especially when compared with the extended LMMLEM (eLMMLEM) algorithm [15] and the so-called LMMaPEM algorithm developed in [12].

The remainder of the paper is organized as follows. On the basis of the Poisson data model, in Section 2, we develop the LMMRFMaP algorithm for MRF-based convex priors. Numerical experiments and comparative studies conducted with hand-held Temporal Compton cameras developed by Damavan Imaging company and punctual 0.2 MBq ^{22}Na sources are next introduced and analyzed in Section 3. Finally, the paper ends with some conclusions and perspectives in Section 4.

2. LMMRFMaP Algorithm

This section is dedicated to formulating a new MaP algorithm for 3D reconstruction from a list-mode Compton data-space $\boldsymbol{L}$. For convenience, we first conduct the formulation under the bin-mode data, as in [16], then derive the model of interest by considering list-mode data as a limiting case of bin-mode data when the bin-widths tend to zero. Thus, we define, for the bin-mode data, a vector of measurements $\mathbf{g}$, where each element g_i, $i \in \mathcal{I}$, describes the random number of counts in a virtual bin i corresponding to the event $\boldsymbol{A}_i$ of $\boldsymbol{L}$. When $g_i \to 1, \forall i \in \mathcal{I}$, then the bin-mode data return to the list-mode ones [17].

We denote $\mathbf{f}$ the source distribution in the considered image volume $\boldsymbol{V}$, where each element f_j, $\forall j \in \mathcal{J}$, characterizes the emission intensity of the voxel v_j in $\boldsymbol{V}$. Under the Bayesian framework, the 3D reconstruction reduces to find the MaP estimate $\hat{\mathbf{f}}_{\text{map}}$ of $\mathbf{f}$, such that

$$\hat{\mathbf{f}}_{\text{map}} = \arg\max_{\mathbf{f}\geq 0}\{\mathcal{L}(\mathbf{g} \mid \mathbf{f}) + \mathcal{L}(\mathbf{f})\}, \tag{1}$$

where $\mathcal{L}(\mathbf{f})$ and $\mathcal{L}(\mathbf{g} \mid \mathbf{f})$ are the log-prior and log-likelihood functions.

Through a MRF model, we introduce the image smoothness as prior knowledge in estimating $\mathbf{f}$. We thus define a *neighborhood system* δ_j of the voxel v_j, $\forall j \in \mathcal{J}$, such that

- v_j cannot be a neighbor of itself (i.e., $v_j \notin \delta_j$);
- If v_j is a neighbor of v_l, then v_l must be a neighbor of v_j (i.e.,$v_j \in \delta_l \Leftrightarrow v_l \in \delta_j$).

Each system δ_j corresponds to a set of *pair-wise* cliques $\mathcal{C}_j \triangleq \{(v_j, v_l) : v_j \in \delta_l\}$ consisting of all *unordered* neighboring voxels pairs (v_j, v_l) satisfying $v_j \in \delta_l$. Using these notions, we can formulate the log-prior function as [18]

$$\mathcal{L}(\mathbf{f}) = -\sum_{j=1}^{J} \sum_{(v_j,v_l)\in\mathcal{C}_j} \beta_{jl} \cdot \rho(f_j - f_l), \tag{2}$$

where β_{jl} denotes the model parameter, and $\rho(\cdot)$ is a *convex* potential function (see [19] for a detailed discussion about different choices of $\rho(\cdot)$, and their meanings and properties). Assuming now the Poisson event-counting model for g_i, we arrive at the log-likelihood function [2]

$$\mathcal{L}(\mathbf{g} \mid \mathbf{f}) = \sum_{i=1}^{I} \left(-\sum_{j=1}^{J} t_{ij} f_j + g_i \ln\left(\sum_{j=1}^{J} t_{ij} f_j \right) - \ln(g_i!) \right), \tag{3}$$

where t_{ij} denotes an element of the *system* response matrix indexed on the event A_i and on the voxel v_j [15]. Obviously, inseparable $f_j, \forall j \in \mathcal{J}$ in $\mathcal{L}(\mathbf{f})$ and $\mathcal{L}(\mathbf{g} \mid \mathbf{f})$ make a direct resolution of (1) mathematically intractable. A well-known technique to overcome this obstacle is to replace $\mathcal{L}(\mathbf{f})$ and $\mathcal{L}(\mathbf{g} \mid \mathbf{f})$ by surrogate functions [12,13]. While an efficient log-likelihood surrogate can be easily found thanks to the common assumption of the Poisson data model, a relevant choice for log-prior surrogate is not always straightforward and closely dependent on the specific adopted prior [20]. That is why we propose substituting only $\mathcal{L}(\mathbf{g} \mid \mathbf{f})$ by an appropriate surrogate, keeping $\mathcal{L}(\mathbf{f})$ as it is, and then developing an iterative scheme to maximize the new objective function.

In [20], Bouman proved that the Q-function in the E-step of the EM algorithm could be an appropriate log-likelihood surrogate for the Poisson data. Sharing their viewpoint, we adapt the Q-function in our eLMMLEM algorithm [15], and obtain a surrogate of $\mathcal{L}(\mathbf{g} \mid \mathbf{f})$ in list-mode (i.e., $g_i \to 1, \forall i \in \mathcal{I}$) as

$$Q_l(\mathbf{f}; \mathbf{f}') = \sum_{i=1}^{I} \sum_{j=1}^{J} \left(\frac{t_{ij} f_j'}{\sum_{s=1}^{J} t_{is} f_s'} \ln(t_{ij} f_j) - t_{ij} f_j \right), \tag{4}$$

where $\mathbf{f}'$ denotes any possible information from earlier iterations, and $f_j', \forall j \in \mathcal{J}$, are elements of $\mathbf{f}'$. Using (2) and (4), equivalences of (1) for the list-mode data are obtained by

$$\hat{\mathbf{f}}_{\text{map}} = \arg\max_{\mathbf{f}\geq 0} \{ \mathcal{L}_{\text{map}}(\mathbf{f}) \}, \tag{5}$$

$$\mathcal{L}_{\text{map}}(\mathbf{f}) = \sum_{i=1}^{I} \sum_{j=1}^{J} \left(\frac{t_{ij} f_j'}{\sum_{s=1}^{J} t_{is} f_s'} \ln(t_{ij} f_j) - t_{ij} f_j \right) - \sum_{j=1}^{J} \sum_{(v_j,v_l)\in\mathcal{C}_j} \beta_{jl} \cdot \rho(f_j - f_l). \tag{6}$$

It is easy to verify that $\mathcal{L}_{\text{map}}(\mathbf{f})$ is a concave function when the potential function is convex. This leads to the line search maximization scheme with simultaneous update as follows.

Theorem 1. *If $\mathcal{L}_{map}(\mathbf{f})$ is concave, then the following iterative schema converges to its maximum*

$$\hat{\mathbf{f}}^{(k+1)} \leftarrow \hat{\mathbf{f}}^{(k)} + \sum_{j=1}^{J} a_j^{(k)} \mathbf{e}_j^{(k)}, \tag{7}$$

where $\mathbf{e}_j^{(k)} \in \mathbb{R}^J$ is vector whose j-th component is 1, remainders are 0, and $a_j^{(k)}$ is given by

$$a_j^{(k)} \leftarrow \arg\max_{a_j\in\mathbb{R}} \left\{ \mathcal{L}_{map}\left(\hat{\mathbf{f}}^{(k)} + a_j \mathbf{e}_j^{(k)} \right) \right\}, \quad \forall j \in \mathcal{J}. \tag{8}$$

Proof. This theorem is easily proved by applying the Taylor expansion to the $\mathcal{L}_{\text{map}}(\mathbf{f})$ concave function. □

The above iterative maximization scheme requires a starting point $\hat{\mathbf{f}}^{(0)}$. As in [21], a potential choice for $\hat{\mathbf{f}}^{(0)}$ is such that $\hat{f}_j^{(0)} = \sum_{i=1}^{I} t_{ij}, \forall j \in \mathcal{J}$.

3. Numerical Experiments and Comparative Studies

To assess the performance of the proposed LMMRFMaP algorithm, we set up an experimental system with a punctual 0.2 MBq ^{22}Na source and a hand-held Temporal Compton camera developed by the Damavan Imaging company.

The camera is sequentially deployed at three different view-angles, as in Figure 1: *the two first views are set perpendicularly on a same horizontal plane* ***H*** *of the punctual source, and the third view is placed vertically above.* The distance between the camera and the source is 500 mm, and the whole system is located 500 mm from the ground. We conducted the acquisition within 15 min to constitute different list-mode data-spaces $\boldsymbol{L}$ for comparative studies among the eLMMLEM, LMMaPEM, and LMMRFMaP algorithms.

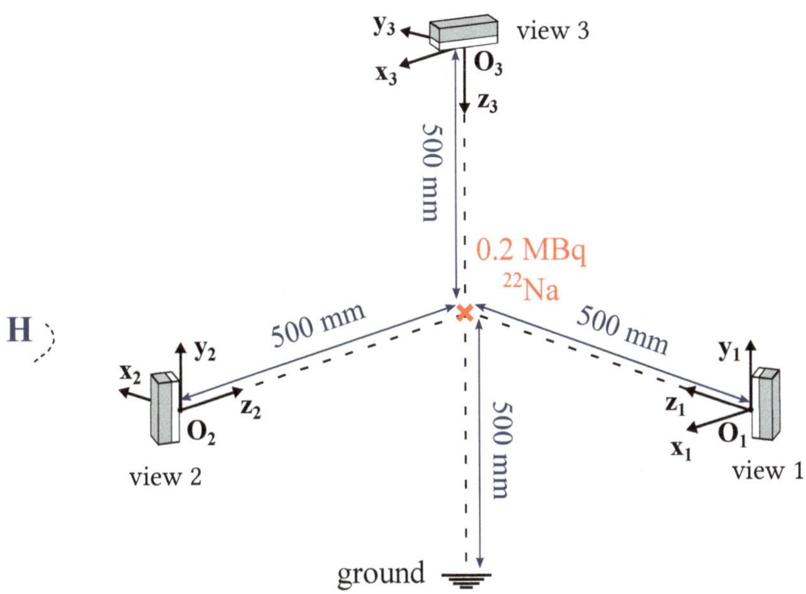

Figure 1. Experiment setting for Compton data acquisition.

Assuming the uniform sensitivity for the camera system, the eLMMLEM algorithm always admits the closed-form update at the $(k+1)$-st iteration as

$$\hat{f}_{\text{mle},j}^{(k+1)} \leftarrow \hat{f}_{\text{mle},j}^{(k)} \sum_{i=1}^{I} \frac{t_{ij}}{\sum_{s=1}^{J} t_{is} \hat{f}_{\text{mle},s}^{(k)}}, \quad \forall j \in \mathcal{J}, k \in \mathbb{N}. \tag{9}$$

Unlike the eLMMLEM algorithm, the update form of the LMMaPEM and LMMRFMaP algorithms depends closely on the choice of prior model. To enable their closed-form update, we choose the zero-mean Gaussian MRF with the potential function $\rho(f_j - f_l) = \frac{1}{2}(f_j - f_l)^2$ for describing the local smoothness of the expected reconstructed image. This choice gives the updates at the $(k+1)$-st iteration as

- For the LMMaPEM algorithm:

$$\hat{f}^{(k+1)}_{\text{mapem},j} \leftarrow \frac{1}{8}\sqrt{\left(1-2\left(\hat{f}^{(k)}_{\text{mapem},j}+\sum_{(v_j,v_l)\in\mathcal{C}_j}\beta_{jl}\hat{f}^{(k)}_{\text{mapem},l}\right)\right)^2+16\sum_{i=1}^{I}\frac{t_{ij}\hat{f}^{(k)}_{\text{mapem},j}}{\sum_{s=1}^{J}t_{is}\hat{f}^{(k)}_{\text{mapem},s}}} - \frac{1}{8}\left(1-2\left(\hat{f}^{(k)}_{\text{mapem},j}+\sum_{(v_j,v_l)\in\mathcal{C}_j}\beta_{jl}\hat{f}^{(k)}_{\text{mapem},l}\right)\right), \quad \forall j\in\mathcal{J}, k\in\mathbb{N}, \quad (10)$$

- For the LMMRFMaP algorithm:

$$\hat{f}^{(k+1)}_{\text{mrfmap},j} \leftarrow \frac{1}{2}\sqrt{\left(1-\sum_{(v_j,v_l)\in\mathcal{C}_j}\beta_{jl}\hat{f}^{(k)}_{\text{mrfmap},l}\right)^2+4\sum_{i=1}^{I}\frac{t_{ij}\hat{f}^{(k)}_{\text{mrfmap},j}}{\sum_{s=1}^{J}t_{is}\hat{f}^{(k)}_{\text{mrfmap},s}}} - \frac{1}{2}\left(1-\sum_{(v_j,v_l)\in\mathcal{C}_j}\beta_{jl}\hat{f}^{(k)}_{\text{mrfmap},l}\right), \quad \forall j\in\mathcal{J}, k\in\mathbb{N}, \quad (11)$$

where $\sum_{(v_j,v_l)\in\mathcal{C}_j}\beta_{jl}=1$, and the pair-wise clique $\mathcal{C}_j$ is chosen as in [22] (i.e., $\mathcal{C}_j$ consists of the eight nearest voxels from the same slice and the single nearest voxels of the upper and lower slices). The updates (9)–(11) also require a probabilistic model for t_{ij}. In this work, we adapt Maxim et al.'s model [23] for Compton data by adding different measurement uncertainties related to the spatial and energy resolution of detectors, as well as the Doppler broadening effect to make the model more flexible.

To see how fast the LMMRFMaP algorithm is when compared to the eLMMLEM and LMMaPEM algorithms, we fix the number of events at $I=20$ events/view and observe the evolution of the associated reconstructed 3D images at the k-th iteration with $k=1,3,5,7$. Visually, we can find from Figure 2 that the point cloud returned by the LMMRFMaP algorithm converges to true punctual source (marked by the red cross) much faster than the one returned by the two other algorithms.

Quantitatively, we propose using the so-called *sum of weighted distances* (SWD) between the reconstructed point cloud and the true source as a measure for the quality assessment at the k-th iteration

$$\text{SWD}^{(k)} = \sum_{j=1}^{J}\hat{f}_j^{(k)}\left\|P_j-P_S\right\|_2, \quad (12)$$

where $\hat{f}_j^{(k)}$ is given either from (9), (10) or (11); P_j and P_S are, respectively, the coordinate vectors of the voxel v_j and the punctual source; and $\|\cdot\|_2$ standards for the Euclidean norm. The SWD exhibits a compromise between the intensity of the voxels and their accuracy with respect to the source location. The lower the value of $\text{SWD}^{(k)}$, the nearer the reconstructed 3D image is to the true source. Fixing the number of events at $I=20$ events/view, we randomly build 100 list-mode data-spaces, reconstruct the images of punctual source following (9)–(11), and compute the associated SWD following (12). Next, we sketch the evolution of the mean and the standard deviation of these SWD with respect to the iteration number k in Figure 3a. Clearly, the proposed LMMRFMaP algorithm exhibits a higher convergence speed and higher reconstruction quality than the LMMaPEM and eLMMLEM algorithms. Similarly, we fix the iteration number at $k=5$, and vary the number of events/view from 4 to 20 with step 4 to evaluate the impact of the events number on the reconstruction performance. Indeed, for each of these configurations, we also build 100 list-mode data-spaces, compute the associated SWD for the considered algorithms, then sketch in Figure 3b the evolution of their mean and standard deviation.

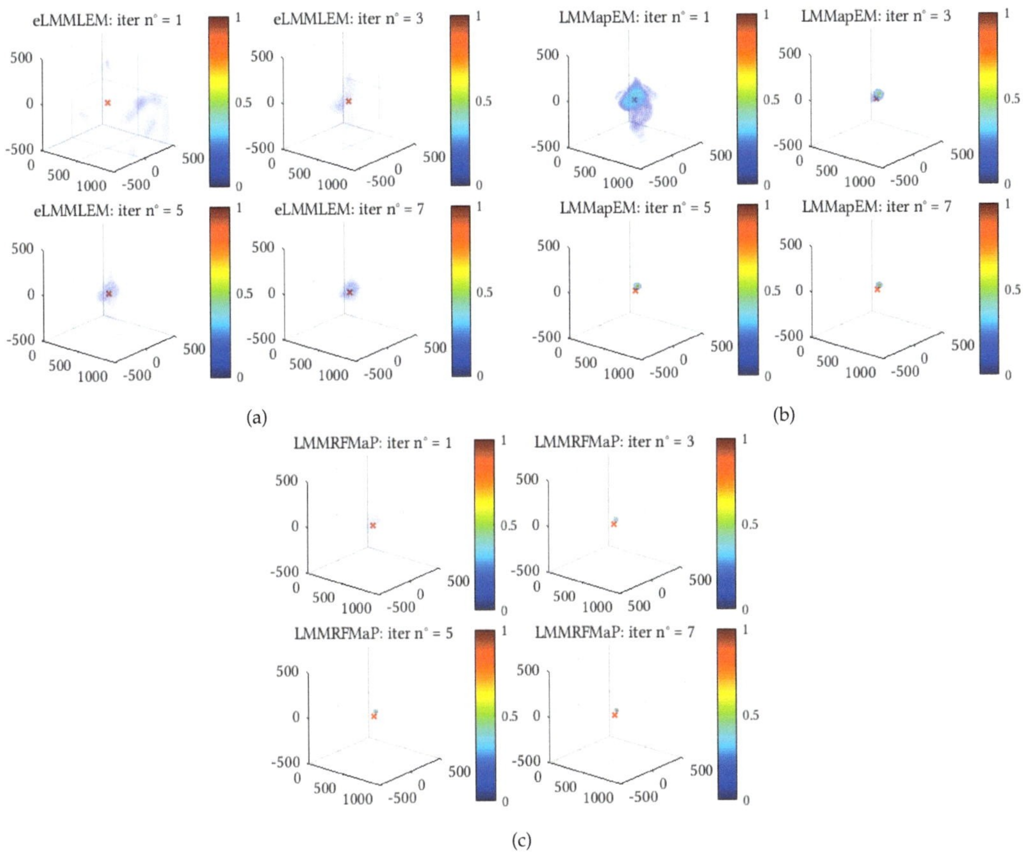

Figure 2. Evolution of point clouds returned by the considered algorithms (20 events/view): (**a**) eLMMLEM algorithm, (**b**) LMMaPEM algorithm, (**c**) LMMRFMaP algorithm.

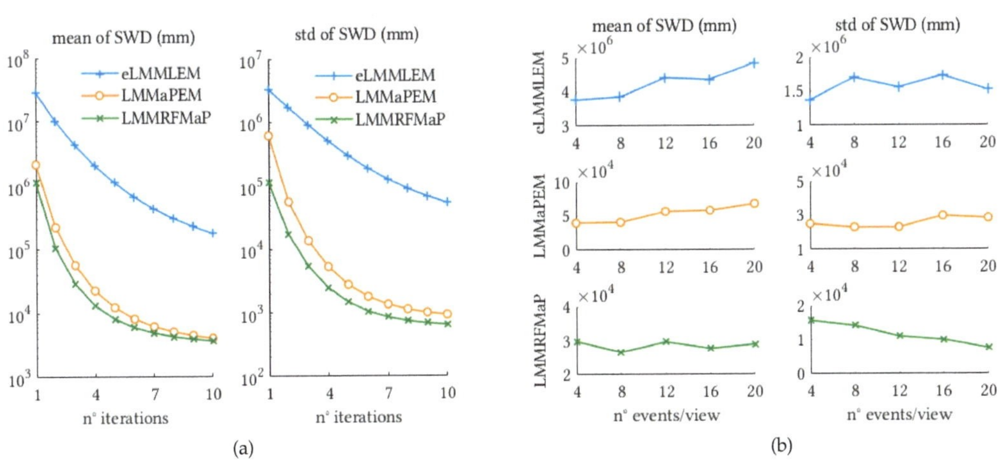

Figure 3. Evolution of the mean and the standard deviation of SWD: (**a**) evolution with respect to the number of iterations, (**b**) evolution with respect to the number of events/view.

We can remark that the mean values of SWD given by the LMMRFMaP algorithm are more-or-less constant when the number of events/view varies, while the ones given by the LMMaPEM and eLMMLEM algorithms present a growing trend. This means that the LMMRFMaP algorithm does not require as large a number of events/view to obtain as good a convergence as the two other algorithms. Of course, as shown by the curve of SWD standard deviation, the higher the number of events/view, the more precise the reconstruction results are.

4. Conclusions and Perspectives

Using the Bayesian framework, in this paper, we proposed the LMMRFMaP algorithm for 3D reconstruction from list-model Compton data. Various comparative studies with more classical statistical reconstruction algorithms on three-view real Compton data confirm that the algorithm can very quickly return an accurate image of a punctual radioactive source with a small number of events/view. These advantages are gained thanks to a novel line search maximization scheme with simultaneous updates developed for MRF-based convex priors. Based on these promising results, ongoing work is devoted to more exhaustive testing of the algorithm performances with other kinds of radioactive sources, such as multiple point sources and extended sources. Another perspective is to extend the proposed algorithm for other kinds of MRF-based priors (e.g., non-convex, mixtures, etc.).

Author Contributions: Conceptualization, N.L., H.S. and A.I.; methodology, N.L. and H.S.; software, N.L.; validation, N.L., H.S. and A.I.; formal analysis, N.L.; investigation, N.L., H.S. and A.I.; resources, N.L., H.S. and A.I.; data curation, N.L. and A.I.; writing—original draft preparation, N.L.; writing—review and editing, N.L., H.S. and A.I.; visualization, N.L.; supervision, H.S.; project administration, H.S. and A.I.; funding acquisition, H.S. and A.I. All authors have read and agreed to the published version of the manuscript.

Funding: This research was funded by the BPI PIA4 DreamScanner project.

Institutional Review Board Statement: Not applicable.

Informed Consent Statement: Not applicable.

Data Availability Statement: The data presented in this study are available on request from the corresponding author. The data are not publicly available due to the confidentiality policy.

Conflicts of Interest: The authors declare no conflict of interest.

References

1. Kinahan, P.E.; Defrise, M.; Clackdoyle, R. Analytic image reconstruction methods. In *Emission Tomography: The Fundamentals of PET and SPECT*; Elsevier Academic Press: Cambridge, MA, USA, 2004; Chapter 20, pp. 421–442.
2. Lalush, D.S.; Wernick, M.N. Iterative image reconstruction. In *Emission Tomography: The Fundamentals of PET and SPECT*; Elsevier Academic Press: Cambridge, MA, USA, 2004; Chapter 21, pp. 443–472.
3. Maxim, V. Filtered backprojection reconstruction and redundancy in Compton camera imaging. *IEEE Trans. Image Process.* **2014**, *23*, 332–341. [CrossRef] [PubMed]
4. Parajuli, R.K.; Sakai, M.; Parajuli, R.; Tashiro, M. Development and Applications of Compton Camera—A Review. *Sensors* **2022**, *22*, 7374. [CrossRef] [PubMed]
5. Shepp, L.; Vardi, Y. Maximum Likelihood Reconstruction for Emission Tomography. *IEEE Trans. Med. Imaging* **1982**, *1*, 113–122. [CrossRef] [PubMed]
6. Parra, L.; Barrett, H.H. List-Mode Likelihood: EM Algorithm and Image Quality Estimation Demonstrated on 2-D PET. *IEEE Trans. Med. Imaging* **1998**, *17*, 228–235. [CrossRef] [PubMed]
7. Sato, Y.; Terasaka, Y.; Utsugi, W.; Kikuchi, H.; Kiyooka, H.; Torii, T. Radiation imaging using a compact Compton camera mounted on a crawler robot inside reactor buildings of Fukushima Daiichi Nuclear Power Station. *J. Nucl. Sci. Technol.* **2019**, *56*, 801–808. [CrossRef]
8. Geman, S.; Geman, D. Stochastic Relaxation, Gibbs distributions, and the Bayesian Restoration of Images. *IEEE Trans. Pattern Anal. Mach. Intell.* **1984**, *PAMI-6*, 721–741. [CrossRef] [PubMed]
9. Li, S.Z. *Markov Random Field Modeling in Image Analysis*, 3rd ed.; Advances in Pattern Recognition; Springer Science & Business Media: Berlin, Germany, 2009.
10. Green, P.J. Bayesian reconstructions from emission tomography data using a modified EM algorithm. *IEEE Trans. Med. Imaging* **1990**, *9*, 84–93. [CrossRef] [PubMed]

11. Lange, K. Convergence of EM image reconstruction algorithms with Gibbs smoothing. *IEEE Trans. Med. Imaging* **1990**, *9*, 439–446. [CrossRef] [PubMed]
12. De Pierro, A.R. A modified expectation maximization algorithm for penalized likelihood estimation in emission tomography. *IEEE Trans. Med. Imaging* **1995**, *14*, 132–137. [CrossRef] [PubMed]
13. Chang, J.H.; Anderson, J.M.; Votaw, J.R. Regularized image reconstruction algorithms for positron emission tomography. *IEEE Trans. Med. Imaging* **2004**, *23*, 1165–1175. [CrossRef] [PubMed]
14. Bouman, C.A.; Sauer, K. A Unified Approach to Statistical Tomography Using Coordinate Descent Optimization. *IEEE Trans. Med. Imaging* **1996**, *5*, 480–492. [CrossRef]
15. Le, N.; Snoussi, H.; Hmissi, M.Z.; Iltis, A.; Lebonvallet, G.; Zeufack, G. An extended list-mode MLEM algorithm for 3D Compton image reconstruction from multi-view data. In Proceedings of the 8th International Conference on Advancements in Nuclear Instrumentation Measurement Methods and Their Applications, Lucca, Italy, 12–16 June 2023.
16. Wilderman, S.J.; Clinthorne, N.H.; Fessler, J.A.; Rogers, W.L. List-mode maximum likelihood reconstruction of Compton scatter camera images in nuclear medicine. In Proceedings of the IEEE Nuclear Science Symposium Conference Record, Toronto, ON, Canada, 8–14 November 1998; Volume 3, pp. 1716–1720.
17. Barrett, H.H.; White, T.; Parra, L.C. List-mode likelihood. *J. Opt. Soc. Am. A* **1997**, *14*, 2914–2923. [CrossRef] [PubMed]
18. Qi, J.; Leahy, R.M. Iterative reconstruction techniques in emission computed tomography. *Phys. Med. Biol.* **2006**, *51*, R541. [CrossRef] [PubMed]
19. Zhang, H.; Wang, J.; Zeng, D.; Tao, X.; Ma, J. Regularization strategies in statistical image reconstruction of low-dose X-ray CT: A review. *Med. Phys.* **2018**, *45*, e886–e907. [CrossRef]
20. Bouman, C.A. *Foundations of Computational Imaging: A Model-Based Approach*; SIAM—Society for Industrial and Applied Mathematics: Philadelphia, PA, USA 2022.
21. Frandes, M.; Magnin, I.E.; Prost, R. Wavelet Thresholding-Based Denoising Method of List-Mode MLEM Algorithm for Compton Imaging. *IEEE Trans. Nucl. Sci.* **2011**, *58*, 714–723. [CrossRef]
22. Villain, N.; Goussard, Y.; Idier, J.; Allain, M. Three-Dimensional Edge-Preserving Image Enhancement for Computed Tomography. *IEEE Trans. Med. Imaging* **2003**, *22*, 1275–1287. [CrossRef] [PubMed]
23. Maxim, V.; Lojacono, X.; Hilaire, E.; Krimmer, J.; Testa, E.; Dauvergne, D.; Magnin, I.; Prost, R. Probabilistic models and numerical calculation of system matrix and sensitivity in list-mode MLEM 3D reconstruction of Compton camera images. *Phys. Med. Biol.* **2016**, *61*, 243. [CrossRef] [PubMed]

physical sciences forum

MDPI

Proceeding Paper

Behavioral Influence of Social Self Perception in a Sociophysical Simulation †

Fabian Sigler [1,2,*], Viktoria Kainz [1,2], Torsten Enßlin [1,2,3], Céline Boehm [4] and Sonja Utz [5,6]

1 Max Planck Institute for Astrophysics, Karl-Schwarzschildstraße 1, 85748 Garching, Germany
2 Faculty of Physics, Ludwig-Maximilians-Universität, Geschwister-Scholl-Platz 1, 80539 Munich, Germany
3 Excellence Cluster ORIGINS, Boltzmannstr. 2, 85748 Garching, Germany
4 School of Physics, The University of Sydney, Physics Road, Camperdown, NSW 2006, Australia
5 Department of Psychology, University of Tübingen, 72076 Tübingen, Germany
6 Leibniz-Institut für Wissensmedien Tübingen, 72076 Tübingen, Germany
* Correspondence: f.sigler@campus.lmu.de
† Presented at the 42nd International Workshop on Bayesian Inference and Maximum Entropy Methods in Science and Engineering, Garching, Germany, 3–7 July 2023.

Abstract: Humans make decisions about their actions based on a combination of their objectives and their knowledge about the state of the world surrounding them. In social interactions, one prevalent goal is the ambition to be perceived to be an honest, trustworthy person in terms of having a reputation of frequently making true statements. Aiming for this goal requires the decision whether to communicate information truthfully or if deceptive lies might improve the reputation even more. The basis of this decision involves not only an individual's belief about others, but also their understanding of others' beliefs, described by the concept of Theory of Mind, and the mental processes from which these beliefs emerge. In the present work, we used the *Reputation Game Simulation* as an approach for modeling the evolution of reputation in agent-based social communication networks, in which agents treat information approximately according to Bayesian logic. We implemented a second-order Theory of Mind based message decision strategy that allows the agents to mentally simulate the impact of different communication options on the knowledge of their counterparts' minds in order to identify the message that is expected to maximize their reputation. Analysis of the communication patterns obtained showed that deception was chosen more frequently than the truthful communication option. However, the efficacy of such deceptive behavior turned out to have a strong correlation with the accuracy of the agents' Theory of Mind representation.

Keywords: sociophysics; information theory; Theory of Mind; agent-based modeling; computational psychology

1. Introduction

Purposeful lying, particularly in a deceptive manner, involves a multifaceted process including understanding the nature of lies itself, producing the lie, and maintaining built-up lies without being spotted as a dishonest person [1]. Therefore, a well-put lie demands significant cognitive abilities, including the capacity to develop a robust *Theory of Mind* (ToM). ToM refers to the ability to infer and deduce the mental states of others by trying to understand what they know, think, belief, plan, etc. only considering informative aspects apart from the others' feelings [2]. The *depth* or *order* of the ToM denotes the number of recursive ToM beliefs that are included in a ToM representation, e.g., a belief of the form "A thinks that B thinks, that C thinks X" corresponds to a second-order ToM belief. Recent sociological studies find that the performance of young children in Theory of Mind–related tasks is not only closely connected to their mere ability to put targeted lies, but also to their tendency to do so in the first place [3]. Although lying may be considered antisocial behavior, this does not necessarily hold when telling white lies or flattering to create a more pleasant social atmosphere [4,5]. Nonetheless, we frequently resort to deceitful lies to gain an advantage for ourselves [6,7], for example in the form of social advancement by attempting to move the own **reputation** in a desired direction. An approach to simulate the dynamics of reputation within social networks was taken by Enßlin, Kainz, and Bœhm (2022) [8,9]. In the "Reputation Game Simulation" (RGS), virtual agents communicate their beliefs to others with the objective of maximizing the own reputation among their fellows. The agents are modeled not to retain every step of communication, but to let them store their information at hand in a compressed way with a fixed number of parameters that are updated according to Bayesian probabilistic reasoning and information theoretical principles.

To send and receive messages, agents use one of many strategies that are implemented in the RGS based on conceivable real-life personality types and communication schemes. In this work, we explore a novel type of reputation-maximization-oriented strategy based on the implementation of a higher-order Theory of Mind.

2. The Reputation Game Simulation

2.1. Basic Principles

In the following, we summarize the most important aspects of the Reputation Game's mechanisms. For the details, we refer to [8].

The game follows the interactions of n agents that exchange information of only one kind, namely, the honesty of other agents. In each round each agent once takes on the role of the speaker(denoted by agent a) and therefore chooses a communication partner b and a topic agent c to transfer a message about to b and then receives a reply from b about c. This means that a and b tell each other how honest they believe c's statements to be in general, which can either be the agents' actual belief in form of an honest statement or a lie. The model is built around this type of information called the **reputation**. The agents use this image of others to decide about the trustworthiness of received messages to properly implement those into their beliefs while striving for maximizing their own reputation in the eyes of their fellows. A higher reputation of an agent also means a stronger influence of his messages on the group. So far, various strategies for achieving this goal are implemented in the RGS, varying in the way lies are used to deceptively influence other agents' beliefs.

The agents' beliefs are represented as the set of two parameters $I_{ab} = (\mu_{ab}, \lambda_{ab})$ that denote the number of honest and dishonest statements, respectively, that a believes b to have told during the game. Thus, following Bayesian logic, they describe a probability distribution over the possible lie frequencies of b in the eyes of a by $P(x_b|I_{ab}) = \mathrm{Beta}(\mu_{ab}+1, \lambda_{ab}+1|I_{ab})$. The mean of a belief $\overline{x}_{ab} \in [0,1]$, calculated from this probability distribution, refers to the reputation of b for agent a and is a measure of the frequency for a's belief for b to tell the truth.

Besides their belief on the behavior of others, the agent's mental states also include the first-order Theory of Mind values of the form I_{abc}, denoting what a thinks b thinks about c. Proper information on the beliefs and plans of others, on the one hand, protects the agents against being manipulated, and, on the other hand, makes it easier the other way round to successfully deceive their counterparts.

After receiving a message, the new belief of an agent ideally would have the form of a superposition of probability distributions for the old belief (that is to be kept in case of being lied to) and one for the new belief, in case the message was honest. The weighting of these two beliefs is according to the assessment of the receiver that the message was a lie or truth, respectively. For the sake of keeping the form of two parameters μ and λ, the superposition is then compressed back into the form of Beta distribution by minimizing the appropriate information distance between those, namely, the Kullback–Leibler divergence.

Whether an agent a lies or tells the truth in a specific situation is not decided by himself but in a random process characterized by the intrinsic, agent-specific lie frequency $1 - x_a$.

If an agent is chosen to lie in the first place, his lies are constructed around first-order ToM values (what a thinks that b thinks about c, i.e., lying positive means transmitting a message that is of the form $(\mu_{abc} + \alpha, \lambda_{abc})$, and for a negative $(\mu_{abc}, \lambda_{abc} + \alpha)$, with an α for that a expects to make the statement credible for b). This means the agents present the topic slightly better or worse than they believe the receiver to currently think about it. Transmitting the exact values of the first-order ToM is what is referred to as a **white lie**. After sending a message, the agents also update their belief on their own honesty according to whether they have just lied or not. This value is also influenced by the information the agents receive by others about themselves.

The agents use several indicators to detect lies in order to protect themselves from deception and adopting false information in their belief state. These include blushing (when agents tell a lie, they run the risk of exposing themselves by a subsequent blushing with frequency $f_b = 0.1$) or unusual surprise. The latter denotes the most crucial part of lie detection and means of evaluating the trustworthiness of a message according to the surprise it generates in the agent's mind when hearing it, i.e., the amount of information gained if the agent updated its initial belief to the information contained in the message. This surprise is calculated by the Kullback–Leibler divergence between the agents' current belief on the topic and the message. The value also is weighted by a factor κ, an agent specific value that denotes the mean of message surprises of the last ten conversations and therefore represents the current "social atmosphere".

Another indicator for honesty is the so-called *confession*. It means that, if a statement of an agent about himself is more negative than the receiver's current belief, the message is seen to be ultimately honest.

Besides the pure reputation-driven dynamics, the RGS contains a second principle that decides whether an agent should be lied positively or negatively about: whether they are regarded as a friend or an enemy. If an agent receives a message about himself wherein the mean reputation is above the perceived average, the sender is added to the receiver's list of friends, or to the list of enemies if the opposite was the case. If an agent decides to communicate false information, they will lie positively about friends, causing the receiver to put more trust in the expected positive subsequent messages of the friend about a. Following the same reasoning, a aims at minimizing the reputation of his enemies.

2.2. Implementation of a Second-Order Theory of Mind

We will now replace the lie direction rule for a new type of agent (called ToM-agent in the following) with the following considerations. The standard mental model of agents in the RGS includes a ToM up to the first order represented by $I_{abc} = (\mu_{abc}, \lambda_{abc})$ describing a's belief about b's zero order belief I_{bc} on c. The ToM-agent's knowledge not solely includes those first-order ToM values but also the very fact that this is the highest order that other agents conceive and an understanding of the mental processes leading to those beliefs. At least we demand the ToM-agents to be able to "imagine" how a hypothetical message affects a hypothetical mind. The rational consequence is to extend the ToM by an extra order, as this allows the ToM-agent to build up a complete picture of his fellows' mental states and, therefore, examine how they are influenced by possible messages and choose the option promising the best result in terms of its expected reputation. The ToM strategy therefore consists of using the extended picture of the surrounding to simulate the impact of different behavioral options on it. Following rational theory, the best response is again to undermine the strategy by increasing one's ToM by one more level and simulate the ToM-agents simulations, respectively, for the next response, resulting in an infinite number of iterations [10]. This is neither reasonable to model nor a process to be expected in real life. Sociology finds that, in practice, people do not apply deductive rational reasoning to the highest level but rather act according to a so-called **bounded rationality** that basically emerges from limitations in human brain capacities and from the resulting belief that others also react according to a bounded rationality [11]. Moreover, in the RGS, the uncertainty about the information of the ToM becomes more uncertain with each order of ToM, making the mental simulations more inaccurate and mostly useless at some point of depth. The proper strategic solution is, therefore, to identify the level of others' reasoning and undermine their strategy by thinking exactly one order deeper than their counterparts, which is exactly what ToM-agents in the RGS do, where they know the others' depth of reasoning to be one.

We also see this approach legitimized by its practical relevance, as sociological studies show that humans in general are able to form and apply second-order ToM beliefs, while higher orders are observed less frequently [1–3,10].

2.3. Implementing a Lie Direction Decision Rule

ToM-agents use the following strategy to decide whether to be honest or lie, and if so, in what direction. The ToM-agents a use their second-order ToM to directly examine the effects of their talk on their counterpart's mental state and how this might influence further conversations and belief changes about himself. Therefore, a ToM-agent imagines a set of virtual agents (*dummies*) holding the mental state $\mathcal{I}_{b'}$ according to a's first- and second-order ToM of his counterpart, agent b. Thus, $\mathcal{I}_{b'c} = I_{abc}$ and $\mathcal{I}_{b'ac} = I_{abac}$, $\mathcal{I}_{b'ca} = I_{abca}$. In the current model, the second-order parameters of the latter shape are hard to extract from conversations as it is only relevant and updated in conversations in which a is not involved. The ToM-agent is therefore not able to have an independent value for the second-order coefficients and his best guess is that c's beliefs were grasped in the equal way by both agents. We therefore assume the ToM-agent to replace $\mathcal{I}_{b'ca} = I_{abca} = I_{aca}$, meaning a assumes b to hold the same first-order belief on c's opinion on a as a himself.

2.4. Finding the Optimal Message

If an agent has chosen communication partner b and topic c, he finds himself with different options for the messages that he is about to be sent, which are a white, a positive, and a negative lie or a's honest opinion. In the first step, the ToM-agent constructs all these possible messages on the basis of his first-order ToM. Then five virtual dummy minds for b are created, of which four are updated under one of the possible messages each, whereas the last is confronted with an arbitrary lie with a subsequent blushing. We denote these options in the following by w, +, −, t, and blush, respectively. a needs to consider the possibility of being exposed as a liar by blushing, as this diminishes the expected gain in reputation by telling a lie. Blushing happens with a fixed frequency $f_b := P(\text{blush}|\text{lie})$ for all agents, which is known to all agents. This procedure is illustrated in Figure 1. The quality of the message is then measured by the expected reputation of a after the conversation. For the honest statement, this is equal to the reputation of a in the eyes of the simulated dummy after the update under the honest information. For the lies, the expected value is calculated as the sum of the reputation after the update under the successful lie, multiplied with the frequency of not blushing, and the update after the lie followed by blushing multiplied with the frequency for that to happen.

The message promising the highest possible expected value is then actually sent to the receiver. As only one positive and one negative lie are simulated, the chosen message is in general close to but not necessarily the ultimate best option.

When talking about the receiver b, b's mind update on a is not influenced by the content of the message but only by the amount of trust that he places in it to be true. Therefore, in this scenario, the white lie option or the truth happens to be the most promising (as currently the ToM-agent does not aim for friendship and its benefits). When talking about oneself, it is not possible to give a generalized answer about what lie, if not the truth, is the optimal message as it arises from a more complex interplay of parameters. Besides the intuitive result of lying positively being the most promising option in some cases, sometimes lying in the downward direction might boost the reputation even more. This is due to the RGS's concept of confession, meaning that if a receiver agent b receives a message from an agent a that presents himself worse than b currently thinks of a, b considers the statement to be ultimately true, as the only purpose of sending a potentially damaging message must lie in the intention of making a true statement. Therefore, if a manages to put a negative lie on his counterpart that lies only slightly below his opinion, the negative impact of the messages content becomes overcompensated by the one certainly true statement that b adds to his belief on a.

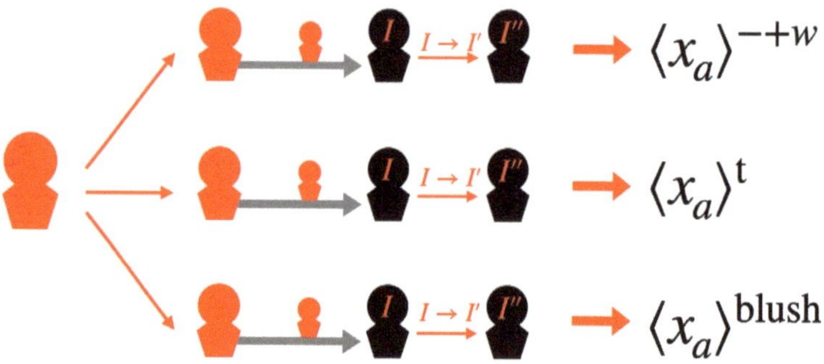

Figure 1. Red agent simulating black's update of possible messages in a conversation about red, who imagines three conversations in which he tells the four options of $+$, $-$, w, and t, as well as the blush option to b and simulates the resulting reputation in the corresponding dummies.

In the case of talking about a third agent c that is not the receiver or the ToM-agent a himself, a not only influences b's opinion on a but also on c and therefore also takes into account the possible subsequent messages from c to b about a. Depending on what a expects c to tell b about himself, it might be reasonable to accept a different effect on his reputation in the first communication with b. If a's statement on c leads to a different reputation of b about c, then c's positive or negative messages about a will be evaluated differently by b and the expected value for the message options changes. In order to investigate the effect of the conversation from c to b about the ToM-agent, he now imagines the four possible messages that c could tell b on the basis of its knowledge about c's mental state, as shown in Figure 2. Those are, again, the honest statement, positive and negative lie, and a lie followed by unintended blushing. The honest statement a expects c to make is given by the first-order ToM I_{aca}, which represents what a thinks c to belief about himself. As c will construct his lies to b around his first-order values I_{cba}, a will construct his expectation for them around his second-order values I_{acba}.

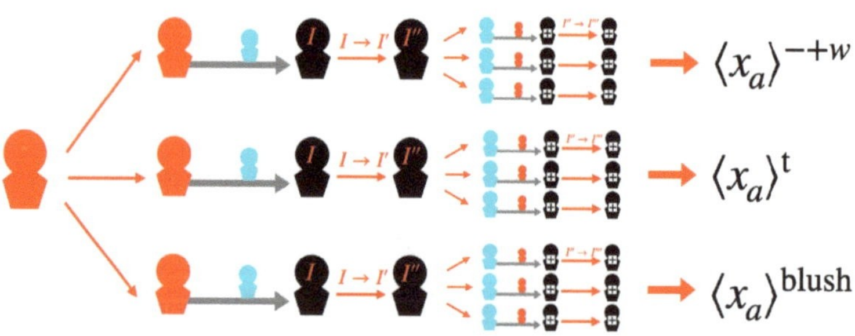

Figure 2. Red agent simulating black's update of possible messages in a conversations about cyan, followed by simulations from cyan to black about red for each of the updated dummies. Each of the five dummies for the options of a is copied three times to simulate the effect of the subsequent message from cyan to black on red for each of the dummies from the first conversation.

For making predictions about c's behavior, a assumes friendships to be symmetrical, meaning, if a sees c as a friend, he automatically assumes that c also considers a as a friend, applying the same logic if c is an enemy of a. This makes a expect c to tell b a positive lie about himself in the first case and a negative in the latter. If c is neither friend nor enemy in the eyes of a, then he expects positive and negative lies with probability of $\frac{1}{2}$ each.

The ToM-agent calculates the expected reputations for each combination of messages from himself about c, followed by a possible message from c to b. Then he evaluates the expected value for each of his options by considering the messages by c and the blushing to happen. The honest message that a expects c to make about a is equal to the first-order parameter I_{aca}. However, when b receives the message, he will extract the new information by comparing the content to his first-order I_{bca}. As we construct the ToM-agent so that he assumes that this value in the form I_{abca} is equal to I_{aca}, the content of the true messages in the subsequent messages in the simulation only influences the belief of b in the trustworthiness of c but not the belief in a, which is of actual interest. Therefore, the analysis of the strategy shows that it is advantageous for the ToM-agent to lie positively about agents that he considers to consider him a friend and negative about agents that he thinks see him as an enemy. Due to the assumption of symmetry, this is equivalent to the current rule of lying positively/negatively about agents the sender himself considers friend/enemy. Only if the first-order value for friendship and the second-order ToM values of I_{acba} were tracked and updated as independent parameters, the rule would become more differentiated. Anyhow, the current structure of the RGS does not provide the mechanism to do so. Moreover, the uncertainty of ToM values increases with the order, which was also a reason to introduce the bounded rationality in the first place.

2.5. ToM-Agent Update Rules

We also introduced additional information update rules that are unique to ToM-agents compared to the agents of the previous version. First, after sending a simulated message to b, the ToM-agent replaces his representation for b's mind with the mind of the dummy that was updated under the chosen message.

We also assume the ToM-agents to be deaf towards messages about themselves. As their lie decision does not depend on random processes but on prudent considerations, we find it reasonable to assume the ToM-agents to be aware of their exact lie and truth count and therefore should not be influenceable in their self reputation. Anyhow, those messages are still used to update the ToM-agents' view on the honesty of the sender.

An agent that understands the general structure of a mind in the RGS so well that he is able to perform the described simulations should also be aware that other agents like himself perform an update on their self reputation after telling a message. Therefore, after receiving a message from a, the ToM-agent b will update his first-order belief I_{baa}. In a two-agent game, the ToM-agent is therefore capable of keeping track of its counterpart's exact mental state at all times, under a known start parameter. Moreover, a lie can be spotted easily if the content of the message does not precisely match the first ToM-agent's first-order ToM. In this constellation, ToM-agents therefore always manage to reach and keep the top reputation after a small number of rounds by effectively outplaying their counterpart. In larger groups in general, the first-order ToM representation is not equal to the real values, which is why we stick to smart lie detection (as described in Enßlin et al. [8]) for more than two agents.

3. Results

Figure 3 shows a reputation statistic for a ToM-agent (red) among two ordinary agents.

Ordinary agents had rather low but fixed lie frequencies, whereas the ToM-agent was free to lie whenever he saw this to be the most promising option according to the rules described above. On average, he showed to tell the truth in only about 14% of the conversations. For an ordinary agent with a comparable, fixed lie frequency, the reputation, as well as his self-reputation, converges against this low value. In the ToM case, the agent's self-reputation (by construction of the ToM update rules) always represents the real value. The way others see the ToM-agent's honesty improved significantly to a value of around 50%, meaning that others held a very distorted image about the ToM-agent by expecting him to tell honest statements in about half of the conversations. We also observe simulations where a very dishonest ToM-agent can even achieve nearly top reputations, but also sometimes expose themselves as deceptive liars and therefore fail completely. As we are dealing with a highly chaotic system, it is often difficult to identify the exact reason for the shape of the reputation dynamic of a game. However, the success of the strategy was found to have a strong correlation with the quality of the ToM-agents' information on the others' beliefs, as shown in Figure 4.

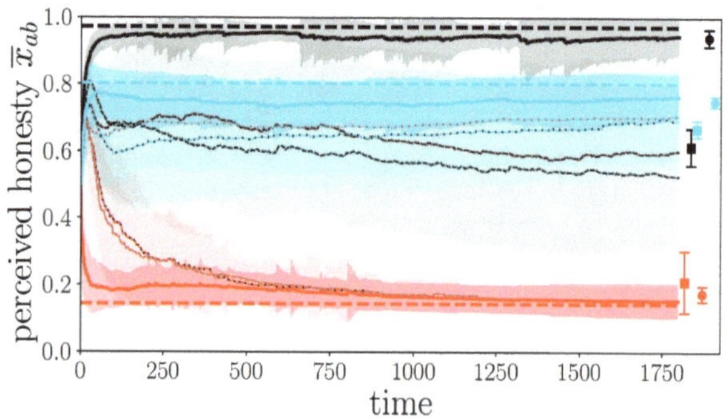

(**a**) Ordinary agent (red) among ordinary agents

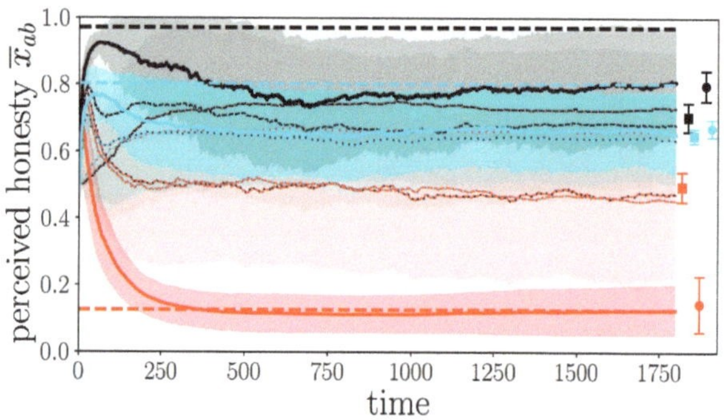

(**b**) ToM-agent (red) among ordinary agents

Figure 3. Comparison of the average achieved reputation and standard deviation for the ToM-agent with the ordinary agent in 100 simulations over the time of 300 rounds. The unicolored lines represent the agents' self-reputations, whereas the red line with black dots denotes the reputation of red in the eyes of black. The dashed lines stand for the honesty frequencies of the agents.

The range around the white lie, in that a lie in general turns out to be better than the truth in the first place, commonly happens to be a lot more narrow than the deviation of the first-order ToM towards the real belief described by it. If this occurs, the deception turns out to be ineffective regardless of the chosen lie option. Dropping a large number of lies close to their counterparts' first-order information and therefore confirming their counterparts' opinion can turn out to be disadvantageous for ToM-agents as future lies then need to be very close to this value. As the ToM-agent's amount of lies is not restricted by an intrinsic lie frequency, he can become very persistent in achieving his goal by permanently dropping lies on a specific agent. Especially in cases where those lies are based on a wrong set of ToM values and agents are confronted with a large variety of different opinions on a specific topic, this can create a high level of social chaos and a high volatility in reputations, especially the self-reputation of the most honest agents.

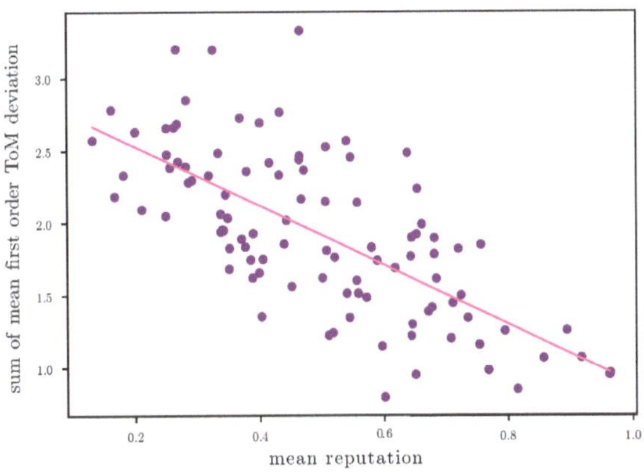

Figure 4. Correlation between the sum of the absolute deviations of first- and second-order Theory of Mind information from the real values and the success of the simulating ToM strategy in terms of the achieved mean reputation in the game.

Ideally, a ToM-agent would identify a statement to be a lie if its content deviates from his first-order ToM values for the sender's belief of the topic. One needs to take into account the possibility that the agent actually intended to send an honest message but the ToM value could be dramatically erroneous and therefore the message's content cannot be used as a measure for its truth. Instead, we take a product of the Kullback-Leibler divergence between the message and the first-order ToM as well as the ToM-agents' belief on the topic. Therefore, the ToM-agents still suffer from the so-called "echo chamber effect" already known from ordinary agents [8]. If a ToM-agent identifies a gain in reputation by manipulating an agent's mind in a certain direction and effectively influencing his counterpart to believe false information, the ToM-agent himself might adopt these false beliefs if the influenced agents themselves start to spread those false beliefs.

4. Discussion and Summary

We have set up a working model of a second-order Theory of Mind agent that uses this mental ability to deliberately deceive others with the goal of maximizing his own reputation, by choosing the communication option that promises the highest expected reputation after simulating different possibilities in the other agents' minds. As it turned out, such ToM-agents showed a strong tendency to tell lies and significantly increased their mean reputation performance compared to the conventional ordinary strategy. It must be noted the success in the individual scenario mostly depends on the quality of the ToM information used for the decision making.

Outlook

We consider the following to be reasonable future extensions of the existing model:

- Include an independent update on I_{abca}: These would give the ToM-agent the opportunity to evaluate the influence of the expected honest statement a third agent might tell to one's counterpart in a subsequent conversation and would lead to more differentiated lie decision rules.
- First-order ToM for friendship: If the ToM-agent holds a picture of their counterparts' friend and enemy lists, he might also have a more accurate belief on the direction of lies from third agents in subsequent conversations, as we found the assumption of symmetrical friendship to be true in many but by far not all cases.
- Choice of topic and conversation partner: Additionally to letting the ToM-agent decide which message about a fixed topic to send to a fixed receiver, one could also let him calculate the expected value for all messages to any agent about any topic in order

to exploit even more options for increasing one's reputation. Moreover, the option to withdraw from conversation could be added, as we found that in some situations none of the possible messages promise an expected reputation that is above the one the ToM-agent expects the receiver to currently hold.

Author Contributions: Conceptualization, F.S., T.E., V.K., S.U. and C.B.; software, F.S., T.E. and V.K.; formal analysis, F.S.; visualization, F.S.; supervision, T.E. and V.K. All authors have read and agreed to the published version of the manuscript.

Funding: This research received no external funding.

Institutional Review Board Statement: Not applicable.

Informed Consent Statement: Not applicable.

Data Availability Statement: Data sharing is not applicable to this article.

Conflicts of Interest: The authors declare no conflict of interest.

References

1. Lee, J.Y.S.; Imuta, K. Lying and Theory of Mind: A Meta-Analysis. *Child Dev.* **2021**, *92*, 536–553. [CrossRef] [PubMed]
2. Böckler-Raettig, A. *Theory of Mind*; utb.: Stuttgart, Germany, 2019.
3. Ding, X.P.; Wellman, H.; Wang, Y.; Fu, G.; Lee, K. Theory-of-Mind Training Causes Honest Young Children to Lie. *Psychol. Sc.* **2015**, *26*, 1812–1821. [CrossRef] [PubMed]
4. Warneken, F.; Orlins, E. Children tell white lies to make others feel better. *Br. J. Dev. Psychol.* **2015**, *33*, 259–270. [CrossRef [PubMed]
5. Biziou-van Pol, L.; Haenen, J.; Novaro, A.; Liberman, O.A.; Capraro, V. Does telling white lies signal pro-social preferences *Judgm. Decis. Mak.* **2015**, *10*, 538–548. [CrossRef]
6. Kathleen, O.M.; Carnevale J., P. A nasty but effective negotiation strategy: Misrepresentation of a common-value issue. *Persona Soc. Psychol. Bull.* **1997**, *23*, 504–515.
7. Steinel, W.; De Dreu, C. Social Motives and Strategic Misrepresentation in Social Decision Making. *J. Personal. Soc. Psychol.* **2004** *86*, 419. [CrossRef] [PubMed]
8. Enßlin, T.; Kainz, V.; Bœhm, C. A Reputation Game Simulation: Emergent Social Phenomena from Information Theory. *Ann Phys.* **2022**, *534*, 2100277. [CrossRef]
9. Enßlin, T.; Kainz, V.; Bœhm, C. Reputation Communication from an Information Perspective. *Phys. Sci. Forum* **2022**, *5*, 15 [CrossRef]
10. Nagel, R. Unraveling in Guessing Games: An Experimental Study. *Am. Econ. Rev.* **1995**, *85*, 1313–1326.
11. Arthur, W.B. Inductive Reasoning and Bounded Rationality. *Am. Econ. Rev.* **1994**, *84*, 406–411.

physical sciences forum

MDPI

Proceeding Paper

Uncertainty Quantification with Deep Ensemble Methods for Super-Resolution of Sentinel 2 Satellite Images †

David Iagaru [1,2,*] and Nina Maria Gottschling [2]

1 CentraleSupélec, Université Paris-Saclay, 91190 Gif-sur-Yvette, France
2 MF-DAS OP-EO Data Science, Deutsches Zentrum für Luft- und Raumfahrt, 82234 Weßling, Germany
* Correspondence: david.iagaru@student-cs.fr

† Presented at the 42nd International Workshop on Bayesian Inference and Maximum Entropy Methods in Science and Engineering, Garching, Germany, 3–7 July 2023.

Abstract: The recently deployed Sentinel 2 satellite constellation produces images in 13 wavelength bands with a Ground Sampling Distance (GSD) of 10 m, 20 m, and 60 m. Super-resolution aims to generate all 13 bands with a spatial resolution of 10 m. This paper investigates the performance of DSen2, a proposed convolutional neural network (CNN)-based method, for tackling super-resolution in terms of accuracy and uncertainty. As the optimization problem for obtaining the weights of a CNN is highly non-convex, there are multiple different local minima for the loss function. This results in several possible CNN models with different weights and thus implies *epistemic uncertainty*. In this work, methods to quantify epistemic uncertainty, termed *weighted deep ensembles* (WDESen2) and its variants), are proposed. These allow the quantification of predictive uncertainty estimates and, moreover, the improvement of the accuracy of the prediction by selective prediction. They involve a consideration of deep ensembles, and each model's importance can be weighted depending on the model's validation loss. We show that weighted deep ensembles improve the accuracy of prediction compared to state-of-the-art methods and deep ensembles. Moreover, the uncertainties can be linked to the underlying inverse problem and physical patterns on the ground. This allows us to improve the trustworthiness of CNN predictions and the predictive accuracy with selective prediction.

Keywords: inverse problems; uncertainty quantification; deep ensemble methods; physics-informed ML; Bayesian inference

1. Introduction

Satellite images of multiple spectral bands acquired from various satellites, such as MODIS, VIIRS, ASTER, Worldview-3, and Sentinel 2, have different spatial resolutions. The resolution varies by a factor of 2–6 based on the instrumentation. This work focuses on the Sentinel 2 satellites, but the proposed methods for super-resolution, WDESen2 and its variants, are not limited to Sentinel 2 data. The constellation consists of two satellites that capture images of the Earth's surface, where each point on the surface is observed for a period of 5 days. Each image contains information from 13 spectral bands, some of which partially overlap, ranging from ultraviolet/visible to infrared. Band B10 is used for atmospheric corrections and was therefore omitted in the data used for this study. The bands are available in several different resolutions: bands B2, B3, B4, and B8 have a GSD of 10 and will be referred to as the *A bands*; B5, B6, B7, B8a, B10, and B11 have a GSD of 20 m and will be referred to as the *B bands*; while bands B1 and B9 have a GSD of 60 m and will be referred to as the *C bands*. The data can be used in various applications, such as hydrology and water resource management [1–3]. Even though there are reasons for the different spatial resolutions, such as storage and transmission bandwidth restrictions, often it is desirable to have all bands available at the highest spatial resolution. The problem of bringing all bands to a GSD of 10 m can be called super-resolution. In the work of Lanaras et al. [4], the authors demonstrated that a CNN-based super-resolution method obtained higher accuracy for all super-resolved bands compared to state-of-the-art algorithms. State-of-the-art algorithms for the super-resolution of multi-spectral satellite data can be divided into three types: pan-sharpening per band, as reviewed in [5,6]; inverting an explicit imaging model, as demonstrated in [7,8], with a reduced-rank method applied for Sentinel 2 data in [9]; and machine learning approaches, such as [4]. The approaches in the second category are often based on the reformulation of super-resolution as a convex optimization problem by re-projecting each image in a subspace determined by the corresponding algorithm, such as in [9]. These methods rely on linear sensing model assumptions and are computationally expensive, as the super-resolution of each image requires a separate optimization. While being computationally more efficient, the CNN-based approach of [4] also better preserves spectral characteristics, achieves a favorable computational speed when run on GPUs, and is globally applicable for Sentinel 2 data without retraining. Thus, our work focuses on the DSen2 method from [4], which combines the information of the bands with lower and higher resolutions for super-resolution. The method relies on the assumption of the *self-similarity* of the upsampling process. This means that upsampling from 20 m to 10 m GSD is assumed to have the same mapping as upsampling from 40 m to 20 m GSD. Based on this assumption, in this work we address the features in B bands that cannot be super-resolved by exploiting the correlation between the bands with higher and lower resolutions. This is achieved in two ways: firstly, by studying extensions of the DSen2 method that predict uncertainty, such as WDESen2, and secondly, by studying the kernel of the inverse problem inherent to the proposed super-resolution method. Specifically, predictive uncertainty is evaluated with deep ensembles, as proposed in [10], and based on the success of weighted versions of deep ensembles for classification problems, as in [11,12], we propose a novel method of weighted deep ensembles for super-resolution. Uncertainty quantification (UQ) methods in DL usually distinguish between epistemic, or model, uncertainty and aleatoric, or data, uncertainty [13]. However, there is an ongoing debate as to whether these two types of uncertainty can be disentangled [14]. Adopting this perspective, we aim at explaining the obtained predictive uncertainties by the underlying physics, i.e., the inverse problem model. Here, we interpret the predictive uncertainties that arise by experimentally relating them to the kernel of the forward operator through projection onto the kernel of the inverse problem model. A different technique based on the generalized kernel has been applied in tomographic image reconstruction to detect hallucinations [15]. Moreover, we evaluate the uncertainties and kernel obtained by selective prediction, which has been applied in classification [16] and regression [17] problems for DL methods.

2. Methods

In this section, a brief introduction to Bayesian methods for uncertainty quantification in CNNs is given, followed by a short overview of deep ensembles as a Bayesian method and the extension thereof for super-resolution that is proposed in this work, WDESen2. Thereafter, the architecture and pipeline for the proposed method and its relation to the underlying inverse problem are described in detail.

2.1. *Bayesian Methods for Uncertainty Quantification in Deep Neural Networks*

2.1.1. Bayesian Methods and Deep Ensembles

We denote as $y \in Y$ the space of the predictions and $x \in X$ the inputs. In Bayesian approaches for uncertainty quantification in CNNs, instead of obtaining a single set of weights θ for the CNN by optimizing a given loss function, the aim is to obtain a posterior distribution of weights, $p(\theta|D)$, given the training data $D = (X, Y)$ [18]. In the following, we give an outline of deep ensembles interpreted as a Bayesian method. Given the posterior distribution, a predictive distribution can be obtained by marginalization over the weights:

$$p(y|x, D) = \int p(y|x, \theta)p(\theta|D)d\theta.$$

Since the posterior distribution, $p(\theta|D)$, is intractable, it is common to either approximate it with a distribution or apply sampling methods [13]. Deep ensembles, as introduced in [10], can be interpreted as a sampling method [18]. The approach consists in training a number $N \in \mathbb{N}$ of ensemble members that are CNNs, $f_{\theta_i} : X \to Y$. Due to the non-convexity of the optimization problem for obtaining the weights and the stochastic gradient descent methods, this results in N different sets of weights, $(\theta_i(D))_{i=1}^N$. Then, the posterior distribution [18] is assumed to be approximated by

$$p(\theta|D) \approx \frac{1}{N}\sum_{i=1}^{N} \delta(\theta = \theta_i). \tag{1}$$

The resulting prediction of the deep ensemble (DE) is the mean of the predictions,

$$\hat{y}_{\mathrm{DE}}(x) = \frac{1}{N}\sum_{i=1}^{N} f_{\theta_i}(x) \tag{2}$$

and the epistemic uncertainty is given by the variance of the predictions,

$$\hat{\sigma}_{\mathrm{DE}}(x)^2 = \frac{1}{N}\sum_{i=1}^{N} f_{\theta_i}(x)^2 - \hat{y}_{\mathrm{DE}}(x)^2. \tag{3}$$

2.1.2. Utilizing the Loss-Weighted Deep Ensembles

In the following, we introduce an extension of deep ensembles, weighted deep ensembles, which assign weight to the prediction by the validation loss of each ensemble member. This approach implicitly approximates the posterior distribution in (1) with weights $(w_i(\theta_i, V))_{i=1}^N$ attributed to each network in the ensemble, as illustrated in Figure 1. The weights $w_i = w(\theta_i, V)$ depend on the validation loss and validation data V and the weights of each ensemble member:

$$p(\theta|D) \approx \sum_{i=1}^{N} w_i \delta(\theta = \theta_i). \tag{4}$$

Correspondingly, the prediction of the weighted deep ensemble (WDE) is given by the weighted mean,

$$\hat{y}_{\mathrm{WDE}}(x) = \sum_{i=1}^{N} w_i f_{\theta_i}(x) \tag{5}$$

and the predictive uncertainty by the variance with respect to the distribution (4),

$$\hat{\sigma}_{\mathrm{WDE}(x)^2} = \sum_{i=1}^{N} w_i f_{\theta_i}(x)^2 - \hat{y}_{\mathrm{WDE}}(x)^2. \tag{6}$$

2.1.3. Derivation of Ensemble Weights: Contribution Ratio Control

As shown in previous work, there are many possibilities for computing the weights $(w_i)_{i=1}^N$. We chose to compute the weights such that the contribution ratio can be controlled by a hyperparameter. The weights depending on the parameters θ_i and the validation loss $L_{\theta_i} = L_{\theta_i}(V)$ are defined as

$$w_i(\theta_i, V) = \left(\frac{1}{L_{\theta_i} + \frac{rL_{\max} - L_{\min}}{1-r}}\right) \Big/ \left(\sum_{j=1}^{N} \frac{1}{L_{\theta_j} + \frac{rL_{\max} - L_{\min}}{1-r}}\right) \quad \forall i \in \{1, \dots, N\},$$

where $L_{\max}(V)$ and $L_{\min}(V)$ are the maximum and the minimum validation loss, respectively. The above definition ensures that the ensemble member with the lowest validation loss is weighted as the most important, i.e., has the highest weight. Moreover, the weights satisfy the following conditions:

$$\frac{min((w_i)_{i=1}^N)}{max((w_i)_{i=1}^N)} = r \in (0,1) \quad 0 < w_i(\theta_i, D) < 1 \quad \forall i \in \{1,\ldots,N\}, \quad \sum_{i=1}^{N} w_i(\theta_i, D) = 1.$$

The last two conditions ensure that (4) is a probability distribution, and the first condition allows one to determine the ratio of contribution, r, a hyperparameter that we fixed at 0.5 in our experiments.

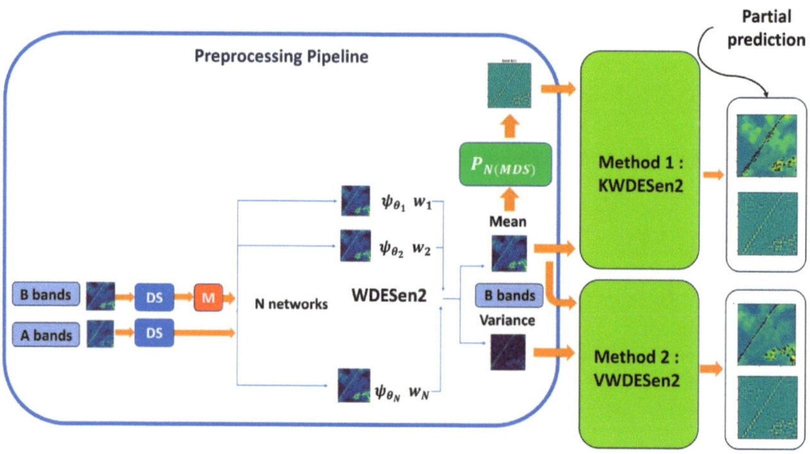

Figure 1. Architecture and pipeline of the KWDESen2 and VWDESen2 methods with N ensemble members.

2.2. Architecture of a Weighted Deep Ensemble for the Super-Resolution of Satellite Data: WDESen2

The weighted deep ensemble (WDESen2) architecture and its extensions, shown in Figure 1, use the architecture of the DSen2 network [4] as a backbone. By ensembling and weighting the predictions, WDESen2 improves the predictive accuracy compared to DSen2. The selective prediction applied to the predictive uncertainties from WDESen2, which are given as variances according to (6), is referred to as variance-based WDESen2 (VWDESen2). The selective prediction applied on the kernel projection is referred to as kernel-based WDESen2 (KWDESen2). The proposed methods are described below. The ensemble members of WDESen2 are the DSen2 networks that super-resolve the B and C bands to 10 m GSD. For this, two networks are used: a CNN, $\mathbb{S}_{\times 6}$, which performs the 6× super-resolution of the C bands from 60 m to 10 m GSD, and another CNN, Ψ_θ, which performs the 2× super-resolution of the B bands from 20 m to 10 m GSD. As the 6× resolution network training procedure is structurally the same as that of the 2× resolution network, we restrict ourselves to describing the latter. For training a CNN to super-resolve the B bands, there do not exist ground-truth data with a 10 m GSD for the B bands. In order to generate training data, we followed the approach of [4]. The A bands and B bands are downsampled to 20 m and 40 m GSD, respectively. For each band and patch of size $W/2 \times H/2$, the linear downsampling operator, $DS : \mathbb{R}^{W \times H} \to \mathbb{R}^{W/2 \times H/2}$, is given below in Figure 2.

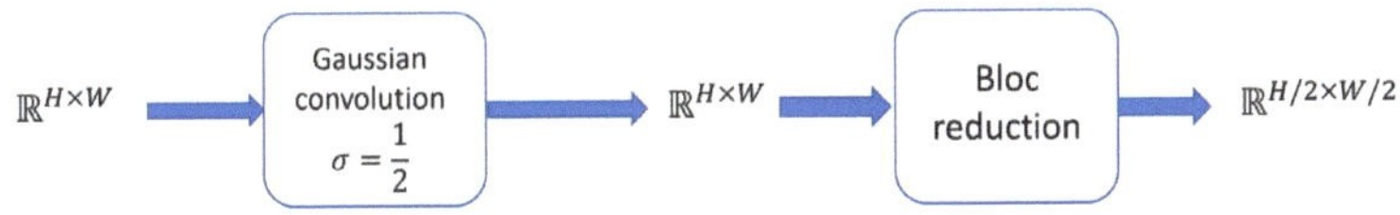

Figure 2. The $DS : \mathbb{R}^{W\times H} \to \mathbb{R}^{W/2\times H/2}$ operator. Here, W and H are the width and the height of a patch.

Then, the downsampled 40 m GSD and 20 m GSD B and A bands are used as inputs, and the 20 m GSD B bands can be used as targets in the training procedure. Based on the assumption of the self-similarity of super-resolution from 40 m GSD to 20 m GSD compared with that from 20 m GSD to 10 m GSD, the resulting network can then be used for 2× super-resolution from 20 m to 10 m GSD. We let $y_A \in \mathbb{R}^{4\times W\times H}$, where $W \times H$ is the patch-size, and $y_B \in \mathbb{R}^{6\times W/2\times H/2}$, where $W/2 \times H/2$ is the patch-size, denote the observed intensities for the A and B bands, respectively. The DSen2 method for 2× super-resolution is given by

$$\Psi_\theta : \mathbb{R}^{4\times W\times H} \times \mathbb{R}^{6\times W/2\times H/2} \to \mathbb{R}^{6\times W\times H},$$
$$(y_A, y_b) \to x_B,$$

where the B bands, as depicted in Figure 1, are upsampled to the same resolution as the A bands with *bi-linear interpolation*. The associated linear operator, $M : \mathbb{R}^{H/2\times W/2} \to \mathbb{R}^{H\times W}$, is applied to each B band.

By retraining N times on the same data splits with the ℓ^1 loss, due to the stochasticity of the training procedure, we obtain N different CNNs, as shown in Table 1. As visualized in Figure 1, the prediction of WDSen2 is composed of a weighted mean, according to (5), and the corresponding variance (6).

The predictions of WDESen2 are post-processed using two different methods, kernel- and variance-based partial prediction, which will be described in the following. For 6× super-resolution to have the same dimension inputs in the CNN Ψ_θ, the B and C bands are upsampled. However, this implies that there is no ground truth with which to compare the predictions.

2.2.1. Kernel Component Exploration

For training, the B band inputs, $y_B \in \mathbb{R}^{6\times W/2\times H/2}$ of Ψ_θ, are given by applying a composition of M and DS, $MDS : \mathbb{R}^{W\times H} \to \mathbb{R}^{W\times H}$, as in Figure 1, band-wise to the training labels $x_B \in \mathbb{R}^{6\times W\times H}$. Thus, Ψ_θ aims to reconstruct x_B^i from $y_B^i = MDSx_B^i$ and the other bands, where x_B^i is one of the B bands at 20 m GSD. As $DS : \mathbb{R}^{W/2\times H/2} \to \mathbb{R}^{W\times H}$ has a higher input dimension than output dimension, the composition MDS, depicted in Figure 3, has a non-trivial kernel. Recall that the kernel of a linear map $A : E \to F$ is defined as $\mathcal{N}(A) := \{x \in E | u(x) = 0\}$. Thus, given an input y_i, different super-resolved solutions x_i may exist depending on the effect of correlation. In order to analyze the kernel component of the super-resolved bands, we use the projection operator $P_{\mathcal{N}(MDS)}$ on the kernel $\mathcal{N}(MDS)$:

$$P_{\mathcal{N}(MDS)} = I - (MDS)^\dagger MDS : \mathbb{R}^{W\times H} \to \mathbb{R}^{W\times H},$$

where $(MDS)^\dagger$ is the Moore–Penrose inverse of MDS [19], and this projection is approximated by a truncated singular value decomposition of MDS. The projection is computed individually per patch and band, as shown on the left-hand side of Figure 4.

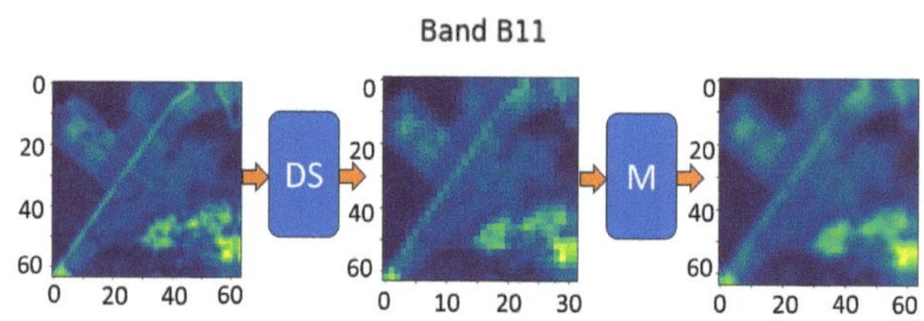

Figure 3. The composition of the DS and M operators, $MDS : \mathbb{R}^{W \times H} \rightarrow \mathbb{R}^{W \times H}$.

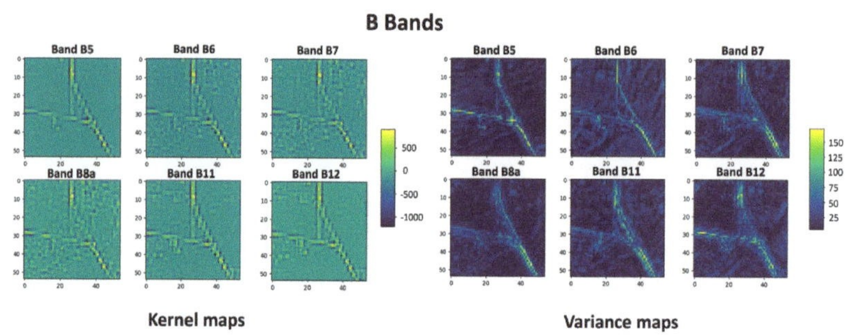

Figure 4. Kernel projection on the mean prediction (left) vs. variance (right) of WDESen2. The variance map visualizes the pixel-wise epistemic uncertainty of WDESen2 with 30 ensemble members. The patch was from the tropical data.

2.3. *Evaluation of the Predicted Variance and Kernel Projection of the Predictions of WDESen2*

In order to utilize the predicted variance of WDESen2 and the kernel projection, as described in Section 2.2.1, we perform selective prediction, as introduced for classification in [16] and applied to DL for regression problems in [17]. The general idea of selective prediction is to make the model abstain from giving a prediction if the uncertainty surpasses a specific threshold. We apply selective prediction to pixel-wise variances, i.e., uncertainties, from WDESen2 and pixel-wise values of the kernel projection. The first method, variance-based WDESen2 (VWDESen2), as visualized in Figure 5, abstains from predicting pixels that have the highest variance. In order to maintain comparability between different methods, we withdraw 5% of the pixels from the mean prediction of WDESen2. The second method is based on the underlying inverse problem, kernel-based WDESen2 (KWDESen2), where the 5% of the pixels from the mean prediction that have the highest values in the kernel projection of the prediction are withdrawn from the prediction.

2.4. *Dataset Splits*

In Table 1, the dataset splits for the training, validation, and test datasets are listed. Inference at test time is performed on patches of two images of size 20×20 km^2. The $2\times$ super-resolution methods are trained and evaluated individually on the *"Tropical landscape"* and *"Mountains landscape"* datasets, as these have different spectral characteristics.

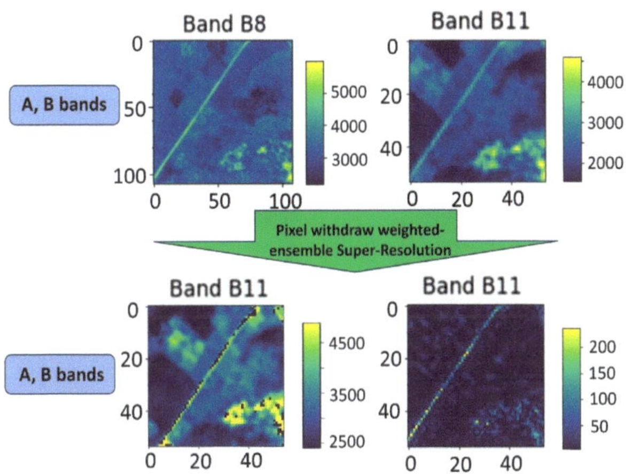

Figure 5. A and B ground-truth bands (**top**), variance-based withdrawal (**bottom left**) and variance map (**bottom right**) for VWDESen2.

Table 1. Data splits for the training, validation, and test datasets. Regarding the images used for inference at test time, the first image will be referred to as "Tropical data" and the second image as "Mountain data".

	Images		Split	Patches
Tropical landscape/farmlands	1	Training	87%	$6500, 64 \times 64$
		Validation	8.5%	$723, 64 \times 64$
		Test	4.5%	$289, 64 \times 64$
Mountains landscape	9	Training	89.5%	$58{,}500, 64 \times 64$
		Validation	10%	$6500, 64 \times 64$
		Test	0.5%	$289, 64 \times 64$

3. Results

3.1. Linking Uncertainties and Physical Patterns on the Ground

Figure 4 visualizes the kernel projections on the mean predictions and the uncertainty of WDESen2. Pixels of high variance correspond to pixels that have high values in the kernel projections. This indicates a possible relation between pixels where the predictions of the ensemble members diverge and the projection onto the kernel of *MDS*. Moreover, comparing Figure 4 to the ground truth shows that the pixels of high variance and the pixels with higher values in the kernel projections correspond to the physical patterns shown in Figure 6.

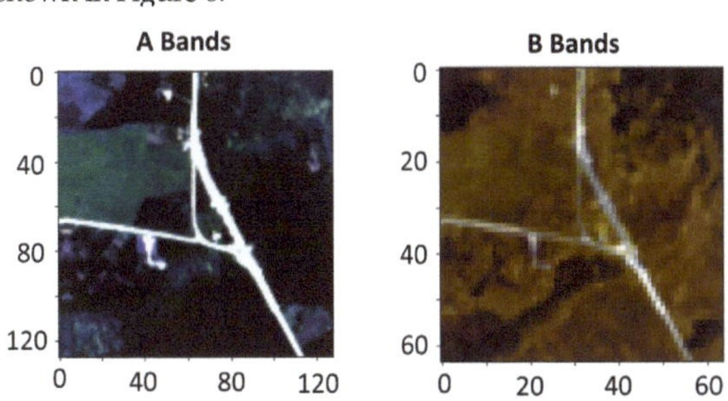

Figure 6. The A and B bands of the patch used in Figure 4.

3.2. Selective Prediction: Improving Accuracy with Uncertainty and Kernel Projections

Figures 7–9 enable one to compare the predictions of WDESen2, KWDESen2, and VWDESen2 with those of deep ensembles, where the predictions are given by (2) and the variance by (3). Selective prediction, as in Section 2.3, was applied to deep ensembles DESen2, resulting in a variance-based DESen2 method (VDESen2) and a kernel-based DESen2 method (KDESen2). Table 2 shows that removing the most uncertain 5% of the total number of pixels could significantly reduce the band-wise RMSE by up to 15% for VDESen2, and removing the pixels with a high value in the kernel projection could reduce the band-wise RMSE by 7% for KDESen2. Moreover, Figures 7–9 show that weighted deep ensembles with and without selective prediction were more accurate in terms of band-wise RMSE than deep ensembles and the baseline model from [4]. Moreover, Figures 8 and 9 indicate that, depending on the dataset, 10–15 ensemble members obtained better accuracy compared to smaller ensembles.

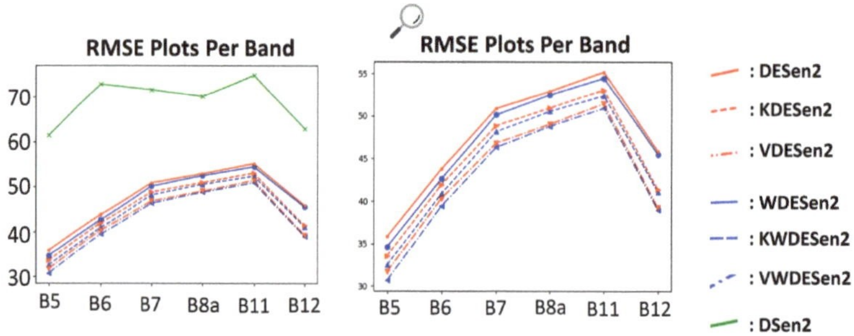

Figure 7. Band-wise RMSE of different deep ensemble methods (DESen2, KDESen2, VDESen2, WDESen2, KWDESen2, and VWDESen2). Each ensemble was composed of 30 members, and the baseline of DSen2 corresponds to 1 ensemble member. The RMSE was computed over 300 patches of tropical data.

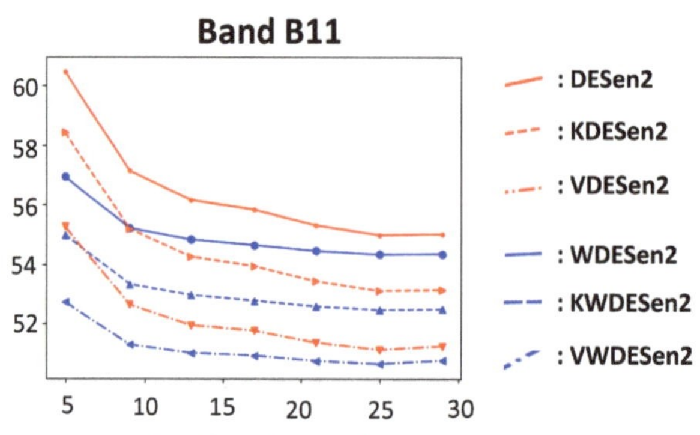

Figure 8. Band B11's RMSE for different numbers of ensemble members on tropical satellite data. For a given number of ensemble members $n \in \{1, \ldots, 30\}$, a deep ensemble with n networks was randomly chosen from the 30 trained models, and this was averaged over 10 times in order to avoid the bias from the network sampling. The results were averaged over the patches.

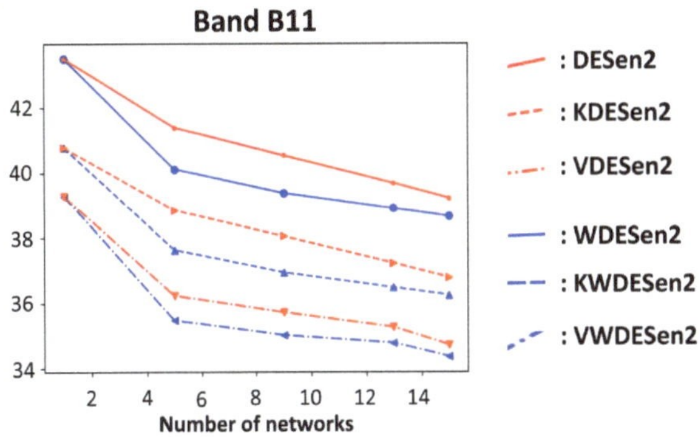

Figure 9. Band B11's RMSE for different numbers of ensemble members on mountain satellite data. For a given number of ensemble members $n \in \{1,\dots,14\}$, a deep ensemble with n networks was randomly chosen from the 14 trained models, and this was averaged over 10 times in order to avoid the bias from the network sampling. The results were averaged over the patches.

Table 2. RMSE of the bicubic method, DSen2 baseline, deep ensemble of DSen2 (DESen2), weighted deep ensemble of DSen2 (WDESen2), weighted deep ensemble of DSen2 with kernel-based pixel withdrawal (KWDESen2), and weighted deep ensemble of DSen2 with variance-based pixel withdrawal (VWDESen2). All deep ensembles had 30 networks. The RMSE was computed over 300 patches of the tropical image.

	B5	B6	B7	B8a	B11	B12	Average
				RMSE			
Bicubic	93.2	105.1	124.6	133.6	105.4	94.8	109.4
DSen2	44.5	54.3	58.9	60.2	62.5	53.0	55.5
DESen2	35.7	43.6	50.8	52.8	55.1	45.8	47.3
WDESen2	34.6	42.6	50.1	52.4	54.4	45.5	46.6
KWDESen2	30.9	39.4	46.2	48.3	49.7	38.2	42.1
VWDESen2	**28.7**	**37.6**	**44.0**	**46.0**	**48.0**	**35.8**	**40.0**

4. Conclusions

This work showed that by utilizing uncertainties in predictions, accuracy and interpretability can be improved compared to when other methods for the super-resolution of satellite data, specifically Sentinel 2 data, are applied. The first proposed method, weighted deep ensembles (WDESen2), improved the accuracy compared to state-of-the-art methods and at the same time provided predictive uncertainties. In the second proposed method, VWDESen2, selective prediction was applied to the uncertainties and yielded a lower RMSE compared to state-of-the-art methods. Moreover, the variance could be visually related to the projection onto the kernel of the underlying inverse problem and physical patterns on the ground. The corresponding pixels could be interpreted as being hard to super-resolve, as the withdrawal of such pixels led to a higher accuracy, resulting in the third proposed method, KWDESen2.

Author Contributions: Conceptualization and methodology, N.M.G. and D.I.; modeling and validation, D.I.; formal analysis, D.I. and N.M.G.; writing—original draft preparation, D.I. and N.M.G.; writing—review and editing, D.I. and N.M.G. All authors have read and agreed to the published version of the manuscript.

Funding: This research received no external funding.

Institutional Review Board Statement: Not applicable.

Informed Consent Statement: Not applicable.

Data Availability Statement: The Sentinel-2 images used can be downloaded from: https://dataspace.copernicus.eu/browser/.

Conflicts of Interest: The authors declare no conflict of interest.

References

1. Mura, M.; Bottalico, F.; Giannetti, F.; Bertani, R.; Giannini, R.; Mancini, M.; Orlandini, S.; Travaglini, D.; Chirici, G. Exploiting th capabilities of the Sentinel-2 multi spectral instrument for predicting growing stock volume in forest ecosystems. *Int. J. App Earth Obs. Geoinf.* **2018**, *66*, 126–134. [CrossRef]
2. Castillo, J.A.A.; Apan, A.A.; Maraseni, T.N.; Salmo, S.G. Estimation and mapping of above-ground biomass of mangrove forest and their replacement land uses in the Philippines using Sentinel imagery. *ISPRS J. Photogramm. Remote Sens.* **2017**, *134*, 70–85 [CrossRef]
3. Toming, K.; Kutser, T.; Laas, A.; Sepp, M.; Paavel, B.; Noges, T. First Experiences in Mapping LakeWater Quality Parameters with Sentinel-2 MSI Imagery. *Remote Sens.* **2016**, *8*, 640. [CrossRef]
4. Lanaras, C.; Bioucas-Dias, J.; Galliani, S.; Baltsavias, E. Super-Resolution of Sentinel-2 Images: Learning a Globally Applicable Deep Neural Network. *ISPRS J. Photogramm. Remote Sens.* **2018**, *146*, 305–319. [CrossRef]
5. Vivone, G.; Alparone, L.; Chanussot, J.; Dalla Mura, M.; Garzelli, A.; Licciardi, G.A.; Restaino, R.; Wald, L. A critical comparison among pansharpening algorithms. *IEEE Trans. Geosci. Remote Sens.* **2014**, *53*, 2565–2586. [CrossRef]
6. Li, S.; Kang, X.; Fang, L.; Hu, J.; Yin, H. Pixel-level image fusion: A survey of the state of the art. *Inf. Fusion* **2017**, *33*, 100–112 [CrossRef]
7. Brodu, N. Super-resolving multiresolution images with band-independent geometry of multispectral pixels. *IEEE Trans. Geosci Remote Sens.* **2017**, *55*, 4610–4617. [CrossRef]
8. Dian, R.; Li, S.; Fang, L.; Bioucas-Dias, J. Hyperspectral image super-resolution via local low-rank and sparse representations In Proceedings of the IGARSS 2018-2018 IEEE International Geoscience and Remote Sensing Symposium, Valencia, Spain, 22–27 July 2018; pp. 4003–4006.
9. Ulfarsson, M.O.; Palsson, F.; Dalla Mura, M.; Sveinsson, J.R. Sentinel-2 sharpening using a reduced-rank method. *IEEE Trans Geosci. Remote Sens.* **2019**, *57*, 6408–6420. [CrossRef]
10. Lakshminarayanan, B.; Pritzel, A.; Blundell, C. Simple and Scalable Predictive Uncertainty Estimation Using Deep Ensembles In Proceedings of the 31st International Conference on Neural Information Processing Systems, Long Beach, CA, USA, 4–9 December 2017; pp. 6405–6416.
11. Omisore, O.M.; Akinyemi, T.O.; Du, W.; Duan, W.; Orji, R.; Do, T.N.; Wang, L. Weighting-Based Deep Ensemble Learning for Recognition of Interventionalists' Hand Motions During Robot-Assisted Intravascular Catheterization. *IEEE Trans. Hum. Mach Syst.* **2022**, *53*, 215–227 [CrossRef]
12. Gupta, R.; Bhatnagar, A.S.; Singh, G. A Weighted Deep Ensemble for Indian Sign Language Recognition. *IETE J. Res.* **2023**, [CrossRef]
13. Gawlikowski, J.; Tassi, C.R.N.; Ali, M.; Lee, J.; Humt, M.; Feng, J.; Kruspe, A.M.; Triebel, R.; Jung, P.; Roscher, R.; et al. A Survey of Uncertainty in Deep Neural Networks. *Artif. Intell. Rev.* **2021**, *56*, 1513–1589. [CrossRef]
14. Gruber, C.; Schenk, P.O.; Schierholz, M.; Kreuter, F.; Kauermann, G. Sources of Uncertainty in Machine Learning—A Statisticians View. *arXiv* **2023**, arXiv:2305.16703.
15. Bhadra, S.; Kelkar, V.A.; Brooks, F.J.; Anastasio, M.A. On hallucinations in tomographic image reconstruction. *IEEE Trans. Med. Imaging* **2021**, *40*, 3249–3260. [CrossRef] [PubMed]
16. El-Yaniv, R. On the Foundations of Noise-free Selective Classification. *J. Mach. Learn. Res.* **2010**, *11*, 1605–1641.
17. Jiang, W.; Zhao, Y.; Wang, Z. Risk-controlled selective prediction for regression deep neural network models. In Proceedings of the 2020 International Joint Conference on Neural Networks (IJCNN), Glasgow, UK, 19–24 July 2020; pp. 1–8.
18. Wilson, A.G.; Izmailov, P. Bayesian Deep Learning and a Probabilistic Perspective of Generalization. In Proceedings of the 34th International Conference on Neural Information Processing Systems, Online, 2–6 December 2020; Curran Associates Inc.: Red Hook, NY, USA, 2020.
19. Courrieu, P. Fast Computation of Moore-Penrose Inverse Matrices. *Neural Inf. Process. Lett. Rev.* **2008**, *8*, 25–29.

physical sciences forum

MDPI

Proceeding Paper

Geodesic Least Squares: Robust Regression Using Information Geometry †

Geert Verdoolaege

Department of Applied Physics, Ghent University, 9000 Ghent, Belgium; geert.verdoolaege@ugent.be

† Presented at the 42nd International Workshop on Bayesian Inference and Maximum Entropy Methods in Science and Engineering, Garching, Germany, 3–7 July 2023.

Abstract: Geodesic least squares (GLS) is a regression technique that operates in spaces of probability distributions. Based on the minimization of the Rao geodesic distance between two probability models of the response variable, GLS is robust against outliers and model misspecification. The method is very simple, without any tuning parameters, owing to its solid foundations rooted in information geometry. Here, we illustrate the robustness properties of GLS using applications in the fields of magnetic confinement fusion and astrophysics. Additional interpretation is gained from visualizations using several models for the manifold of Gaussian probability distributions.

Keywords: robust regression; information geometry; nuclear fusion; energy confinement; astrophysics; Tully–Fisher scaling

1. Introduction

Geodesic least squares is a relatively new technique for regression analysis in a regular Euclidean data space, which takes into account uncertainty estimates (error bars) on all regression variables [1]. It is a minimum distance technique that uses the geodesic distance (Rao distance) on a statistical manifold, as studied in the field of information geometry [2,3]. Crucially, the method is easily implemented, yet it exhibits good robustness properties in the presence of outliers and model uncertainty.

With this contribution, on the one hand, we aim to demonstrate the robustness of the technique with recent and new applications. On the other hand, we intend to stress the considerable value, in some domains of application, of analysis techniques that are easy to implement, requiring little tuning or expert knowledge, while addressing shortcomings of overly simple data analysis techniques that are nevertheless commonly employed. Hence, we aim to promote the use of geometric methods in machine learning by showing that proper treatment of the complex, nonlinear character of certain data can greatly improve the performance of even the simplest analysis techniques. We focus on the estimation of power laws, which are ubiquitous in various domains, such as astronomy, biology and geology, reflecting scale invariance and details of the underlying dynamical processes.

Two applications are targeted in this work. The first is situated in the domain of magnetic confinement fusion (MCF), specifically the critical scaling law for the energy confinement time. MCF is a field where the methods of modern data science have only recently been introduced and where the ease of implementation and interpretation are likely to accelerate the widespread adoption of new analysis techniques. The second application concerns a key scaling law in astrophysics: the baryonic Tully–Fisher relation. This is a remarkably tight relation between the total baryonic mass of disk galaxies and their rotational velocity, which is of great practical and theoretical significance in astrophysics and cosmology.

The exposition starts with a description of the GLS regression technique in Section 2. We then move to the application in fusion science in Section 3, followed by the astrophysical application in Section 4. Section 5 concludes the paper.

2. Geodesic Least Squares Regression

GLS relies on the concept of a statistical manifold, which is a Riemannian manifold that uses the Fisher information as a metric tensor. As such, it is a nonlinear space wherein the points themselves represent probability distributions. Geodesics, i.e., curves of a locally minimal distance between points on the manifold, can be constructed using the Levi–Civita connection, from which a geodesic distance can be obtained. Given a parametric probability model $p(\{x_m\}|\{\theta^j\})$ for variables x_m ($m = 1, \ldots, M$) with parameters θ^j ($j = 1, \ldots, n_{\text{par}}$), the Fisher information (covariance of the score) provides a unique metric tensor, with elements given by

$$g_{ij} = -\mathbb{E}\left[\frac{\partial^2}{\partial\theta^i\partial\theta^j} \ln p(\{x_m\}|\{\theta^k\})\right], \qquad i,j,k = 1, \ldots, n_{\text{par}}.$$

For instance, in the case of a univariate normal distribution $p(x|\mu,\sigma)$, parametrized by its mean μ and standard deviation σ, the Fisher information matrix is [4]

$$I(\mu,\sigma) = \frac{1}{\sigma^2}\begin{pmatrix} 1 & 0 \\ 0 & 2 \end{pmatrix}.$$

To explain the principle of GLS regression, let us assume an exact functional relationship $\eta = f(\{\xi_m\}, \alpha_j)$, parametrized by the set α_j ($j = 1, \ldots, n_{\text{par}}$), between a set of hidden predictor variables $\xi_1, \ldots, \xi_{n_{\text{pred}}}$ and a single hidden response variable η (in the remainder of this paper, we no longer make a notational distinction between covariantly and contravariantly transforming quantities). Let us also consider additive uncertainty on all variables, yielding the respective measurable variables $y, x_1, \ldots, x_{n_{\text{pred}}}$:

$$y = \eta + \epsilon_y, \quad x_1 = \xi_1 + \epsilon_{x_1}, \quad \ldots$$

For simplicity, here we consider zero-mean normal error distributions, but other models can be used as well:

$$\epsilon_y \sim \mathcal{N}(0,\sigma_y), \quad \epsilon_{x_1} \sim \mathcal{N}(0,\sigma_{x_1}), \quad \ldots$$

Importantly, it is assumed that an estimate of the measurement uncertainties is available, determined by the experimentalist. This will be used to construct a modeled distribution for the data. Furthermore, we consider the general case, which will be useful in Section 3, where the data consist of $n_{\text{obs},k}$ independent observations of the response and predictor variables from a class k (e.g., a fusion device), out of a total of n_{class} classes (e.g., multiple fusion devices). The unobserved values $\xi_{m,i_k,k}$ ($m = 1, \ldots, n_{\text{pred}}$, $i_k = 1, \ldots, n_{\text{obs},k}$, $k = 1, \ldots, n_{\text{class}}$) need to be marginalized out. If f is linear,

$$\eta = \alpha_0 + \sum_{j=1}^{n_{\text{pred}}} \alpha_j \xi_j, \qquad (n_{\text{par}} = n_{\text{pred}} + 1)$$

and then the joint likelihood is again Gaussian:

$$p(\{y_{i_k,k}\}, \{x_{j,i_k,k}\}|\{\alpha_0, \alpha_j\}) = \prod_{k=1}^{n_{\text{class}}} \prod_{i_k=1}^{n_{\text{obs},k}} \frac{1}{\sqrt{2\pi\sigma^2_{\text{mod},i_k,k}}} \exp\left\{-\frac{1}{2}\frac{\left[y_{i_k,k} - f(\{x_{j,i_k,k}, \alpha_0, \alpha_j\})\right]^2}{\sigma^2_{\text{mod},i_k,k}}\right\}, \tag{1}$$

where [5]

$$\sigma^2_{\text{mod},i_k,k} = \sigma^2_{y,i_k,k} + \sum_{j=1}^{n_{\text{pred}}} \alpha_j^2 \sigma^2_{x_j,i_k,k}, \tag{2}$$

and the standard deviations $\sigma_{x_j,i_k,k}$ may in general be different for each data point. Note that, in Equation (1), we have plugged the *measured* data points $x_{j,i_k,k}$ into f. The marginalization causes their uncertainty to propagate through the functional relation f, hence turning up in Equation (2). In fact, due to the linearity of f, this result agrees with standard Gaussian error propagation. By analogy, for a power-law model,

$$\eta = \alpha_0 \prod_{j=1}^{n_{\text{pred}}} \xi_j^{\alpha_j},$$

we approximate the likelihood by using a Gaussian distribution with the standard deviation obtained by the Gaussian error propagation rules:

$$\sigma^2_{\text{mod},i_k,k} = \sigma^2_{y,i_k,k} + \eta^2_{i_k,k} \sum_{j=1}^{n_{\text{pred}}} \alpha_j^2 \left(\frac{\sigma_{x_j,i_k,k}}{x_{j,i_k,k}} \right)^2.$$

Although the method has general applicability, in the remainder of this paper, we only consider linear and power-law functional models and a Gaussian statistical model.

We call $\sigma_{\text{mod},i_k,k}$ the *modeled standard deviation*, as the expression (2) hinges on the model assumptions, and the likelihood (1) is the *modeled distribution*. In reality, data outliers and model uncertainty may cause the spread of the data around the fitted hypersurface to be significantly larger. Therefore, in addition to the modeled distribution, an empirical likelihood is considered, which again for simplicity we assume to be Gaussian and centered on each measured data point:

$$p(y|y_{i_k,k}, \sigma_{\text{obs},i_k,k}) = \frac{1}{\sqrt{2\pi\sigma^2_{\text{obs},i_k,k}}} \exp\left[-\frac{1}{2} \frac{(y - y_{i_k,k})^2}{\sigma^2_{\text{obs},i_k,k}} \right]. \tag{3}$$

The standard deviations $\sigma_{\text{obs},i_k,k}$, which we call the *observed standard deviations*, model the actual dispersion of the data and are learned from the data. This *observed distribution* is designed with minimal assumptions in mind (in the context of the generalized linear model, this is referred to as the *saturated model* [6]).

The next step is to minimize the geodesic distance between the modeled distributions (Equation (1)) and observed distributions (Equation (3)) for all the data points, thereby estimating the model parameters and the observed standard deviations. Because of the independence assumption, this can be written as a minimization of the sum of squared geodesic distances (GDs):

$$\{\hat{\alpha}_0, \hat{\alpha}_j, \hat{\sigma}_{\text{obs},i_k,k}\} = \underset{\alpha_0,\alpha_j \in \mathbb{R}, \sigma_{\text{obs},i_k,k} \in \mathbb{R}^+}{\text{Arg Min}} \sum_{k=1}^{n_{\text{class}}} \sum_{i_k=1}^{n_{\text{obs},k}} \text{GD}^2\Big[p(\{y_{i_k,k}\}, \{x_{j,i_k,k}\} | \{\alpha_0, \alpha_j\}) \Big| p(y|y_{i_k,k}, \sigma_{\text{obs},i_k,k}) \Big].$$

In this form, the problem is severely underdetermined, which we solve by assuming a fixed observed standard deviation $\sigma_{\text{obs},k}$ for all points from class k or a fixed relative error. For the application in fusion science, this is a reasonable assumption, as the error bars for all the measurements from a single device are fixed for each particular variable, although they can differ from one machine to another. In the astrophysics application, this is trivially the case, as only a single class is involved.

Hence, GLS acquires its robustness properties by allowing the modeled and observed distributions to differ, as demonstrated on synthetic and real data in [1]. Incidentally, it is interesting to note that by forcing the equality of the modeled and observed standard deviations (along the entire geodesic), GLS becomes equivalent to the maximum likelihood estimation (MLE), as the corresponding geodesic distance is the Mahalanobis distance [7]. Nevertheless, we stress that the technique is easily implemented, in particular owing to the closed-form expression for the geodesic distance between two univariate normal distributions with the parameters μ_1, σ_1 and μ_2, σ_2 [4]:

$$\text{GD}(\mu_1, \sigma_1 | \mu_2, \sigma_2) = \sqrt{2} \ln \frac{1+\delta}{1-\delta} = 2\sqrt{2} \tanh^{-1} \delta, \qquad \delta \equiv \left[\frac{(\mu_1 - \mu_2)^2 + 2(\sigma_1 - \sigma_2)^2}{(\mu_1 - \mu_2)^2 + 2(\sigma_1 + \sigma_2)^2} \right]^{1/2}.$$

GLS is similar to the minimum distance estimation (MDE), where the Hellinger distance is a popular similarity measure [8,9], although there are several differences. First and foremost, GLS calculates the geodesic distance between each *individual* pair of modeled and observed distributions of the response variable, corresponding to an individual measurement point. As such, each individual data point acquires the status of a probability distribution in its own right. Consequently, GLS performs regression between probability distributions on a probabilistic manifold. In contrast, MDE usually considers a distance between a kernel density estimate of the distribution of residuals, on the one hand, and the parametric model, on the other hand, based on the entire data sample. Secondly, in GLS, all parameters of the modeled distribution are explicitly modeled, which is similar to the ideas behind the link function in the generalized linear model [6]. In the present work, this is accomplished by explicitly modeling both the mean and standard deviation of the Gaussian-modeled distribution. Additionally, a final difference is that the Rao geodesic distance is used as a similarity measure, which offers the appeal of intuitive geometric interpretation.

3. GLS for Energy Confinement Scaling in Tokamaks

To introduce the application in MCF, we first briefly sketch the context and then we show the regression results, as well as a number of visualizations.

3.1. Energy Confinement in Magnetic Confinement Fusion

The field of magnetic confinement fusion aims at the development of nuclear fusion as a clean, safe and as good as inexhaustible source of energy. In MCF, a mixture of hydrogen isotopes (deuterium and tritium) is heated to thermonuclear temperatures ($\sim$15 keV, or about 1.5×10^8 K) and the resulting plasma is confined using magnetic fields [10]. The most advanced configuration, shaped as a torus, is referred to as the *tokamak*. The large-scale international ITER project, the construction of which in Southern France is nearing completion, represents the next step in this endeavor.

One of the primary remaining challenges of MCF is to improve the confinement of heat in the plasma in order to maximize fusion performance. The thermal *energy confinement time* $\tau_{\mathrm{E,th}}$ is a critical figure of merit in this regard, characterizing the time scale of the energy loss from the plasma through transport processes [11]. However, the detailed physics of the energy transport in fusion plasmas, which is of a predominantly turbulent nature, is not fully understood. Therefore, a semi-empirical approach is often taken to estimate the dependencies of $\tau_{\mathrm{E,th}}$ on the plasma conditions, by means of regression analysis in a database of measurements from multiple fusion devices. Here, a subset will be used of the recently updated international confinement database DB5.2.3-STD5, accessible through [12], containing 6233 data points from 18 tokamaks [13]. This allows for extrapolating the confinement performance to new machines, such as ITER and future fusion demonstration reactors (DEMO). In addition, these semi-empirical laws provide a reference for assessing the quality of global confinement in present-day experiments and can guide the development of theoretical models for heat transport in tokamaks.

The dependencies of the energy confinement time are usually written in the form of a power law:

$$\tau_{\mathrm{E,th}} = \alpha_0\, I_{\mathrm{p}}^{\alpha_I}\, B_{\mathrm{t}}^{\alpha_B}\, \bar{n}_{\mathrm{e}}^{\alpha_n}\, P_{\mathrm{l,th}}^{\alpha_P}\, R_{\mathrm{geo}}^{\alpha_R}\, (1+\delta)^{\alpha_{1+\delta}} \kappa_{\mathrm{a}}^{\alpha_\kappa}\, \epsilon^{\alpha_\epsilon}\, M_{\mathrm{eff}}^{\alpha_M}. \tag{4}$$

The predictor variables are the plasma current I_{p} (MA), the toroidal magnetic field B_{t} (T), the line-averaged electron density $\bar{n}_{\mathrm{e}}$ (10^{19} m^{-3}), the thermal power lost due to transport $P_{\mathrm{l,th}}$ (MW), the major radius R_{geo} (m) of the torus, the plasma shape variables δ (triangularity), κ_{a} (elongation) and ϵ (inverse aspect ratio) and, finally, the effective atomic mass M_{eff} (or isotope mass) of the plasma. The details of these variables are not essential for the remainder of the present paper, except for M_{eff}. Indeed, most data in the database were obtained from plasmas made up of hydrogen ($M_{\mathrm{eff}} = 1$) or deuterium ($M_{\mathrm{eff}} = 2$), while the ITER and fusion reactors will eventually work with a $50 - 50$ mixture of deuterium and tritium (DT; $M_{\mathrm{eff}} = 2.5$).

With the GLS regression technique, we aim to exploit information that is available about the uncertainty on the measured data. Indeed, estimates of the uncertainty on the experimentally measured database variables are provided in the confinement database in terms of percentage errors (i.e., a constant absolute error on a logarithmic scale). This includes statistical and possibly systematic uncertainty arising purely from the measurement. For the purposes of the present paper, all the measurements of the database variables are assumed to arise from an underlying normal distribution, with the standard deviation given by the absolute error derived from the corresponding measurement and its percentage error. For the effective mass M_{eff}, it was deemed better to assume a fixed absolute error for the data used in the scaling, with a typical value of 0.20.

3.2. Regression Results

The parameters of the confinement time in Equation (4) are now estimated using GLS and compared with those obtained with MLE (or the maximum a posteriori method with flat priors), the latter corresponding to GLS with a constant standard deviation along the geodesic. In this way, the advantage of allowing the standard deviation to differ between the modeled and observed distributions can clearly be shown. The modeled standard deviations, obtained from the percentage errors provided in the database, are known to vary among the different machines. However, due to the complexity of heat transport, engineering details of the devices and the relative simplicity of the regression model, it is expected that the actual uncertainty of the confinement data is significantly larger [13]. Therefore, the flexibility offered by GLS is expected to be beneficial in this case, as it allows for the observed distribution to deviate from the modeled distribution.

At this point, we note that Equation (4) is often considered on a logarithmic scale, leading to a linear relation and rendering the data approximately homoscedastic. However, as the logarithm alters the data distribution, the inference results using MLE or simple least squares may be different on the logarithmic scale vs. the original scale, if in both cases a normal error distribution is assumed [14]. In [1], it was shown that GLS is less sensitive to these statistical model assumptions than MLE. We will come back to this point in Section 4.

In the application currently under discussion, nonlinear power-law regression was carried out on the original scale. Table 1 shows the parameter estimates of the confinement scaling law, using MLE and GLS. A prediction of the confinement time for a standard ITER baseline scenario in a DT plasma is also shown [13]. It is important to note that the error bars (one standard deviation) for the parameter estimates were derived based on the assumptions of the regression model (for GLS, a bootstrapping technique was used). A more thorough uncertainty analysis has revealed that more realistic error bars are likely to be at least a factor of 10 larger [13]. This is partly due to interdependencies between the predictor variables (multicollinearity on a logarithmic scale). The analysis also suggests that, within these more realistic error bars, some of the predictor variables have no significant impact on the confinement. A subset of predictor variables thus seems more suitable for the regression, with the following emerging as the most salient features: I_p, $P_{l,th}$, R_{geo} and M_{eff}. Therefore, Table 2 shows the regression results with this reduced set of predictor variables.

From these regression experiments, a crucial difference between the results from the MLE and GLS becomes apparent. The estimated dependence of confinement on the isotope mass M_{eff} is (much) stronger with GLS than MLE. That this can potentially have important consequences for the expected confinement in ITER can clearly be seen in the predictions. In the most outspoken case in Table 2, the predicted confinement time in ITER is 30% higher according to GLS than with MLE. Caution is required here, as it has been shown in the past that the power-law model may not be sufficiently flexible to capture all the trends of the confinement time [13]. Nevertheless, a clear favorable effect of the isotope mass ('isotope effect') has been noted in earlier studies [15].

Table 1. Estimates of the parameters (intercept α_0 and exponents) and prediction toward ITER in power-law scalings with MLE and GLS of the energy confinement time using subsets of the DB5.2.3-STD5 database. All error bars correspond to one standard deviation.

	α_0	α_I	α_B	α_n	α_P	α_R	$\alpha_{1+\delta}$	α_κ	α_ϵ	α_M	$\hat{\tau}_{E,th,ITER}$
MLE	0.0432 ±0.0014	1.143 ±0.020	0.060 ±0.017	0.1914 ±0.0097	−0.7007 ±0.0075	1.511 ±0.031	0.342 ±0.046	0.785 ±0.033	−0.083 ±0.049	0.317 ±0.014	2.893 ±0.034
GLS	0.0273 ±0.0024	1.187 ±0.024	−0.008 ±0.029	0.166 ±0.015	−0.7511 ±0.0089	1.485 ±0.042	0.509 ±0.060	0.823 ±0.083	−0.365 ±0.087	0.767 ±0.096	2.996 ±0.057

Table 2. Estimates of the parameters (intercept α_0 and exponents) and prediction toward ITER in power-law scalings with MLE and GLS of the energy confinement time with a reduced set of predictor variables.

	α_0	α_I	α_P	α_R	α_M	$\hat{\tau}_{E,th,ITER}$
MLE	0.0952 ±0.0029	1.270 ±0.012	−0.6194 ±0.0079	1.182 ±0.015	0.570 ±0.026	2.717 ±0.037
GLS	0.0280 ±0.0028	1.124 ±0.016	−0.6605 ±0.0097	1.389 ±0.019	2.41 ±0.13	3.535 ±0.098

3.3. Visualizations

One of the main advantages of geometric approaches to statistics and machine learning is the possibility to visualize methods and results and to interpret the results geometrically. We here explore several techniques to visualize or approximate the manifold of univariate Gaussian distributions, which is well known to possess constant negative scalar curvature, corresponding to hyperbolic geometry [16].

The first visualization is the Poincaré half-plane, pictured in Figure 1 (a number of data points at a high confinement time have been excluded to improve the visualization). The geodesics are half-circles ending on the horizontal axis (as well as vertical lines with constant μ). Hence, a characteristic of the geodesics is their route through a region of larger standard deviation, relative to that of the end points.

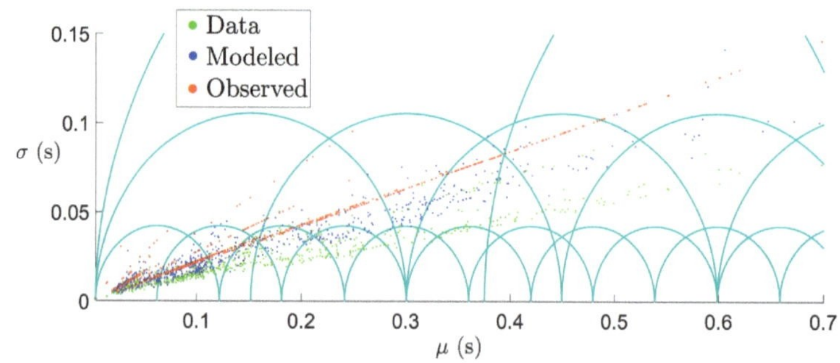

Figure 1. Illustration of the Poincaré half-plane with several half-circle geodesics in the background. The confinement time data (value and error bar) are plotted in the half-plane, as well as the modeled and predicted values obtained by GLS.

The downside of the Poincaré half-plane is that the Euclidean metric in that plane is, of course, different from the metric on the manifold, which causes visual discrepancies when judging the distance between points. For this reason, we present another visualization using a partial immersion of the manifold in three-dimensional Euclidean space, as shown in Figure 2a (in the interest of the visualization, the surface has been rescaled by a factor of 10 in the μ direction and some data points with low confinement times have been excluded). A rendering is shown of one blade of a particular pseudosphere, namely, the *tractroid*, which is locally isometric to the actual manifold for $\sigma > 1$. The parallels of the tractroid are lines of constant standard deviation σ, while the meridians (the tractrices) are lines of constant mean μ. This representation of the normal manifold is periodic in the μ direction. The generally higher standard deviation of the observed distributions, compared to the measured confinement times, is a sign of additional sources of uncertainty.

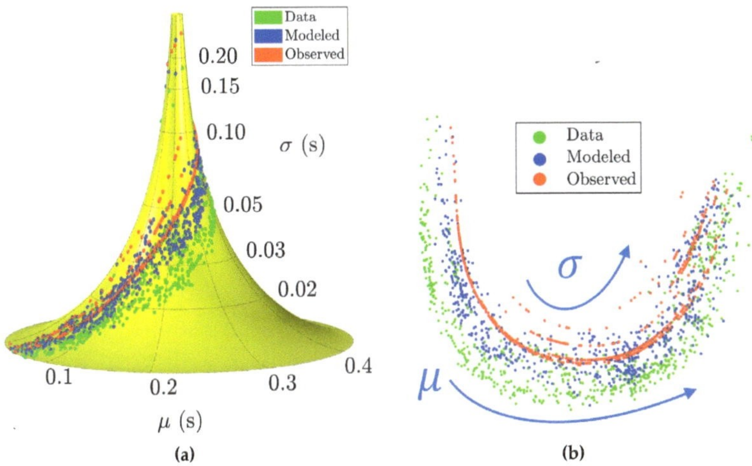

Figure 2. (**a**) Visualization of the confinement time data on a pseudosphere, as well as the modeled and observed distributions obtained by GLS. (**b**) Projection of the confinement data using MDS, roughly indicating the directions of increasing μ and σ.

A visualization in three-dimensional space clearly also has its disadvantages, which is addressed by the final visualization in Figure 2b. This is a projection on a two-dimensional Euclidean plane that is locally approximately isometric to the original manifold. It has been obtained using multidimensional scaling (MDS), which aims at preserving the pairwise distances between all points: geodesic distances on the manifold are approximated by Euclidean distances in the plane.

4. GLS for Tully–Fisher Scaling in Astrophysics

The second application that we consider aims to give a very simple additional illustration of the robustness of GLS and the interpretation thereof.

4.1. The Baryonic Tully–Fisher Relation

The baryonic Tully–Fisher relation (BTFR) between the total (stellar + gaseous) baryonic mass M_b of disk galaxies and their rotational velocity V_f is of fundamental importance in astrophysics and cosmology [17]. It is a remarkably simple and tight empirical relation of the form

$$M_\mathrm{b} = \alpha_0 V_\mathrm{f}^{\alpha_V}. \tag{5}$$

The BTFR serves as a tool for determining cosmic distances, provides constraints on galaxy formation and evolution models and serves as a test for the Lambda cold dark matter paradigm (ΛCDM) in cosmology. In this scaling problem, we use data from 47 gas-rich galaxies, as detailed in [17]. The data also contain estimates of the measurement errors (38% on M_b and 10% on V_f), which again we treat here as a single standard deviation.

4.2. Regression Results

In this application, we demonstrate the robustness of GLS against outliers, by comparing between the loglinear and nonlinear regression of the BTFR. The scalings obtained with GLS are shown in Figure 3, reflecting consistent estimates for both regressions. In particular, whereas the data point corresponding to the largest V_f and M_b does not have the characteristics of an outlier on the logarithmic scale, it may be considered as such on the original scale. Nevertheless, the trend estimated by GLS seems not significantly affected by this outlier.

Furthermore, the relative error derived from the observed standard deviation estimated by GLS amounts to 63%, which is considerably larger than the value of ca. 40% predicted by the model. Again, this is an indication that the scatter on the scaling law is not due to measurement error alone, which has particular significance in this application, as it may provide evidence for the ΛCDM vs. so-called modified Newtonian dynamics (MOND) cosmological models.

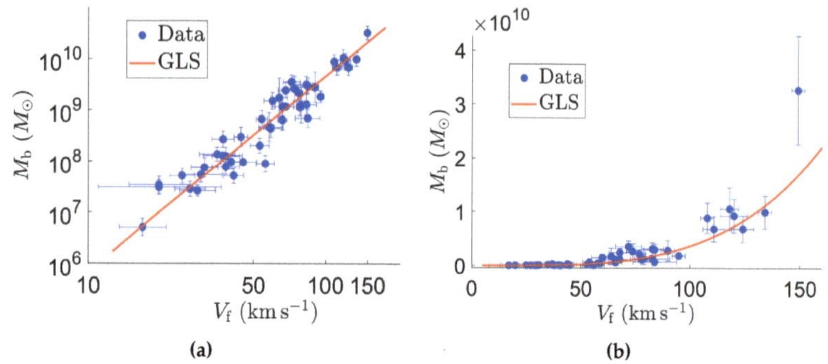

Figure 3. Fit of the BTFR by GLS (**a**) on a logarithmic scale and (**b**) on the original scale of the data.

4.3. Visualization

It is interesting to visualize the data and the fit by GLS, as this provides insight into the way GLS succeeds in ignoring the outlier. This is illustrated in Figure 4, showing the Poincaré half-plane with the BTFR data, as well as the modeled and observed distributions obtained by GLS. The geodesic between the modeled outlier and the data point is shown (dashed curve), as well as that between the modeled outlier and the corresponding observed distribution (full curve). The latter is (slightly) shorter, because the increased observed standard deviation leads to a decrease in the geodesic distance.

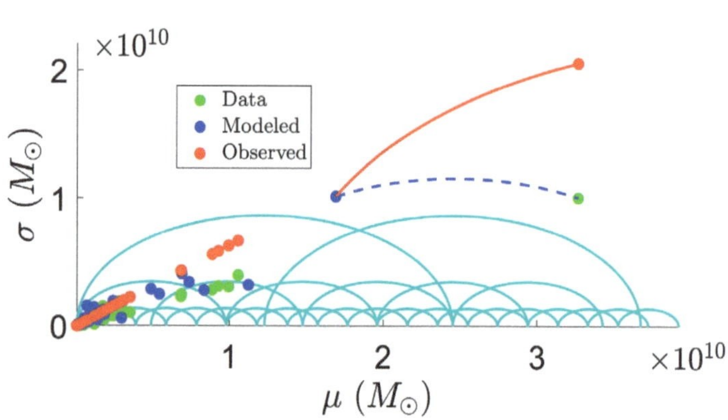

Figure 4. The BTFR data in the Poincaré half-plane, with geodesics considered by GLS shown for the case of the outlier.

5. Conclusions

Geodesic least squares regression has been presented, demonstrating the robustness of the method in two applications in the physical sciences. Essentially a least squares technique, it is easily implemented, which, as we have stressed, is a key asset for the adoption of the method in scientific disciplines where advanced methods of statistics and machine learning are not yet widely used. The data and results can be visualized in several ways, offering additional options for interpretation.

We have applied the GLS technique to the energy confinement scaling in fusion devices of the tokamak design based on an international multi-machine database. A new regression by GLS with a reduced set of predictor variables suggests an isotope effect that is much more favorable than predicted by regular MLE. This could lead to significantly improved confinement in the next-step fusion device ITER. A second application of GLS has been aimed at demonstrating the robustness against outliers, with a corresponding intuitive geometrical interpretation.

As GLS combines simplicity with good robustness properties, it can be of particular use for practitioners in various application domains. Requiring little to no tuning or learning of hyperparameters, it can be readily employed without expert knowledge about the details of the method.

Funding: This research received no external funding

Institutional Review Board Statement: Not applicable.

Informed Consent Statement: Not applicable.

Data Availability Statement: The International Global H-mode Confinement Database is openly available here: https://osf.io/drwcq (accessed on 12 July 2023). The data used in this study to fit the Tully-Fisher scaling are available in [17].

Conflicts of Interest: The author declares no conflict of interest.

References

1. Verdoolaege, G. A New Robust Regression Method Based on Minimization of Geodesic Distances on a Probabilistic Manifold: Application to Power Laws. *Entropy* **2015**, *17*, 4602–4626. [CrossRef]
2. Amari, S.; Nagaoka, H. *Methods of Information Geometry*; American Mathematical Society: Providence, RI, USA, 2000.
3. Nielsen, F. An elementrary introduction to information geometry. *Entropy* **2020**, *22*, 1100. [CrossRef] [PubMed]
4. Burbea, J.; Rao, C. Entropy differential metric, distance and divergence measures in probability spaces: A unified approach. *J. Multivar. Anal.* **1982**, *12*, 575–596. [CrossRef]
5. Von Toussaint, U. General Hyperplane Prior Distributions Based on Geometric Invariances for Bayesian Multivariate Linear Regression. *Entropy* **2015**, *17*, 3898–3912. [CrossRef]
6. McCullagh, P.; Nelder, J. *Generalized Linear Models*; Chapman & Hall/CRC: Boca Raton, FL, USA, 1989.
7. Rao, C.; Nagaoka, H. Differential metrics in probability spaces. In *Differential Geometry in Statistical Inference*; Amari, S.I., Barndorff-Nielsen, O.E., Kass, R.E., Lauritzen, S.L., Rao, C.R., Eds.; Institute of Mathematical Statistics: Hayward, CA, USA, 1987; Chapter 5, pp. 217–240.
8. Beran, R. Minimum Hellinger distance estimates for parametric models. *Ann. Stat.* **1977**, *5*, 445–463. [CrossRef]
9. Pak, R. Minimum Hellinger distance estimation in simple regression models; distribution and efficiency. *Stat. Probab. Lett.* **1996**, *26*, 263–269. [CrossRef]
10. Freidberg, J. *Plasma Physics and Fusion Energy*; Cambridge University Press: Cambridge, UK, 2008.
11. ITER Physics Expert Group on Confinement and Transport; ITER Physics Expert Group on Confinement Modelling and Database; ITER Physics Basis Editors. Chapter 2: Plasma confinement and transport. *Nucl. Fusion* **1999**, *39*, 2175–2249. [CrossRef]
12. H-Mode Database Working Group. International Global H-Mode Confinement Database. 2021. Available online: https://osf.io/drwcq (accessed on 12 July 2023).
13. Verdoolaege, G.; Kaye, S.M.; Angioni, C.; Kardaun, O.J.W.F.; Maslov, M.; Romanelli, M.; Ryter, F.; Thomsen, K.; the ASDEX Upgrade Team; the EUROfusion MST1 Team. The updated ITPA global H-mode confinement database: Description and analysis. *Nucl. Fusion* **2021**, *61*, 076006. [CrossRef]
14. Xiao, X.; White, E.P.; Hooten, M.B.; Durham, S.L. On the use of log-transformations vs. nonlinear regression for analyzing biological power laws. *Ecology* **2011**, *92*, 1887–1894. [CrossRef] [PubMed]
15. Maggi, C.; Weisen, H.; Hillesheim, J.C.; Chankin, A.; Delabie, E.; Horvath, L.; Auriemma, F.; Carvalho, I.S.; Corrigan, G.; Flanagan, J.; et al. Isotope effects on L-H threshold and confinement in tokamak plasmas. *Plasma Phys. Control. Fusion* **2018**, *60*, 014045. [CrossRef]
16. Nielsen, F.; Nock, R. Visualizing hyperbolic Voronoi diagrams. In Proceedings of the 30th Annual Symposium on Computational Geometry (SOCG'14), Kyoto, Japan, 8–11 June 2014; pp. 90–91.

17. McGaugh, S. The baryonic Tully-Fisher relation of gas-rich galaxies as a test of ΛCDM and MOND. *Astron. J.* **2012**, *143*, 4 [CrossRef]

physical sciences forum

MDPI

Proceeding Paper

Magnetohydrodynamic Equilibrium Reconstruction with Consistent Uncertainties †

Robert Köberl [1,2,*], Robert Babin [1,3] and Christopher G. Albert [3]

1 Max-Planck-Institute for Plasma Physics, 85748 Garching, Germany; robert.babin@ipp.mpg.de
2 School of Computation, Information and Technology (CIT), Technical University of Munich (TUM), 85748 Garching, Germany
3 Fusion@ÖAW, Institut für Theoretische Physik Computational Physics (ITPcp), Technische Universität Graz (TU Graz), 8010 Graz, Austria; albert@tugraz.at
* Correspondence: robert.koeberl@ipp.mpg.de

† Presented at the 42nd International Workshop on Bayesian Inference and Maximum Entropy Methods in Science and Engineering, Garching, Germany, 3–7 July 2023.

Abstract: We report on progress towards a probabilistic framework for consistent uncertainty quantification and propagation in the analysis and numerical modeling of physics in magnetically confined plasmas in the stellarator configuration. A frequent starting point in this process is the calculation of a magnetohydrodynamic equilibrium from plasma profiles. Profiles, and thus the equilibrium, are typically reconstructed from experimental data. What sets equilibrium reconstruction apart from usual inverse problems is that profiles are given as functions over a magnetic flux derived from the magnetic field, rather than spatial coordinates. This makes it a fixed-point problem that is traditionally left inconsistent or solved iteratively in a least-squares sense. The aim here is progressing towards a straightforward and transparent process to quantify and propagate uncertainties and their correlations for function-valued fields and profiles in this setting. We propose a framework that utilizes a low-dimensional prior distribution of equilibria, constructed with principal component analysis. A surrogate of the forward model is trained to enable faster sampling.

Keywords: inverse problem; fixed-point problem; Bayesian analysis; dimensionality reduction; polynomial chaos expansion; uncertainty quantification; application

1. Introduction

A frequent starting point for calculations in magnetic confinement fusion is the magnetohydrodynamic equilibrium. Ideal magnetohydrodynamics (MHD) [1] describes plasma as a single quasi-neutral fluid under the assumptions of infinite electrical conductivity, small ion gyroradius and negligible electron momentum. An MHD equilibrium is a solution to the static MHD equations (Equations (1)–(3)), describing the magnetic field ($\boldsymbol{B}$) and its relation to the current density ($\boldsymbol{J}$) and the pressure (p):

$$\boldsymbol{J} \times \boldsymbol{B} = \nabla p, \tag{1}$$

$$\nabla \times \boldsymbol{B} = \mu_0 \boldsymbol{J}, \tag{2}$$

$$\nabla \cdot \boldsymbol{B} = 0. \tag{3}$$

The solutions to Equations (1)–(3) are typically obtained in magnetic flux coordinates [2], curvilinear coordinates that greatly simplify calculations, for example, by aligning in such a way that $\boldsymbol{B}$ appears straight. A large part of a typical MHD solution is, therefore, the mapping from flux coordinates to real-space coordinates. Inferring an MHD equilibrium from experimental diagnostics, a task known as equilibrium reconstruction, is of great interest for fusion experiments, to gain insight into unmeasurable or hard-to-measure quantities [3], such as flux surface geometry.

This task is routinely performed for tokamaks [4,5], where the toroidal symmetry further simplifies the MHD equations but proves to be more complicated for three dimensional (3D) systems, like stellarators. Typically, MHD solvers take the pressure and toroidal current profile as functions of a radial flux coordinate ρ as input and solve for the optimal geometry of the magnetic flux surfaces. From the resulting equilibrium configuration, synthetic diagnostic signals can be calculated and compared to the corresponding real physical measurements. This makes 3D equilibrium reconstruction a fixed-point problem that is traditionally solved iteratively in a least-squares sense [3,6,7]. While the least-squares approach can result in suitable reconstructed parameters and uncertainty estimates [8], evaluations of the forward model are computationally demanding (a single equilibrium reconstruction can take up to several hours [9,10]), and common error estimates might fail to capture relevant uncertainties when high levels of noise are present for the equilibrium diagnostics [11]. Existing Bayesian frameworks [12] are also limited by the high computational costs of the forward model. Therefore, we aim to build a Bayesian framework for 3D equilibrium reconstruction that allows for fast sampling from the posterior distribution of equilibria while maintaining relevant physical constraints. In this work, we present, as a proof of concept, such a framework for a configuration of the Wendelstein 7-X stellarator (W7-X).

2. Methods

For the proposed equilibrium reconstruction approach using the Bayesian formalism, two major components are required: a prior distribution of equilibria and a fast evaluation of the likelihood function, both of which are not readily accessible. A schematic depiction of the steps taken to obtain these major components is shown in Scheme 1. To define the prior distribution of equilibria, a physically motivated prior distribution over equilibrium-defining parameters is formulated (Section 2.1). Samples from this prior distribution are then propagated into the space of equilibria using the Variational Moments Equilibrium Code (VMEC) [13] (Section 2.2) and subsequently into the space of synthetic equilibrium diagnostic signals with the code DIAGNO [14]. Since the space of equilibria is very high-dimensional, dimensionality reduction is performed on the equilibria using principal component analysis (PCA) (Section 2.3). In this way, a low-dimensional distribution over equilibrium configurations, which enables the training of polynomial chaos expansion (PCE) [15] surrogate models, can be defined (Section 2.4). These models map low-dimensional parameters c_i onto the space of synthetic equilibrium diagnostic signals and thus circumvent high computational costs arising from the DIAGNO forward model. Through the combination of the surrogates with the low-dimensional equilibrium prior, the low-dimensional posterior distribution of equilibrium parameters can be inferred using Markov Chain Monte Carlo (MCMC) sampling in a Bayesian formalism, based on a set of (synthetic) measurement signals (Section 3).

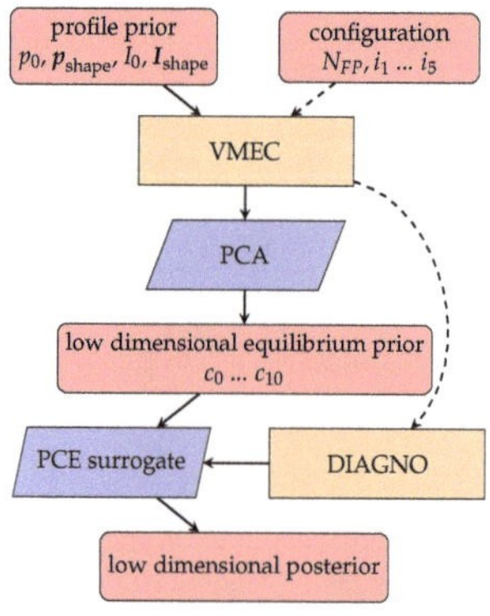

Scheme 1. Overview of the proposed framework.

2.1. Prior Distribution of Current and Pressure Profiles

We define a prior distribution for pressure $p(s)$ and toroidal current $I(s)$ profiles. They are functions of normalized toroidal flux $s = \frac{\psi}{\psi_0}$ and define an MHD equilibrium together with the total toroidal flux (ψ_0). Here, ψ denotes the toroidal flux. Often, reconstruction methods utilize simple parameterized functions [3,7,8], for example, two-power profiles, for describing equilibrium profiles. In order not to restrict the method to only favorable profiles, a broader prior is desired. Following similar concepts already realized in the MINERVA framework [10,12,16], we propose the use of a Gaussian Process [17] (GP) as a function generator for the profile gradients. This allows us to more easily specify the length scales and smoothness constraints based on physics information. For the pressure profile, the correlation length scales are larger at the core of the plasma than at the boundary, which is encoded in the GP kernel used in the MINERVA framework for W7-X [10]. We further constrain the pressure profile to be monotonic and 0 at the plasma boundary and enforce these constraints with exponentiation and integration, as well as a renormalization. For the current profile GP, a smooth rational quadratic kernel [17] is used, since detailed information on the current profile shape can be hard to capture with magnetic diagnostics alone [11] and the toroidal current is expected to be comparatively low for standard W7-X equilibria. Again, exponentiation and integration ensure monotonicity, but in contrast to the pressure profile, the current is 0 at the magnetic axis, and the scaling factor can be positive or negative. Using exponentiation as a positivity transform for the profile gradients allows for steep gradients but discourages many local plateaus that are found when using the absolute value as the transform.

$$\tilde{p}(s) \sim GP_{\text{Minerva}} \qquad p(s) = \frac{p_0}{\int_0^1 \exp(\tilde{p})\, ds'} \int_s^1 \exp(\tilde{p})\, ds' \tag{4}$$

$$\tilde{I}(s) \sim GP_{\text{RQ}} \qquad I(s) = \frac{I_0}{\int_0^1 \exp(\tilde{I})\, ds'} \int_1^s \exp(\tilde{I})\, ds' \tag{5}$$

Finally, the GP is approximated with a PCA to allow for the space-filling latin hypercube sampling of the parameter space. Six shape parameters for the pressure profile and three shape parameters for the current profile correspond to an explained variance ratio of 99% each. Combined with a parameter ψ_0 for the total toroidal magnetic flux, this gives us 12 equilibrium-defining parameters, as detailed in Table 1. Several realizations of the equilibrium profile shapes obtained with the described procedure are visualized in Figure 1. From the prior defined in Table 1, 4000 samples are drawn and propagated into the space of MHD equilibria using VMEC.

Table 1. The prior parameters and their distribution.

Symbol	Distribution	Description
ψ_0	Uniform in $[-2.5\ \text{Wb}, -1.6\ \text{Wb}]$	Total toroidal magnetic flux
p_0	Uniform in $[0, 200\ \text{kPa}]$	Pressure at the magnetic axis
p_{shape}	$\mathcal{GP}_{\text{Minerva}}\begin{pmatrix} \sigma_f & = 1 \\ \lambda_{\text{core}} & = 0.2 \\ \lambda_{\text{edge}} & = 0.1 \\ s_0 & = 0.9 \\ s_w & = 0.1 \end{pmatrix}$	Pressure profile shape factors reduced to 6 principal components
I_0	Uniform in $[-10\ \text{kA}, 10\ \text{kA}]$	Total toroidal current
I_{shape}	$\mathcal{GP}_{\text{RQ}}\begin{pmatrix} \sigma & = 2 \\ \lambda & = 0.2 \\ \alpha & = 2 \end{pmatrix}$	Current profile shape factors reduced to 3 principal components

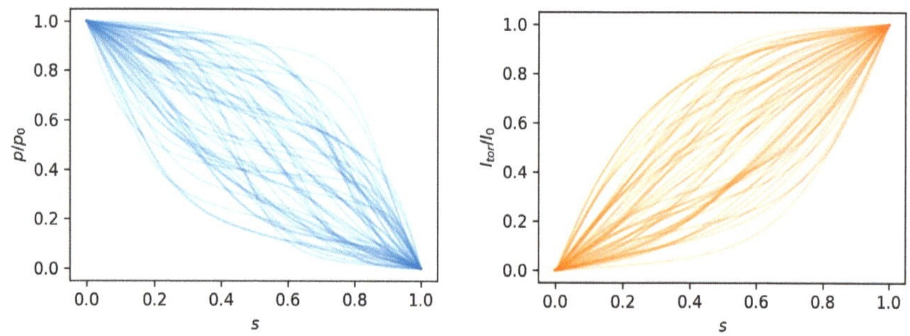

Figure 1. Samples drawn from the prior distribution of (**left**) the pressure profile shape and (**right**) the current profile shape.

2.2. VMEC

VMEC solves Equations (1)–(3) using a double-Fourier basis for the inverse coordinate representation $x(\rho,\theta,\zeta) = (R, Z, \phi)$. Here, $\rho = \sqrt{s}$ denotes the magnetic flux surfaces, and θ and ζ are poloidal and toroidal angle-like flux coordinates. VMEC uses an energy functional in place of the local formulation, which is guaranteed to converge to an equilibrium in the absence of magnetic islands. This is not guaranteed for 3D geometries like stellarators [2], but VMEC can provide a suitable approximation of magnetic flux surfaces. In contrast to many other equilibrium codes, VMEC does not require a fixed plasma boundary but can calculate equilibrium flux surfaces, including the last closed flux surface, given a coil current configuration $[i_1 ... i_5]$, as well as ψ_0, $p(s)$ and $I(s)$.

A VMEC equilibrium can be fully described by the Fourier series for cylindrical coordinates $R(\rho,\theta,\zeta)$, $Z(\rho,\theta,\zeta)$ and potential $\lambda(\rho,\theta,\zeta)$, rotational transform profile $\iota(\rho) = \frac{d\chi}{d\psi}$ and total toroidal flux $\psi_0 = \psi(\rho = 1)$. Here, χ denotes the poloidal flux, and the cylindrical angle (ϕ) is a periodic continuation of the toroidal angle (ζ) with the number of field periods (N_{FP}). The straight-field-line angle ($\theta^\star = \theta + \lambda(\rho,\zeta,\theta)$) can be calculated from λ. With stellarator symmetry, the Fourier series for R only contains cosine components, while Z and λ only contain sine components. For the radial direction, VMEC uses a finite difference scheme and a linear spacing in the normalized toroidal flux $s = \rho^2 = \frac{\psi}{\psi_0}$.

The pressure profile ($p(\rho)$) can be calculated with a weak formulation of the radial force balance [13] under the assumption that Equation (1) is fulfilled:

$$p(\rho) = \int_\rho^1 -\frac{1}{\mu_0 \langle \mathcal{J} \rangle} \frac{d\psi}{d\rho} \left(\frac{d}{d\rho} \langle B_\zeta \rangle + \iota \frac{d}{d\rho} \langle B_\theta \rangle \right) d\rho \tag{6}$$

where $\langle Q \rangle = \frac{1}{4\pi^2} \iint Q \, d\theta d\zeta$ denotes an average over a flux surface and $\mathcal{J}$ is the determinant of the Jacobian.

2.3. Dimensionality Reduction

With $S = 99$ radial points, $M = 25$ poloidal modes and $N = 12$ toroidal modes, a VMEC equilibrium can be described using $S \cdot M \cdot N$ parameters for R, Z and λ, respectively; S parameters for ι; and a single parameter for ψ_0. With a total of 85,636 non-zero parameters, this space is difficult to handle with Bayesian methods. To reduce the dimensionality, principal component analysis (PCA) was applied on the VMEC equilibria.

To compensate for the incomparable units of the five quantities, they were all rescaled such that the total variance of each group of parameters was 1. For the total explained variance ratio of 99.9%, it was found that 11 principal components were sufficient to represent the equilibrium prior. The comparatively low dimensionality now facilitates Bayesian inference directly on the equilibrium parameters and also significantly reduces the number of data needed to represent the posterior distribution or a collection of samples.

For each of the five quantities R, Z, λ, ι and ψ_0, the reconstruction error due to dimensionality reduction was evaluated using the L^2 metric (Figure 2). The relative error stayed below 1% for all quantities except λ. This lead to lower quality of the direction of $\boldsymbol{B}$ and derived quantities like p. The distribution of the low-dimensional parameters (c_i) was approximated with kernel density estimation (KDE) and assumed to be independent. This assumption held well enough and marginally increased the width of the total prior.

In Figure 3, the pressure profile is shown for a validation sample drawn from the prior distribution, together with a pressure profile calculated with Equation (6), and the same calculation applied after the validation sample was pushed through the dimensionality reduction. Indeed, the pressure profile can be calculated from the five quantities R, Z, λ, ι and ψ_0, which determine $\boldsymbol{J}$ and $\boldsymbol{B}$, with only a small deviation near the magnetic axis, where VMEC is known to be inaccurate. However, dimensionality reduction significantly reduces the accuracy of this calculation.

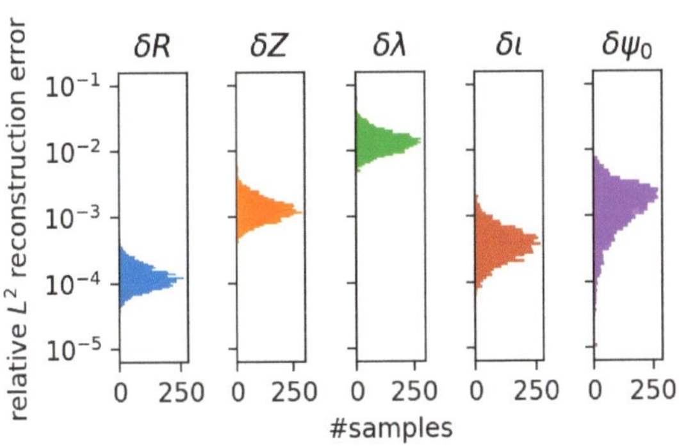

Figure 2. Relative L^2 error of the low-dimensional equilibrium representation, evaluated for the samples from the prior.

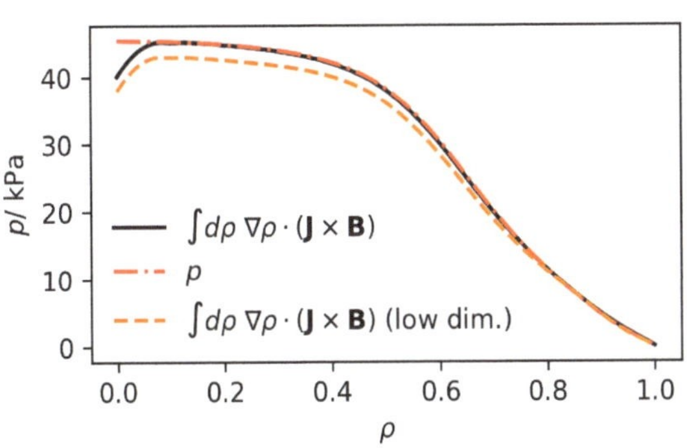

Figure 3. Input pressure profile p compared with the pressure derived from $\nabla p = \boldsymbol{J} \times \boldsymbol{B}$ for a validation sample drawn from the prior and its low-dimensional representation.

2.4. Synthetic Diagnostics Surrogate Model

We trained polynomial chaos expansion (PCE) surrogate models that map from equilibrium coefficients c_i to magnetic equilibrium diagnostics of W7-X. The aim of these surrogate models was to circumvent prohibitively high computational costs arising from DIAGNO forward model evaluations, which occur during sampling. The magnetic equilibrium diagnostics considered were saddle coils, diamagnetic loops and segmented Rogowski coils [18]. PCEs were built using the linear regression methods provided in the `chaospy` [19] Python package. The polynomials used in the expansions were chosen through hyperbolic truncation. In this truncation scheme, all multivariate polynomials, where the associated degrees of the univariate polynomials α_i fulfill [20]

$$d \geq \left(\sum_i \alpha_i^h \right)^{1/h}, \tag{7}$$

are retained. Here, d denotes the maximum polynomial degree occurring in the expansion, and $h \in (0,1]$ determines the number of retained polynomials. While several other methods for choosing the basis polynomials in a sparse sense exist [21], we found that the hyperbolic truncation scheme with $d = 6$ and $h = 0.8$ worked well in the considered case. To estimate the performance of the resulting PCEs, the relative leave-one-out error ε_{loo}^R [22] was calculated for each surrogate model. The resulting ε_{loo}^R for these PCEs are visualized in Figure 4, grouped in terms of the synthetic diagnostic signal that is being approximated. Note that ε_{loo}^R is given relative to the variance present in the training data.

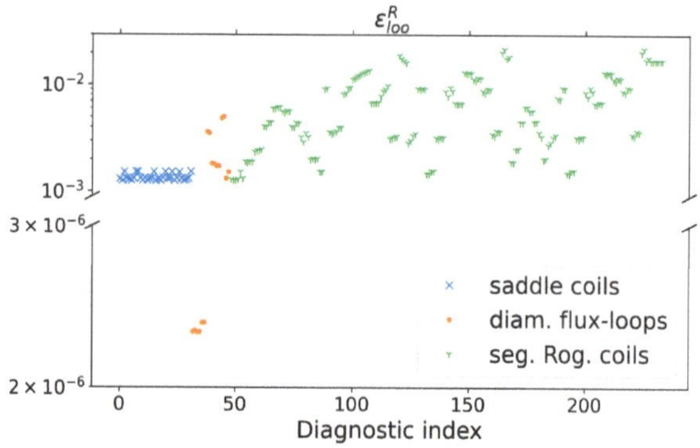

Figure 4. Relative leave-one-out error ε_{loo}^R of the polynomial chaos expansion surrogate models, mapping from the low-dimensional equilibrium parameters to the synthetic diagnostics.

3. Results

With the previously discussed prior on the low-dimensional parameters (c_i) and the synthetic diagnostic surrogate, we tried to reconstruct a W7-X standard configuration equilibrium from a set of corresponding synthetic magnetic equilibrium diagnostic signals.

A set of equilibrium-defining parameters from the distribution discussed in Section 2.1 were propagated using the VMEC/DIAGNO forward model, providing a known ground truth. The resulting synthetic magnetic equilibrium diagnostic signals were then used together with a Gaussian likelihood function and the low-dimensional equilibrium prior to estimate the posterior distribution of c_i. For the likelihood function, the synthetic diagnostics were assumed to be uncorrelated with a 5% error. Two methods for estimating the posterior were used: Laplace's approximation and MCMC sampling. In both methods the evaluation of the synthetic magnetic equilibrium diagnostics was performed with the PCE surrogate models. For the MCMC method, samples were drawn using NUTS [23]. Figure 5 depicts one-dimensional projections of the posterior estimates for the parameters c_i with respect to the associated prior distribution. Seen with respect to the prior width, there is a high uncertainty in the posterior distribution of parameters c_4 to c_{10}. However, they contribute little to the explained variance of the equilibrium PCA. To visualize how this estimated uncertainty appears in the physically relevant space, the maximum a posteriori estimate (MAP) and several samples were transformed back into the space of VMEC equilibria. The flux surface geometry, total toroidal flux and rotational transform profile are visualized in Figures 6–8, respectively. In addition to the reconstructed equilibria, the ground-truth evaluation of VMEC is visualized. One can observe that the MAP estimate is comparatively close to the ground-truth values and that for all three quantities, the ground-truth value is covered within the area spanned by the posterior samples.

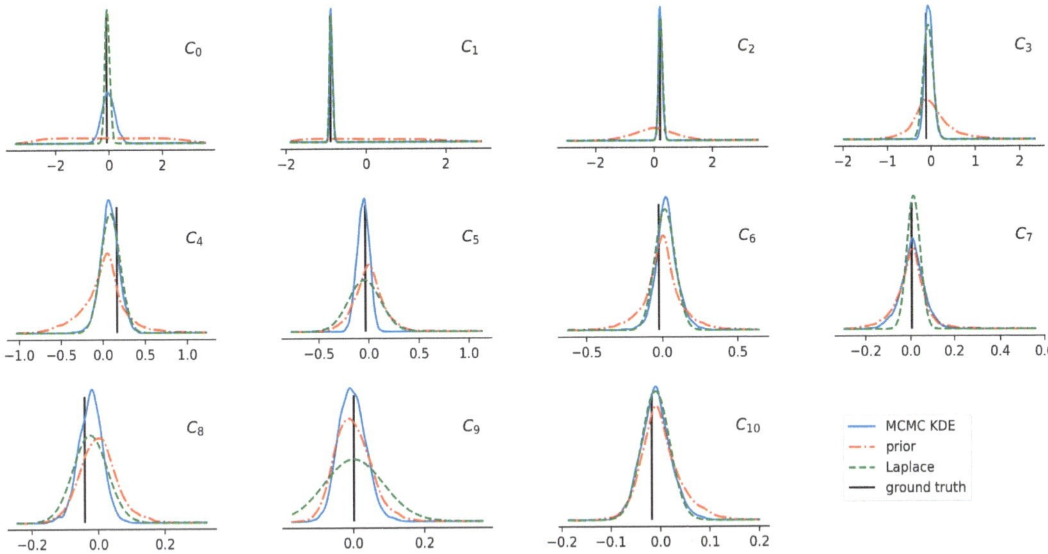

Figure 5. One-dimensional projections of the low-dimensional equilibrium parameter prior distribution, kernel density estimation (KDE) and Laplace's approximation of the posterior distribution.

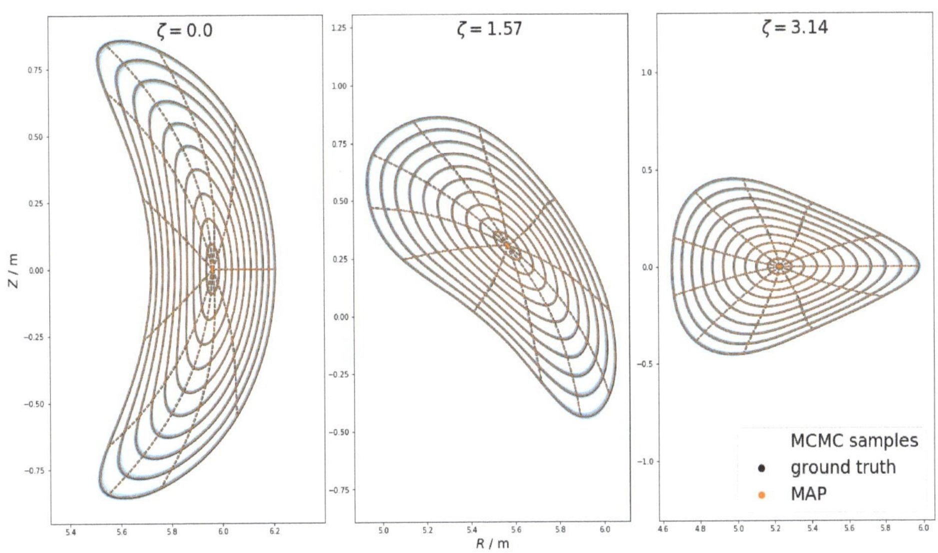

Figure 6. Posterior samples of the flux surface geometry in comparison to the ground truth. The solid lines are linearly spaced contours of constant ρ, and the dashed lines are constant-$\theta^\star$ contours.

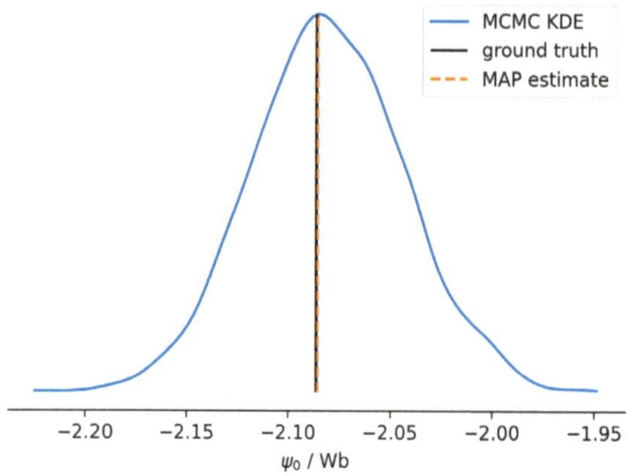

Figure 7. Posterior kernel density estimation (KDE) of the total toroidal flux (ψ_0).

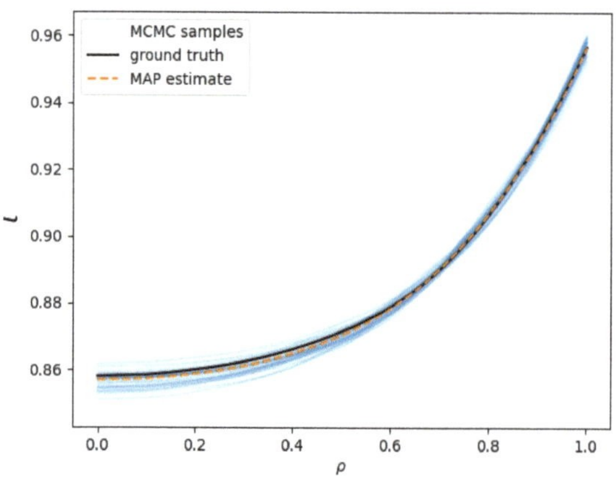

Figure 8. Rotational transform profiles $\iota(\rho)$ drawn from the posterior distribution in comparison to the ground truth.

With Equation (6), the radial pressure profile of the posterior samples can be estimated (Figure 9). Significant differences between the MAP estimate and the ground-truth profile can be observed. However, the ground-truth profile is still covered within the space spanned by the posterior samples, and the magnetic diagnostics used for this reconstruction are rather insensitive to the pressure profile. The systematic discrepancy close to the magnetic axis can be attributed to the use of VMEC, which is known to have problems in this region. Additionally, the profile shapes expose nonphysical values near the last closed flux surface, as well as non-monotonicity, which was not present in the pressure profile prior. One reason for this, as shown above in Figure 3, is that PCA dimensionality reduction affects the derived pressure profile more than magnetic quantities, like the rotational transform profile (Figure 2).

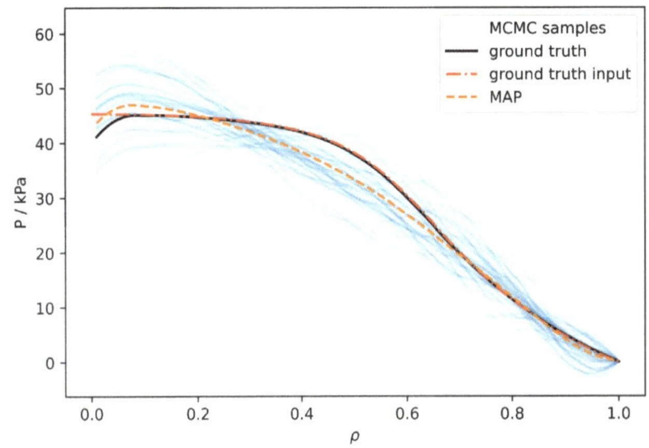

Figure 9. Pressure profiles calculated from the posterior using $\nabla p = \boldsymbol{J} \times \boldsymbol{B}$ together with the ground-truth pressure profile.

4. Discussion

The results presented in Section 3 show that the presented framework is able to reconstruct an equilibrium configuration drawn from the prior distribution. That is, the ground truth lies within the region of high probability of the estimated posterior distributions, and the MAP estimate closely matches the true flux surface geometry and rotational transform profile. While the MAP estimate and posterior mean are found to be equivalent, Laplace's approximation is not sufficient to model the posterior. Quantities that are derived from the magnetic quantities R, Z, λ, ι and ψ_0, such as pressure profile $p(\boldsymbol{J}, \boldsymbol{B})$ are not reconstructed well, partially due to dimensionality reduction, but also because the considered magnetic diagnostics are not very sensitive to the pressure profile. While the derived pressure profiles fulfill the weak force balance by construction, they lack the physically motivated constraints that are encoded into the profile prior distribution in Section 2.1. In particular, they can violate the positivity constraint. However, the true profile does lie within the posterior distribution, and a pressure-sensitive diagnostic could be used to determine it, particularly as the mapping between real-space and flux coordinates is reconstructed well.

The presented framework offers a flexible prior distribution of profiles but directly reconstructs the flux surface geometry from magnetic diagnostics, allowing samples of the flux surface geometry to be drawn from the posterior. The equilibrium prior has to be constructed once for each device configuration, e.g., coil currents, but it offers reusability from this point onward and can be well described with only a few principal components. This not only allows Bayesian methods to be applied effectively but also significantly reduces the number of data needed to describe a posterior distribution or collection of samples. The PCE surrogate model similarly reduces the computational cost of 3D equilibrium reconstruction and offers fast access to uncertainties.

With the presented methods, we provide a proof of concept for a fast Bayesian equilibrium reconstruction framework. The accurate reconstruction of the derived quantities and thus the physical consistency of the posterior samples warrant further investigation. This could be achieved with different dimensionality reduction methods, for example, using a variational autoencoder with additional physics constraints, or different surrogate models, as well as more accurate MHD solvers.

Author Contributions: Conceptualization, R.K., R.B. and C.G.A.; methodology, R.K., R.B. and C.G.A.; software, R.K. and R.B.; investigation, R.K. and R.B.; data curation, R.K. and R.B.; writing—original draft preparation, R.K. and R.B.; writing—review and editing, R.K., R.B. and C.G.A.; visualization, R.K. and R.B.; supervision, R.K. and C.G.A.; project administration, C.G.A.; funding acquisition, C.G.A. All authors have read and agreed to the published version of the manuscript.

Funding: The present contribution was supported by the Helmholtz Association of German Research Centers under the joint research school HIDSS-0006 'Munich School for Data Science - MUDS'. This work was carried out within the framework of the EUROfusion Consortium, funded by the European Union via the Euratom Research and Training Programme (Grant Agreement No. 101052200-EUROfusion). Views and opinions expressed are, however, those of the authors only and do not necessarily reflect those of the European Union or the European Commission. Neither the European Union nor the European Commission can be held responsible for them.

Institutional Review Board Statement: Not applicable.

Informed Consent Statement: Not applicable.

Data Availability Statement: The data presented in this study are openly available via Zenodo at https://doi.org/10.5281/zenodo.8063360 (accessed on 21 June 2023).

Acknowledgments: The authors thank Udo von Toussaint, Florian Hindenlang and Omar Maj for insightful discussions, and the reviewers for their helpful remarks.

Conflicts of Interest: The authors declare no conflict of interest.

References

1. Freidberg, J.P. *Ideal MHD*; Cambridge University Press: New York, NY, USA, 2014.
2. D'haeseleer, W.D.; Hitchon, W.N.G.; Callen, J.D.; Shohet, J.L. *Flux Coordinates and Magnetic Field Structure*; Springer Berlin/Heidelberg, Germany, 1991. [CrossRef]
3. Lazerson, S.; The DIII-D Team. Three-Dimensional Equilibrium Reconstruction on the DIII-D Device. *Nucl. Fusion* **2015**, *55*, 023009. [CrossRef]
4. Fischer, R.; Fuchs, C.J.; Kurzan, B.; Suttrop, W.; Wolfrum, E.; ASDEX Upgrade Team. Integrated Data Analysis of Profile Diagnostics at ASDEX Upgrade. *Fusion Sci. Technol.* **2010**, *58*, 675–684. [CrossRef]
5. Fischer, R.; Giannone, L.; Illerhaus, J.; McCarthy, P.J.; McDermott, R.M.; ASDEX Upgrade Team. Estimation and Uncertainties of Profiles and Equilibria for Fusion Modeling Codes. *Fusion Sci. Technol.* **2020**, *76*, 879–893. [CrossRef]
6. Andreeva, T.; Alonso, J.; Bozhenkov, S.; Brandt, C.; Endler, M.; Fuchert, G.; Geiger, J.; Grahl, M.; Klinger, T.; Krychowiak, M.; et al. Equilibrium Evaluation for Wendelstein 7-X Experiment Programs in the First Divertor Phase. *Fusion Eng. Des.* **2019**, *146*, 299–302. [CrossRef]
7. Sanders, M.; Ida, K.; Yoshinuma, M.; Suzuki, C.; Yoshimura, Y.; Seki, R.; Emoto, M.; Yoshida, M.; Kobayashi, T. Analysis of the Motional Stark Effect (MSE) Diagnostic to Measure the Rotational Transform and Current Profile in the Large Helical Device. *Rev. Sci. Instrum.* **2021**, *92*, 053503. [CrossRef] [PubMed]
8. Cianciosa, M.R.; Hanson, J.D.; Maurer, D.A. Uncertainty Analysis in 3D Equilibrium Reconstruction. *Fusion Sci. Technol.* **2018**, *74*, 1–12. [CrossRef]
9. Seal, S.K.; Cianciosa, M.R.; Hirshman, S.P.; Wingen, A.; Wilcox, R.S.; Unterberg, E.A. Parallel Reconstruction of Three Dimensional Magnetohydrodynamic Equilibria in Plasma Confinement Devices. In Proceedings of the 2017 46th International Conference on Parallel Processing (ICPP), Bristol, UK, 14–17 August 2017; pp. 282–291. [CrossRef]
10. Merlo, A.; Böckenhoff, D.; Schilling, J.; Höfel, Z.; Kwak, S.; Svensson, J.; Pavone, A.; Lazerson, S.A.; Pedersen, T.S. Proof of Concept of a Fast Surrogate Model of the VMEC Code via Neural Networks in Wendelstein 7-X Scenarios. *Nucl. Fusion* **2021**, *61*, 096039. [CrossRef]
11. Köberl, R.; von Toussaint, U.; Bungartz, H.J.; Schilling, J.; Albert, C.G.; W7-X Team. Uncertainty Quantification in Three Dimensional Magnetohydrodynamic Equilibrium Reconstruction via Surrogate-Assisted Bayesian Inference. *Contrib. Plasma Phys.* **2023**, *63*, e202200173. [CrossRef]
12. Svensson, J.; Werner, A. Large Scale Bayesian Data Analysis for Nuclear Fusion Experiments. In Proceedings of the 2007 IEEE International Symposium on Intelligent Signal Processing, Alcala de Henares, Spain, 3–5 October 2007; pp. 1–6. [CrossRef]
13. Hirshman, S.P. Steepest-Descent Moment Method for Three-Dimensional Magnetohydrodynamic Equilibria. *Phys. Fluids* **1983**, *26*, 3553. [CrossRef]
14. Lazerson, S.; Sakakibara, S.; Suzuki, Y. A Magnetic Diagnostic Code for 3D Fusion Equilibria. *Plasma Phys. Control. Fusion* **2013**, *55*, 025014. [CrossRef]
15. Xiu, D.; Karniadakis, G.E. The Wiener–Askey Polynomial Chaos for Stochastic Differential Equations. *SIAM J. Sci. Comput.* **2002**, *24*, 619–644. [CrossRef]
16. Pavone, A.; Svensson, J.; Langenberg, A.; Höfel, U.; Kwak, S.; Pablant, N.; Wolf, R.C. Neural Network Approximation of Bayesian Models for the Inference of Ion and Electron Temperature Profiles at W7-X. *Plasma Phys. Control. Fusion* **2019**, *61*, 075012. [CrossRef]
17. Rasmussen, C.E.; Williams, C.K.I. *Gaussian Processes for Machine Learning*; Adaptive Computation and Machine Learning; MIT Press: Cambridge, MA, USA, 2006.

8. Endler, M.; Brucker, B.; Bykov, V.; Cardella, A.; Carls, A.; Dobmeier, F.; Dudek, A.; Fellinger, J.; Geiger, J.; Grosser, K.; et al. Engineering Design for the Magnetic Diagnostics of Wendelstein 7-X. *Fusion Eng. Des.* **2015**, *100*, 468–494. [CrossRef]
9. Feinberg, J.; Langtangen, H.P. Chaospy: An Open Source Tool for Designing Methods of Uncertainty Quantification. *J. Comput. Sci.* **2015**, *11*, 46–57. [CrossRef]
0. Blatman, G.; Sudret, B. Adaptive Sparse Polynomial Chaos Expansion Based on Least Angle Regression. *J. Comput. Phys.* **2011**, *230*, 2345–2367. [CrossRef]
1. Lüthen, N.; Marelli, S.; Sudret, B. Sparse Polynomial Chaos Expansions: Literature Survey and Benchmark. *SIAM/ASA J. Uncertain. Quantif.* **2021**, *9*, 593–649. [CrossRef]
2. Schobi, R.; Sudret, B.; Wiart, J. Polynomial-Chaos-Based Kriging. *Int. J. Uncertain. Quantif.* **2015**, *5*, 171–193. [CrossRef]
3. Hoffman, M.D.; Gelman, A. The No-U-Turn Sampler: Adaptively Setting Path Lengths in Hamiltonian Monte Carlo. *arXiv* **2011**, arXiv:1111.4246. [CrossRef]

physical sciences forum

MDPI

Proceeding Paper

Improving Inferences about Exoplanet Habitability †

Risinie D. Perera and Kevin H. Knuth *

Physics Department, University at Albany (SUNY), Albany, NY 12222, USA; rperera@albany.edu
* Correspondence: kknuth@albany.edu

† Presented at the 42nd International Workshop on Bayesian Inference and Maximum Entropy Methods in Science and Engineering, Garching, Germany, 3–7 July 2023.

Abstract: Assessing the habitability of exoplanets (planets orbiting other stars) is of great importance in deciding which planets warrant further careful study. Planets in the habitable zones of stars like our Sun are sufficiently far away from the star so that the light rays from the star can be assumed to be parallel, leading to straightforward analytic models for stellar illumination of the planet's surface. However, for planets in the close-in habitable zones of dim red dwarf stars, such as the potentially habitable planet orbiting our nearest stellar neighbor, Proxima Centauri, the analytic illumination models based on the parallel ray approximation do not hold, resulting in severe biases in the estimates of the planetary characteristics, thus impacting efforts to understand the planet's atmosphere and climate. In this paper, we present our efforts to improve the instellation (stellar illumination) models for close-in orbiting planets and the significance of the implementation of these improved models into EXONEST, which is a Bayesian machine learning application for exoplanet characterization. The ultimate goal is to use these improved models and parameter estimates to model the climates of close-in orbiting exoplanets using planetary General Circulation Models (GCM). More specifically, we are working to apply our instellation corrections to the NASA ROCKE-3D GCM to study the climates of the potentially habitable planets in the Trappist-1 system.

Keywords: exoplanet characterization; Bayesian inference; EXONEST; ROCKE-3D GCM; General Circulation Models; astrophysics; instellation; Trappist-1

1. Introduction

Typically, the illumination of exoplanets is determined under the assumption that the incident stellar radiation reaches the exoplanet as plane parallel rays. In this situation, half of the planet is fully illuminated and the other half is dark. For extremely close-in exoplanets for which the star-planet separation is on the order of only a few stellar radii, it is particularly important to take into account the finite angular size of the host star. Doing so reveals that two additional zones are necessary to describe the illumination of the exoplanet. Previous work has shown that the planet will have four distinct illumination zones: the fully-illuminated zone, the first penumbral zone, the second penumbral zone, and the un-illuminated (or night) zone [1] (Figure 1).

The fully-illuminated zone (FZ) is the region on the planet in which the entire apparent disk of the star is visible in the sky. However, due to the close-in geometry, the surface area of the visible star is less than one-half of the total area. The first penumbral zone (PZ_1) is the region on the planet in which a portion of the star (less than one-half of the potentially visible area) is occluded by the horizon; that is, the star is on the verge of setting. The second penumbral zone (PZ_2) is the region on the planet in which more than one-half of the potentially visible area of the star is occluded by the horizon and the un-illuminated zone (UZ) is the region on the planet from which the star cannot be seen.

The result is that the rather abrupt terminator between the day-side and night-side of a distant planet is, in the case of a close-in planet, spread out into the two penumbral zones, which impinge on both the fully illuminated and the un-illuminated zones. So while a smaller surface area of the close-in planet is fully-illuminated, more of the planet's surface receives light because the un-illuminated zone is also smaller. For example, in the case of the close-in orbiting hot Jupiter, Kepler-91b, about 30% of the surface is in full daylight, 40% of the surface is in partial penumbral light, and 30% of the surface is in full darkness, so that approximately 70% of the planet is receiving some light at any point in time [2]. These differences in illumination are expected to impact the planet's climate.

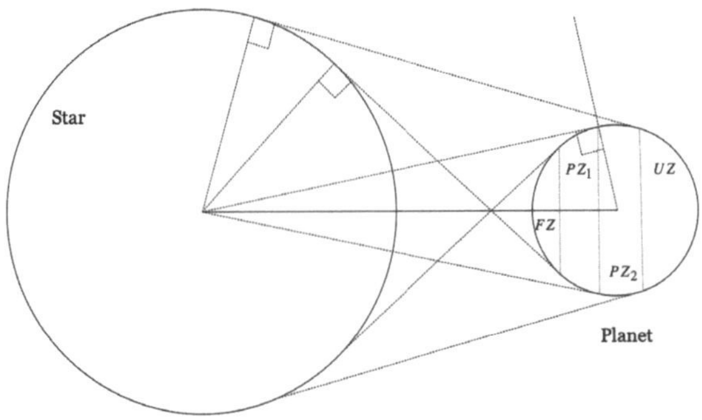

Figure 1. Extremely close-in exoplanets exhibit four zones of illumination: a fully illuminated zone (FZ) defined by the inner tangents, two penumbral zones (PZ_1, PZ_2) defined by whether more than half or less than half of the star is visible, and a dark un-illuminated zone (UZ) defined by the outer tangents.

The main goal of this paper is to construct analytic expressions for the flux incident on an exoplanet to be implemented in EXONEST, a Bayesian machine learning application developed for exoplanet characterization [2,3].

2. Methods

In this section we build analytic expressions for the light received by a point on the exoplanet taking into account the finite angular size of the host star.

2.1. Instellation Setup

Suppose that the star emits P photons per unit time. Integrating the photons over the blackbody spectrum will give the amount of emitted energy, and therefore this P is related to the power emitted by the star. If each area element of the star, $d\sigma_S$, emits dp photons per unit time, the total number of photons emitted by the star in unit time, P, is given by:

$$P = \int dp. \tag{1}$$

If p is the number of photons emitted by an area element in a unit time, then

$$dp = p\, d\sigma_S, \tag{2}$$

and

$$P = \iint p\, d\sigma_S. \tag{3}$$

Let the radius of the star be R_S. Assuming the star to be a sphere, we can write $d\sigma_S$ in terms of the surface area element as

$$P = \iint pR_S^2 \sin\theta \, d\theta d\omega. \tag{4}$$

Integrating over the whole star yields

$$P = p(4\pi R_S^2), \tag{5}$$

which allows us to write

$$p = \frac{P}{4\pi R_S^2}. \tag{6}$$

Some of these photons will hit the planet.

Next, let us consider photons emitted from the area element on the stellar surface at point A, and determine how many of these photons will be incident at a point B on the planet. Let $df_{A\to B}$ denote the number of photons emitted by A and arriving at B in a unit time. Then, from Figure 2:

$$df_{A\to B} = \left[\frac{(\hat{m}\cdot -\hat{\rho'})(\hat{n}\cdot\hat{\rho'})}{\rho'^2}\right] dp. \tag{7}$$

Substituting $\hat{\rho'} = \frac{\vec{\rho'}}{\rho'}$ and dp using Equations (2)–(6) gives:

$$df_{A\to B} = -\frac{P}{4\pi}\left[\frac{(\hat{m}\cdot\vec{\rho'})(\hat{n}\cdot\vec{\rho'})}{\rho'^4}\right] \sin\theta\, d\theta d\omega. \tag{8}$$

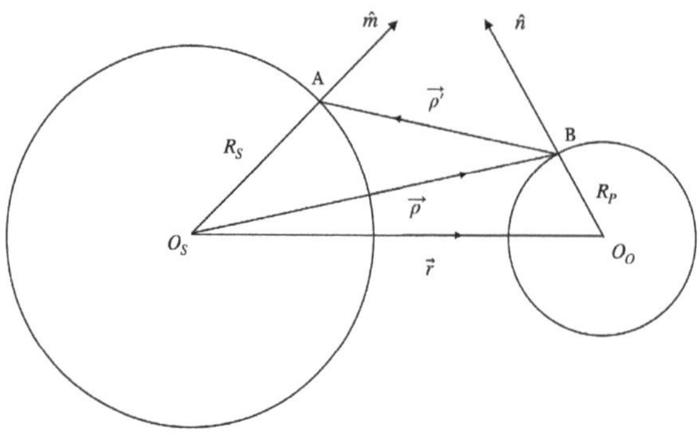

Figure 2. Star–planet geometry involving light emission at point A on the star and incident on the planet at point B: The star with radius R_S is centered at O_S and the exoplanet with radius R_P is centered at O_O. The star–planet separation is denoted by the vector $\vec{r}$.

2.2. Coordinate Systems

When we have a star–planet dynamical system, it is convenient to introduce three coordinate systems: stellar, planetocentric, and overall. For the scope of this paper, it suffices to use just two, stellar and overall, by choosing the origin of the overall coordinate system at the center of the exoplanet such that it would coincide with the planetocentric coordinate system (Figure 3). Furthermore, the axial symmetry of stellar illumination about the axis joining the centers of the planet and the star can be exploited by orienting the z-axis of the overall coordinate system in this direction. With this coordinate system, the problem simplifies to having to determine the illumination along one semicircular arc.

The stellar and overall Cartesian coordinate systems are related to one another as follows:

$$(x_O, y_O, z_O)_O = (-x_O, y_O, r - z_O)_S \tag{9}$$

and

$$(x_S, y_S, z_S)_S = (-x_S, y_S, r - z_S)_O. \tag{10}$$

Now that we have chosen appropriate coordinate systems, next let us derive the vectors required for the illumination in (8).

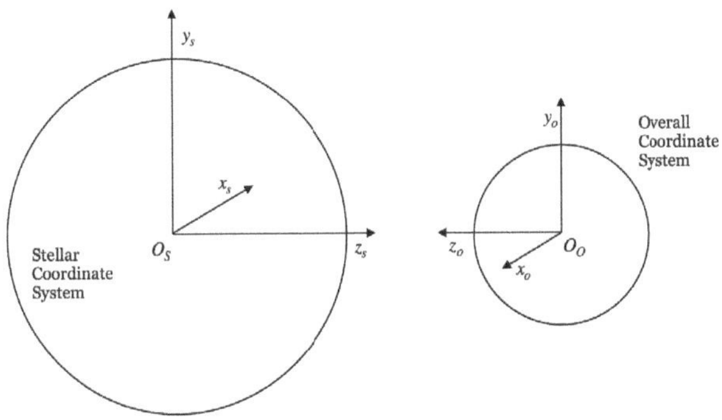

Figure 3. Orientations of the stellar and overall coordinate systems with origins located at O_S and O_O, respectively. The stellar coordinate system, which employs spherical coordinates (θ, ω), is used to define points on the star, while the overall coordinate system employs spherical coordinates (η, ϕ).

2.3. *The Vector $\vec{\rho}$*

The vector $\vec{\rho}$ is given by

$$\vec{\rho} = \vec{r} + \overrightarrow{O_P B}, \tag{11}$$

where in the overall frame

$$\vec{r} = -r\,\hat{z_o}, \tag{12}$$

and $\overrightarrow{O_P B}$ is the position vector of the point B in the overall frame:

$$\overrightarrow{O_P B} = R_P \sin\eta \cos\phi\ \hat{x_o} + R_P \sin\eta \sin\phi\ \hat{y_o} + R_P \cos\eta\ \hat{z_o}. \tag{13}$$

The sum then gives

$$\vec{\rho} = R_P \sin\eta \cos\phi\ \hat{x_o} + R_P \sin\eta \sin\phi\ \hat{y_o} + (R_P \cos\eta - r)\ \hat{z_o}, \tag{14}$$

which has a magnitude of

$$\rho = \left(r^2 - 2rR_P \cos\eta + {R_P}^2\right)^{1/2}. \tag{15}$$

We can now verify that if the point B is located at $(\eta, \phi) = (0, \pi/2)$, then

$$\rho(\eta = 0, \phi = \pi/2) = \left(r^2 - 2rR_P + {R_P}^2\right)^{1/2} = r - R_P, \tag{16}$$

as expected.

2.4. *The Vector $\vec{\rho'}$*

Let us consider the triangle $O_S AB$ in Figure 2.

$$\overrightarrow{BA} = \overrightarrow{BO_S} + \overrightarrow{O_S A}, \tag{17}$$

$$\vec{\rho'} = -\vec{\rho} + \overrightarrow{O_S A}. \tag{18}$$

The vector $\overrightarrow{O_S A}$ in the stellar coordinate system is given by

$$\overrightarrow{O_S A} = R_S \sin\theta \cos\omega\ \hat{x}_s + R_S \sin\theta \sin\omega\ \hat{y}_s + R_S \cos\theta\ \hat{z}_S. \tag{19}$$

Constructing the vector $\overrightarrow{O_S A}$ in the overall coordinate system requires some care. We can use the fact that

$$\overrightarrow{O_S A} = \overrightarrow{O_S O_O} + \overrightarrow{O_O A}, \tag{20}$$

where by the relation in Equation (10),

$$\overrightarrow{O_O A} = -R_S \sin\theta \cos\omega\ \hat{x}_o + R_S \sin\theta \sin\omega\ \hat{y}_o + (r - R_S \cos\theta)\ \hat{z}_o \tag{21}$$

and

$$\overrightarrow{O_S O_O} = -r\ \hat{z}_o, \tag{22}$$

such that

$$\overrightarrow{O_S A} = -R_S \sin\theta \cos\omega\ \hat{x}_o + R_S \sin\theta \sin\omega\ \hat{y}_o - R_S \cos\theta\ \hat{z}_o, \tag{23}$$

which simplifies to

$$\begin{aligned}\vec{\rho}' = -(R_P \sin\eta \cos\phi + R_S \sin\theta \cos\omega)\ \hat{x}_o \\ - (R_P \sin\eta \sin\phi - R_S \sin\theta \sin\omega)\ \hat{y}_o \\ + (r - R_P \cos\eta - R_S \cos\theta)\ \hat{z}_o.\end{aligned} \tag{24}$$

The squared-magnitude of ρ' is

$$\begin{aligned}\rho'^2 = R_P^2 + R_S^2 + r^2 - 2r(R_P \cos\eta + R_S \cos\theta) + \\ 2R_P R_S(\sin\eta \cos\phi \sin\theta \cos\omega + \cos\eta \cos\theta - \sin\eta \sin\phi \sin\theta \sin\omega).\end{aligned} \tag{25}$$

The distance ρ' is then found by taking the square root of the result above.

We can check this result by looking at points $B = (\eta = 0, \phi = \pi/2)$ and $A = (\theta = 0, \omega = 0)$ for which

$$\rho'^2 = R_P^2 + R_S^2 + r^2 - 2r(R_P + R_S) + 2R_P R_S, \tag{26}$$

such that

$$\rho' = r - (R_P + R_S), \tag{27}$$

as expected.

2.5. The Unit Vectors

Now let us obtain the relevant unit vectors, $\hat{m}$ and $\hat{n}$, in the overall coordinate system. Note that $\hat{m}$ is the unit vector in the $\overrightarrow{O_S A}$ direction and therefore

$$\hat{m} = \frac{\overrightarrow{O_S A}}{|\overrightarrow{O_S A}|}. \tag{28}$$

Noting that $|\overrightarrow{O_S A}| = R_S$, by Equation (23) we have

$$\hat{m} = -\sin\theta \cos\omega\ \hat{x}_o + \sin\theta \sin\omega\ \hat{y}_o - \cos\theta\ \hat{z}_o. \tag{29}$$

Similarly noting that $\hat{n}$ is the unit vector in the $\overrightarrow{O_P B}$ direction and using Equation (13) gives

$$\hat{n} = \sin\eta \cos\phi\ \hat{x}_o + \sin\eta \sin\phi\ \hat{y}_o + \cos\eta\ \hat{z}_o. \tag{30}$$

With an understanding of how to construct the relevant vectors and express them in the overall coordinate system, we can now write out the necessary integrals to obtain the total instellation.

2.6. Total Instellation

To compute the total instellation, we must integrate over the visible surface of the star, which is described by the spherical coordinates θ and ω. The total instellation at a point B on the planet is given by integrating over all possible points, A, on the star that is visible to point B on the planet:

$$f_B = -\int_{\omega_1}^{\omega_2}\int_{\theta_1}^{\theta_2} \frac{P}{4\pi}\left[\frac{(\hat{m}\cdot\vec{\rho}')(\hat{n}\cdot\vec{\rho}')}{\rho'^4}\right]\sin\theta\, d\theta\, d\omega \qquad (31)$$

where θ_1, θ_2, ω_1 and ω_2 are the limits of integration that depend on the zone in which point B is located.

Upon substitution of $\vec{\rho}'$, its magnitude and the unit vectors $\hat{m}$ and $\hat{n}$, into the total instellation in (31), it simplifies to 5 distinct double integrals with respect to θ and ω, which must be integrated over the limits of each zone to obtain an analytic model.

3. Discussion

In this section, we discuss the applications and implications of improved instellation models to illustrate the benefits of having accurate models for stellar illumination in order to determine the habitability of close-in exoplanets.

3.1. Bayesian Exoplanet Characterization and Model Testing

The EXONEST software suite performs Bayesian model-based exoplanet characterization using photometric and radial velocity data [2–5]. EXONEST relies on predictive photometric models that generate a predicted stellar light curve based on a set of hypothesized model parameters. Bayesian probability theory [6–9] uses this predicted stellar light curve to compute the posterior probability that the set of hypothetical model parameters could have been responsible for the recorded light curve. This allows detailed astrophysical models to be evaluated by the data analysis algorithm, as well as statistical testing among competing models [3,8,10,11]. EXONEST allows the user to employ one of several variants of the Nested Sampling Algorithm, such as Nested Sampling [8,12], MultiNest [13,14], and the state-of-the-art Nested Sampling with Galilean Monte Carlo [15–17]. These Nested Sampling algorithms provide estimates of the model parameter values as well as the Bayesian evidence, which allows for statistical model testing. Using evidence-based Bayesian techniques [10], exoplanetary models can be ranked based on the Bayesian evidence providing the capability for scientific hypothesis testing. One can test various photometric effects and compute the evidence to determine the probability that these effects are present in any given system. Relying on data from the NASA Exoplanet Archive, we can utilize EXONEST and its ability to perform Bayesian model testing to evaluate the efficacy of our improved instellation models by computing the Bayesian evidence of the parallel-ray approximation models versus our proposed models.

In any problem involving model parameter estimation, there are legitimate concerns about potential degeneracies among the model parameters, especially as the number of parameters increases. Degeneracies can be expected when more than one parameter affects the predicted light curve in a similar way. In reality, this is a warning that additional, different data are needed to effectively estimate the model parameters. Examining the cross-correlations between the parameter estimates is a very effective way to identify degeneracies. Model testing, as described above, can be used to detect degeneracies as well, since a degenerate set of model parameters will not significantly increase the likelihood function, but it will significantly decrease the prior probability, resulting in lower evidence.

3.2. Implications for Potentially Habitable M-Dwarf Worlds

The exoplanet TRAPPIST-1d ($R_{star} = 0.00056\,\text{AU} = 0.121 R_{SUN}$, $a = 0.022\,\text{AU} = 39 R_{star}$), which closely orbits its host M-dwarf star in the habitable zone (HZ), has a penumbral zone that covers about 2.5% of its surface or 8.25 million km^2. While this is a small percentage of the surface of the planet, this is greater than the area of Central and South America combined, which contains several important biogeographical regions on Earth. (e.g., [18–21]). Our estimates reveal that planets orbiting M-dwarf stars on the inner edge of the HZ will have penumbral zones ranging in area from around 1 to 3% of the surface area of the planet. While this may seem like a small value, it may have its greatest climatological impact at the poles of low obliquity worlds where the ice–albedo feedback manifests itself consistently and tends to drive mean surface temperatures lower, which can affect the fractional habitability metrics like that proposed in [22]. A detailed understanding of the incident stellar radiation in these penumbral regions is expected to have significant relevance to understanding habitability on these worlds.

3.3. Climate Modeling of Close-In Orbiting Planets

As mentioned before, corrections are required to the illumination models used in planetary General Circulation Models (GCM) in order to properly model the climate of a close-in exoplanet. The analytical expressions for the incident stellar flux can be applied to the illumination model in the ROCKE-3D GCM [23]. Not only have such corrections never been applied in a GCM, but their effects have never been quantified. For example, differences in illumination at the poles of zero obliquity worlds due to these corrected illumination models could have large effects on ice coverage at the poles. Recent work [24] has shown that polar supercontinents and ocean circulation can affect global mean surface temperatures by several degrees because of the extent of polar ice and the resulting ice-albedo feedback. We expect corrected illumination models to have similar effects. Using the modified ROCKE-3D code for close-in planets, we could study the climates of potentially habitable planets. Critically, we can make one-to-one comparisons of GCM results between penumbral and parallel ray (the current default) illuminations and derive statistical measures of difference (e.g., systematic latitudinal—and global—surface temperature differences, ice-cover, precipitation patterns). We could look at a variety of topographies, land/sea masks, and surface water content used in previous ROCKE-3D publications such as Proxima Centauri b [25], Future Earth [24], and ancient Venus [26].

4. Conclusions

Developing accurate analytical stellar instellation models for close-in orbiting exoplanets will increase both the accuracy and precision of inferred model parameter estimates obtained from data, thereby improving exoplanet characterization and potentially the detection of habitable exoplanets. This is a multidisciplinary project, founded on astrophysics with machine learning. Studying the implications of these improved illumination models in the context of exoplanetary climate modeling will provide a better understanding of the climates and atmospheric physics of close-in orbiting exoplanets.

Author Contributions: The authors contributed equally to the project. All authors have read and agreed to the published version of the manuscript.

Funding: This research received no external funding.

Institutional Review Board Statement: Not applicable.

Informed Consent Statement: Not applicable.

Data Availability Statement: Data sharing is not applicable to this article.

Acknowledgments: We wish to thank Jennifer L. Carter, Michael J. Way, Oleg Lunin and Jonathan Petruccelli for helpful advice as well as the organizers, Udo von Toussaint, Roland Preuss, Svenja Eichler and Anja Bauer for their generous hospitality.

Conflicts of Interest: The authors declare no conflict of interest.

References

1. Carter, J.L. *Estimation of Planetary Photometric Emissions for Extremely Close-in Exoplanets*; University at Albany: Albany, NY, USA, 2018.
2. Knuth, K.H.; Placek, B.; Angerhausen, D.; Carter, J.L.; D'Angelo, B.; Gai, A.D.; Carado, B. Exonest: The bayesian exoplanetary explorer. *Entropy* **2017**, *19*, 559. [CrossRef]
3. Placek, B.; Knuth, K.H.; Angerhausen, D. EXONEST: Bayesian model selection applied to the detection and characterization of exoplanets via photometric variations. *Astrophys. J.* **2014**, *795*, 112. [CrossRef]
4. Placek, B. *Bayesian Detection and Characterization of Extra-Solar Planets via Photometric Variations*; University at Albany: Albany, NY, USA, 2014.
5. Placek, B.; Knuth, K.H.; Angerhausen, D.; Jenkins, J.M. Characterization of Kepler-91b and the investigation of a potential Trojan companion using EXONEST. *Astrophys. J.* **2015**, *814*, 147. [CrossRef]
6. Jaynes, E.T. *Probability Theory: The Logic of Science*; Cambridge University Press: Cambridge, UK, 2003.
7. Gregory, P. *Bayesian Logical Data Analysis for the Physical Sciences: A Comparative Approach with Mathematica® Support*; Cambridge University Press: Cambridge, UK, 2005.
8. Sivia, D.; Skilling, J. *Data Analysis: A Bayesian Tutorial*; OUP Oxford: Oxford, UK, 2006.
9. Knuth, K.H.; Skilling, J. Foundations of inference. *Axioms* **2012**, *1*, 38–73. [CrossRef]
10. Knuth, K.H.; Habeck, M.; Malakar, N.K.; Mubeen, A.M.; Placek, B. Bayesian evidence and model selection. *Digit. Signal Process.* **2015**, *47*, 50–67. [CrossRef]
11. Gai, A.D.; Knuth, K.H. Bayesian model testing of ellipsoidal variations on stars due to hot Jupiters. *Astrophys. J.* **2018**, *853*, 49. [CrossRef]
12. Skilling, J. Nested sampling. *AIP Conf. Proc.* **2004**, *735*, 395–405.
13. Feroz, F.; Hobson, M.P. Multimodal nested sampling: An efficient and robust alternative to Markov Chain Monte Carlo methods for astronomical data analyses. *Mon. Not. R. Astron. Soc.* **2008**, *384*, 449–463. [CrossRef]
14. Feroz, F.; Hobson, M.; Bridges, M. MultiNest: An efficient and robust Bayesian inference tool for cosmology and particle physics. *Mon. Not. R. Astron. Soc.* **2009**, *398*, 1601–1614. [CrossRef]
15. Skilling, J. Bayesian computation in big spaces-nested sampling and Galilean Monte Carlo. *AIP Conf. Proc.* **2012**, *1443*, 145–156.
16. Feroz, F.; Skilling, J. Exploring multi-modal distributions with nested sampling. *AIP Conf. Proc.* **2013**, *1553*, 106–113.
17. Skilling, J. Galilean and Hamiltonian Monte Carlo. *Multidiscip. Digit. Publ. Inst. Proc.* **2019**, *33*, 19.
18. Dickinson, R.E.; Kennedy, P. Impacts on regional climate of Amazon deforestation. *Geophys. Res. Lett.* **1992**, *19*, 1947–1950. [CrossRef]
19. Yang, Y.; Saatchi, S.S.; Xu, L.; Yu, Y.; Choi, S.; Phillips, N.; Kennedy, R.; Keller, M.; Knyazikhin, Y.; Myneni, R.B. Post-drought decline of the Amazon carbon sink. *Nat. Commun.* **2018**, *9*, 3172. [CrossRef] [PubMed]
20. Molina, R.; Salazar, J.; Martínez, J.; Villegas, J.; Arias, P. Exponential growth of precipitation along a "forest-fed moisture convey or belt" above the Amazon. *J. Geophys. Res. Atmos* **2019**, *124*, 2589–2599. [CrossRef]
21. Pereira, J.C. Reducing catastrophic climate risk by revolutionizing the Amazon: Novel pathways for Brazilian diplomacy. In *Climate Change and Global Development: Market, Global Players and Empirical Evidence*; Springer: Berlin/Heidelberg, Germany, 2019; pp. 189–218.
22. Spiegel, D.S.; Menou, K.; Scharf, C.A. Habitable climates. *Astrophys. J.* **2008**, *681*, 1609. [CrossRef]
23. Way, M.J.; Aleinov, I.; Amundsen, D.S.; Chandler, M.; Clune, T.; Del Genio, A.D.; Fujii, Y.; Kelley, M.; Kiang, N.Y.; Sohl, L.; et al. Resolving orbital and climate keys of Earth and extraterrestrial environments with dynamics (ROCKE-3D) 1.0: A general circulation model for simulating the climates of rocky planets. *Astrophys. J. Suppl. Ser.* **2017**, *231*, 12. [CrossRef]
24. Way, M.J.; Davies, H.S.; Duarte, J.C.; Green, J. The climates of Earth's next supercontinent: Effects of tectonics, rotation rate, and insolation. *Geochem. Geophys. Geosyst.* **2021**, *22*, e2021GC009983. [CrossRef]
25. Del Genio, A.D.; Way, M.J.; Amundsen, D.S.; Aleinov, I.; Kelley, M.; Kiang, N.Y.; Clune, T.L. Habitable climate scenarios for Proxima Centauri b with a dynamic ocean. *Astrobiology* **2019**, *19*, 99–125. [CrossRef] [PubMed]
26. Way, M.J.; Del Genio, A.D. Venusian habitable climate scenarios: Modeling Venus through time and applications to slowly rotating Venus-like exoplanets. *J. Geophys. Res. Planets* **2020**, *125*, e2019JE006276. [CrossRef]

physical sciences forum

MDPI

Proceeding Paper

Bayesian Model Selection and Parameter Estimation for Complex Impedance Spectroscopy Data of Endothelial Cell Monolayers †

Franziska Zimmermann [1,2], Frauke Viola Härtel [1,3], Anupam Das [1], Thomas Noll [1] and Peter Dieterich [1,*]

[1] Institut für Physiologie, Medizinische Fakultät Carl Gustav Carus, TU Dresden, Fetscherstr. 74, 01307 Dresden, Germany; franziska-zimmermann@web.de (F.Z.); frauke.haertel@physiologie.med.uni-giessen.de (F.V.H.); anupam.das@tu-dresden.de (A.D.); thomas.noll@tu-dresden.de (T.N.)

[2] Carl-von-Basedow-Klinikum Saalekreis gGmbH, Weiße Mauer 52, 06217 Merseburg, Germany

[3] Physiologisches Institut, Fachbereich Medizin der Justus-Liebig-Universität, Aulweg 129, 35392 Giessen, Germany

* Correspondence: peter.dieterich@tu-dresden.de

† Presented at the 42nd International Workshop on Bayesian Inference and Maximum Entropy Methods in Science and Engineering, Garching, Germany, 3–7 July 2023.

Abstract: Endothelial barrier function can be quantified by the determination of the transendothelial resistance (TER) via impedance spectroscopy. However, TER can only be obtained indirectly based on a mathematical model. Models usually comprise a sequence of a resistance in parallel with a capacitor (RC-circuit), each for the cell layer (including TER) and the filter substrate, one resistance (R) for the medium, and a constant phase element (CPE) for the electrode–electrolyte interface. We applied Bayesian data analysis on a variety of model variants. Phase and absolute values of impedance data were acquired over time by a commercial device for measurements of pure medium, medium and raw filter, and medium with cell-covered filter stimulated with different agents. Medium and raw filter were best described by a series of four and three RC-circuits, respectively. Parameter estimation of the TER showed a concentration-dependent decrease in response to thrombin. Model comparison indicated that even high concentrations of thrombin did not fully disrupt the endothelial barrier. Insights in the biophysical meaning of model parameters were gained through complemental cell-free measurements with sodium chloride. In summary, Bayesian analysis allows for valid parameter estimation and the selection of models with different complexity under various experimental conditions to characterize endothelial barrier function.

Keywords: impedance spectroscopy; transendothelial resistance; Bayesian data analysis

1. Introduction

The endothelium forms the essential barrier between the intravascular and the tissue space. It is indispensable for physiological and pathophysiological processes like the selectivity of the blood–brain barrier or the formation of oedema. An important parameter for describing this barrier is permeability. A method for quantifying the endothelial barrier function is impedance spectroscopy. Here, an alternating current is applied and the resulting complex resistance, the impedance $Z(f)$, is measured for different frequencies f. The defining parameters are the frequency-dependent course of absolute value $|\mathbf{Z}|$ and the phase **Ph** of the complex impedance. For the measurements, isolated cell monolayers are cultivated on a filter insert. For characterizing the monolayer, the transendothelial resistance *TER* is often used. This parameter cannot be measured directly, and therefore it will be obtained by mathematical modeling with an equivalent circuit, which describes the measuring system (see Figure 1). The model parameters have to be optimized to obtain the best fit of the model to data values. Commercially available impedance spectroscopes allow for the easy experimental access of **|Z|** and **Ph** and often a direct data analysis with preset models. However, these models are neither exchangeable nor expandable. The standard deviation of the resulting model parameters is often not provided and should therefore be calculated via repeated measurements. Further, a quantitative model comparison is not possible. Hence, we used Bayesian Data analysis for evaluating impedance spectroscopy data.

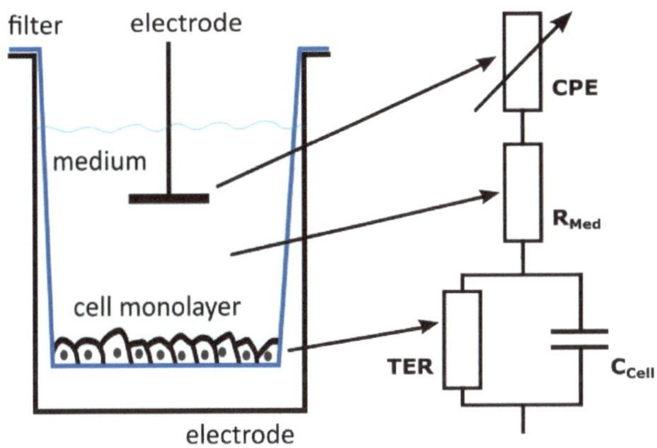

Figure 1. Scheme of the measurement setup: Isolated cells are cultivated on a filter and inserted in a so-called well, which is embedded in tissue medium. **|Z|** and **Ph** are measured via a two-electrode setup. The transendothelial resistance, TER, as a quantity of endothelial barrier function, is determined by an equivalent circuit model. One typically used model consists of a constant phase element (CPE), describing the non-ideal capacity behavior of the electrode–electrolyte interface, an ohmic resistance for the medium, and an RC-circuit (in this case with R = TER and C = C_{Cell}) for the cell monolayer cultivated on the filter support.

2. Materials and Methods

Measurements were performed with a commercially available impedance spectroscope (CellZscope®—nanoAnalytics GmbH, Münster, Germany). Three different conditions were measured—only medium, medium and filter, and medium with cell-covered filter. We used human umbilical vein endothelial cells (HUVEC), which were isolated and cultivated according to established protocols [1,2]. ECGM (Endothelial Cell Growth Medium, PromoCell®, Heidelberg, Germany) was used as cell medium. The filter inserts were composed of polyester (Transwell® Permeable Support, Corning, NY, USA) with an area of 0.33 cm^2, a pore size of 0.4 µm, and 4×10^6 pores/cm^2. Stimulations were carried out with thrombin, which leads to a concentration-dependent increase in the macromolecule permeability via the contraction of endothelial cells and intercellular gap formation [3,4]. Four different concentrations (0.1 U/mL, 0.2 U/mL, 1.0 U/mL, and 2.0 U/mL) were used. Furthermore, cells were exposed to triton X-100 0.1%, which destroys the lipid layer of the cell membrane [5]. Cell medium was used as a reference for stimulation. Measurements were performed at 42 frequencies in the range from 3 Hz to 100 kHz; this whole spectrum is the data set per time point. The time period of 0.5 h before to 1.5 h after stimulation was analyzed. The temporal resolution was about 5 min, which resulted in 54 time points for each stimulation.

Bayesian data analysis is based on the Bayesian Theorem:

$$\mathrm{P}(H|D,I) = \frac{\mathrm{P}(D|H,I) \cdot \mathrm{P}(H|I)}{\mathrm{P}(D|I)} \tag{1}$$

It is a logically consistent method for determining model parameters given a model hypothesis H by taking into account the measured data, their errors D, and previous information

I [6,7]. The result is the posterior $P(H|D, I)$, which represents the probability density functions of the parameters. It is based on the posterior mean, and its standard deviation can be calculated. Previous knowledge of parameters can be included as a prior $P(H|I)$, and both data sets **|Z|** and **Ph** can be incorporated in the likelihood $P(D|H, I)$ for the analysis. The determination of the evidence $P(D|I)$ allows for quantitative model comparisons ([8] (chapter 3.5) and [9]) and therefore valid decisions on the most appropriate model based on the respective data. Bayesian data analysis was performed with a MultiNested-Sampling algorithm based on [10] and the Python interface of [11].

As all parameters of the models are positive and may vary over several orders of magnitude, the Jeffreys' prior was used. The implementation within nested sampling was performed as in the procedure of [9].

Since data uncertainties and possible correlations of **|Z|** and **Ph** of the CellZscope are not known, an algorithm was implemented to estimate those for each measured frequency for every single well during steady state measurements at the end of the experiment. Analyzing the frequency course of the estimated uncertainties indicated that **|Z|** and **Ph** are not measured directly by the commercial device, and therefore their uncertainties may be correlated. This was included by a correlation matrix in the likelihood.

Different equivalent circuits were analyzed for the three experimental conditions. The commonality of all of the models chosen for the analysis was a constant phase element CPE, describing the electrode–electrolyte interface, combined with a series connection of several RC-circuits (the parallel connection of a resistance R and a capacitor C). The impedance of such an RC-circuit is calculated by $Z_{RC} = \frac{R}{1+\mathrm{i}\cdot\omega\cdot C\cdot R}$. The exact number n_{RC} of the RC-circuits was determined as a parameter via model comparison. The particular R and C are denominated with the index e and the number 1 to n_{RC}. The impedance of the CPE is calculated by $Z_{CPE} = \frac{1}{A\cdot(\mathrm{i}\cdot\omega)^n}$, so it is characterized by two parameters A and n, which are denoted by $\mathrm{CPE_A}$ and $\mathrm{CPE_n}$ in this work. The exponent n is limited between 0 and 1, describing for the lower limit a pure ohmic and for the higher limit a pure capacitative performance, respectively.

For measurements with medium only, this equivalent circuit was compared to a model with an additional ohmic resistance in series (as in [12]).

For the filter, 6 different elements added in series were tested: an ohmic resistance, an RC-circuit, an RC-circuit with a definite capacitance calculated from the physical properties of the filter, an R| |CPE-circuit, and two more complex models, which attempted to include the pores of the filter. The cell monolayer was modeled by an RC-circuit. The capacitance of cells was already determined experimentally in 1925 by [13] for erythrocytes with $C_{Ce} = 0.81\ \mu\mathrm{F/cm^2}$. This magnitude was also verified for other cell types [14]. Thus, this capacitance of cells could be separated via specifying the corresponding prior limits.

For obtaining deeper insight into the biophysical meaning of the model components and parameters, further experiments with deionised water ($\mathrm{dd_{H_2O}}$) and sodium chloride solutions were carried out. A 1 M stock solution was diluted in 9 different concentrations ranging from ratio 1:10^{12} to 1:10. For each concentration, several data sets were measured, and these data sets were numbered consecutively since there is not such a relevant time course like for the stimulation experiments.

3. Results

After an extensive and successful validation with self simulated test data (data not shown), the algorithm was applied to the experimental data. For each time point, the impedance spectrum was analyzed with different models and the model probabilities and parameters were estimated. For measurements with medium only, a model consisting of the CPE and $n_{\mathrm{RC}} = 3$ RC-circuits had the best results. This is in contrast to the common assumption that the medium should be described as a simple ohmic resistance. It is noteworthy that the capacitance of one RC-circuit resulted in the low range of 10^{-9} F, whereas the others were estimated within the range of 10^{-5}–10^{-6} F. This low capacitance will be denominated as $\mathrm{C_{e1}}$.

Analyzing the wells with medium and raw filter showed that also the model containing a CPE and n_{RC} RC-circuits without further elements proved to be the most appropriate model. For up to 20% of all time points with n_{RC}= 4, no valid parameter estimation was possible; therefore, it was decided to except n_{RC} = 3 as the optimal number of RC-circuits for the whole time course. To compare the values of the capacities for the different measurement conditions, estimated capacities were sorted according to magnitude and the mean value of all time points was calculated. There was high equality for two of the capacities of medium only and medium with raw filter. Especially, the formerly mentioned C_{e1} with the range of 10^{-9} F was in good accordance between the two measuring conditions. However, for C_{e3} there was a discrepancy of one magnitude. That difference is not due to the missing RC-circuit. For some time points, even n_{RC} = 4 could be estimated, and there C_{e3} remained stable and C_{e4} was again in good accordance to medium measurements (see Figure 2a). Therefore, this RC-circuit describes filter specific behavior.

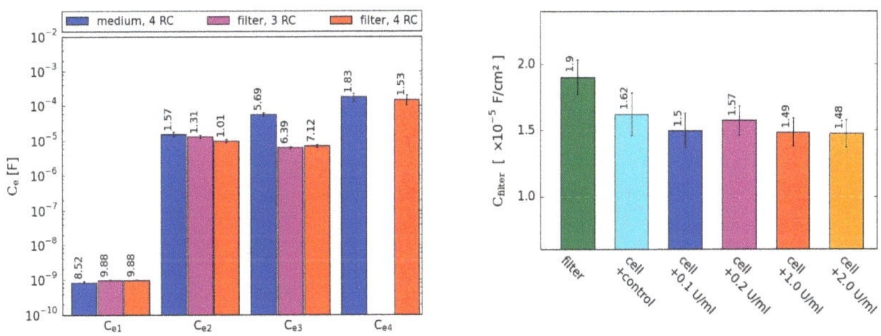

(a) (b)

Figure 2. Comparison of the mean of resulting capacities of the RC-circuits of different measurement conditions. Values of the bars have to be multiplied with the respective order of the magnitude indicated by the y-axis in (**a**) or the label of the y-axis in (**b**), respectively. (**a**) Resulting capacities of the RC-circuits for medium only and medium with filter. Medium only resulted in n_{RC} = 4 (blue), whereas most of the data of medium and raw filter resulted in n_{RC} = 3 (magenta) and only for some time points n_{RC} = 4 (red). There was good accordance within parameter estimates of C_{e1} and C_{e2}, respectively, between the two measuring conditions. For C_{e3}, a difference of about one magnitude is noticeable, which is stable for n_{RC} = 3 and n_{RC} = 4. Hence, the divergent C_{e3} is characteristic for the filter. (**b**) Comparison of the filter specific capacitance from measurements with raw filter and those with cell-covered filter stimulated with different concentrations of thrombin. This resulted in a stable difference, indicating that the cells have an influence on filter properties.

Analyzing the wells with medium and cell-covered filter, the model probability indicated that again a CPE with n_{RC} = 4 RC circuits is best for describing the data. As mentioned above, the range of the capacitance of the monolayer is well known from the literature; therefore, it can be separated via the prior. Comparing the filter describing capacitance C_{e3} from the measurements of the raw filter and the cell-covered filter, one can see that there is quite a difference indicating that the cells have an influence on filter properties. This may be due to the migration of small cellular extensions into the filter pores (see Figure 2b). Thus, parameters extracted from measurements with only medium or medium and raw filter should not be used as direct references for those with cells.

The resulting TER showed the well-known concentration-dependent decrease after thrombin stimulation. The parameter estimation was robust and with small uncertainties (see Figure 3).

To further analyze and quantify the damage of the monolayer caused by high concentrations of thrombin or triton X-100, a model comparison was executed for a model of a CPE with n_{RC} RC-circuits with or without the cell-specific RC-circuit, which can be strictly separated by the prior as explained above. After adding thrombin 2.0 U/mL, there is a high increase in the evidence of the cell-free models (with n_{RC} = 3 to 5), but the cell model still remains with the highest evidence (see Figure 4a). When adding triton X-100, the evidence of a filter with n_{RC} = 4 is much higher, resulting in a shift of model probabilities, indicating that there is no intact monolayer remaining, which has to be incorporated in the model (see Figure 4b).

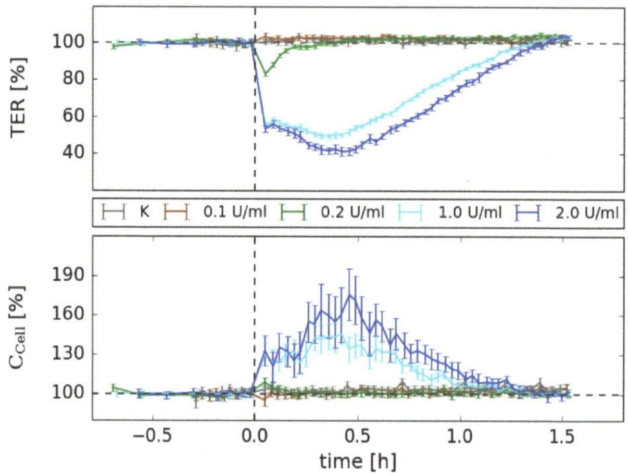

Figure 3. Time course of TER and C_{Cell} after stimulation with different concentrations of thrombin. The model analysis (CPE and n_{RC} = 4 RC-circuits) showed a concentration-dependent decrease in the TER. Magnitude and time needed for regeneration were also dependent of thrombin concentration. For all measured concentrations, the initial levels are reached within the observation time of 1.5 h. C_{Cell} shows an inversely concentration-dependent increase after stimulation with thrombin.

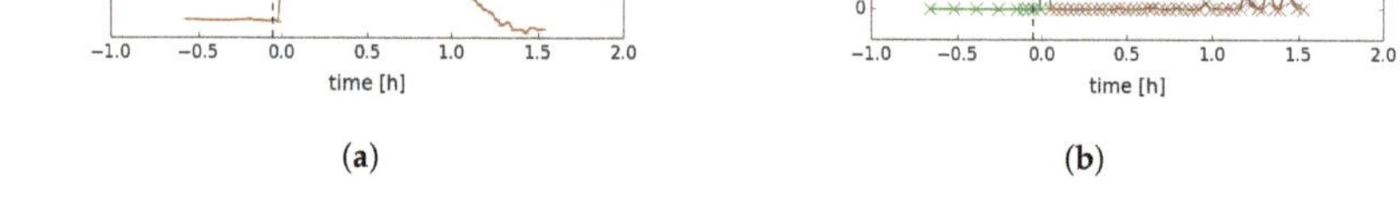

Figure 4. Resulting evidence and model probability for a model with a cell-specific RC-circuit (grey) and a filter model with n_{RC} = 3 to 5 RC-circuits (brown, green, and light blue) but without a cell-specific one. The dashed line indicates the addition of the stimulation agents (high concentration of thrombin or triton X-100) with disrupting effect on endothelial barrier function. (**a**) After adding high concentration of thrombin, there is a high rise of the evidence of the filter models (without the cell-specific RC-circuit), but the cell model (with the cell-specific RC-circuit) still remains as the one with the highest evidence. (**b**) After adding triton X-100, there is a shift in model probability to a cell-free model, indicating that the monolayer is not intact any more.

The measurements with sodium chloride and deionised water (dd_{H_2O}) were also analyzed with the model of a CPE and n_{RC} RC-circuits. The analysis revealed the low capacitance C_{el} mentioned above, which also resulted in an equivalent range without dependency on sodium chloride concentration (see Figure 5). There is only a fluctuation of less than 1 magnitude for higher concentrations. The other capacities showed an interesting course with low values for low concentrations and an abrupt rise after a kind of edge concentration to magnitudes comparable to those estimated with cell medium (data not shown). Ref. [15] compared a circuit model of a series connection of a capacitance and an RC-circuit for a two-electrode setup with a dynamic density functional theory equivalent to the Nernst–Planck–Poisson theory of two sorts of ions in a fluid. The estimated parameters were compared to the equations of this simple model. There was good accordance of the magnitude of this low, concentration-independent capacitance, as illustrated with the blue line in Figure 5. This indicated that this is mainly determined by the setup properties like the distance and size of the electrodes. The slight divergence for higher concentrations may be due to additional correlation effects that were included in the theoretical model of [15].

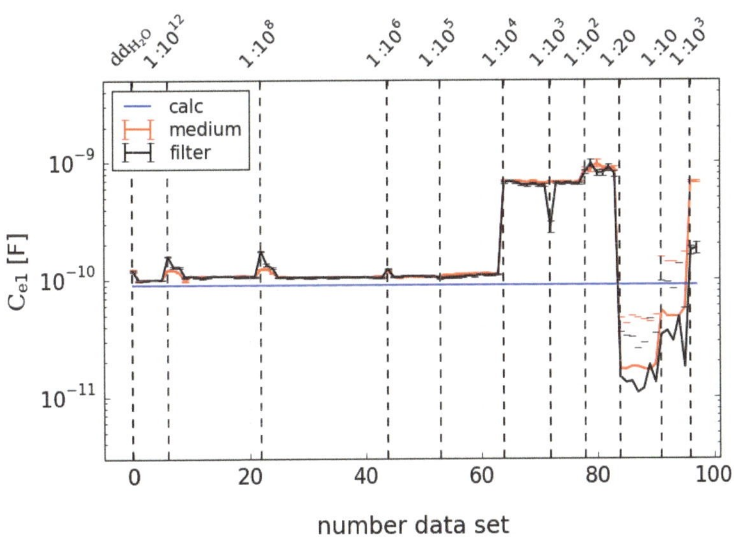

Figure 5. Comparison of estimated lowest capacitance C_{el} as a function of sodium chloride concentration of the medium (as a ratio of the 1 M stock solution) of a well with pure medium (red) and a well with medium and raw filter (black). The capacitance, which already attracted attention in the results of measurements with cell medium because of the really small magnitude, is independent of concentration of the medium and of the presence of a filter. It was compared to calculated values (blue) according to a model for a similar setup published by [15]. There is really high accordance in the magnitude and the concentration-independent course. As it can be seen by the equations in the cited publication, this capacitance mainly depends on setup criteria like electrode size and distance.

By contrast, the corresponding resistance is clearly concentration-dependent and also showed high similarity to the calculated values, especially for higher concentrations (Figure 6). This means that this resistance is determined mainly by the medium. Differences in low concentrations are probably due to the discrepancy of the theoretical assumed concentration of sodium chloride and real values since there was no control with additional measurements (like conductivity of the medium). Accordance was optimized through consideration of a factor for real sodium chloride concentration, as indicated in blue and red; the green line shows the calculated values without this factor.

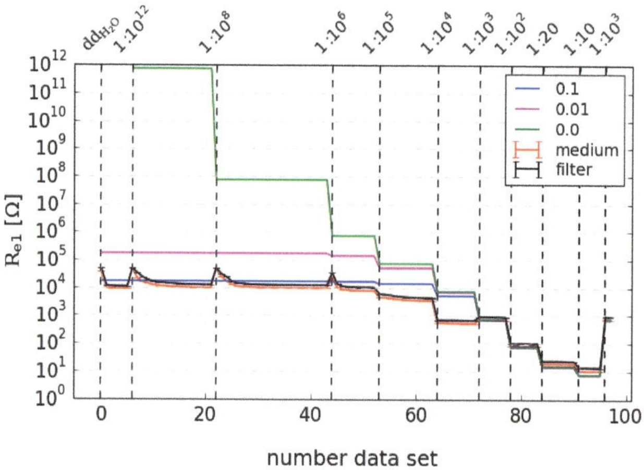

Figure 6. Course of the estimated resistance corresponding to the lowest capacitance as a function of sodium chloride concentration of the medium (as a ratio of the 1 M stock solution) of a well with pure medium (red) and a well with medium and raw filter (black). There is a clear dependence on concentration. Results are compared to calculated values according to a model for a similar setup published by [15], considering an extra factor for real concentration of sodium chloride (green, purple, and blue), which is presumptively underestimated due to basic experimental protocol. Especially with this factor there is really good congruence to the calculated values. The underlying equations show that this resistance is mainly determined by characteristics of the medium.

4. Discussion

We have shown that Bayesian analysis is suitable for analyzing impedance spectroscopy data. By applying the MultiNested sampling algorithm to the extracted data of absolute value **| Z |** and phase **Ph** of the complex impedance, we could benefit from the easy handling of a commercial device and still enhance parameter estimation. The algorithm was verified extensively with self-simulated test data. Model replacement and extensions can be easily carried out. Here, models consisting basically of a CPE and various RC circuits extended with other elements have been analyzed. Regarding the model parameters, there was only limited prior knowledge, except for the RC circuit describing the cell monolayer. Implementing the Jeffreys prior even enabled us to analyze models with a higher number of parameters. That shows that the correct choice of priors is crucial for parameter estimation without distorting the results.

The model comparison indicated that the description of the medium, known from the literature as pure ohmic resistance, is inadequate. A model of a CPE and $n_{RC} = 4$ RC-circuits had the best results. With further basic experiments with sodium chloride as the medium, it could be shown that the very low capacitance of one RC-circuit mainly depends on setup characteristics like electrode size, whereas the corresponding resistance depends on ion concentration. For the capacities of the RC-circuits, the course indicated that they are determined by ion concentration since they showed an abrupt rise after a border concentration. We suppose that this is due to the formation of thin ionic layers operating as capacitors. For low concentrations, the ionic layer is thicker and therefore capacitance is lower, and for concentrations above a definite border, this leads to a thinner ionic layer, resulting in higher capacitance. To further evaluate this, additional experiments with definite medium composition could be carried out.

Another important insight is that cells have an influence on filter properties. This might be due to cellular migration processes within the pores or again the formation of thin ionic layers between cells and filter. Therefore, filter parameters must be estimated via an adequate model for each measuring condition and should not be subtracted by reference measurements of medium with raw filter as carried out by the Cellzscope algorithm. Like for all setup elements, there is a high risk of incorrect determination of cell parameters if other elements like the cell medium are not respected properly.

The concentration-dependent course of TER after thrombin stimulation was estimated in agreement with literature data. The decrease in TER is discussed well in the literature and occurs due to cell contraction and intercellular gap formation. The parallel increase in cell capacitance is rarely discussed. We assume that there may be processes of vesicle formation and pinozytosis due to intracellular calcium rise, which can have an influence on cell capacitance. To verify this hypothesis, additional microscopic studies could be carried out.

A model comparison was used to characterize the influence of agents known to challenge endothelial barrier function. In our experiments, the highest used concentration of thrombin, which led to a decrease in TER of about 40%, still did not lead to a change of the model. Data still were described best with a model with a cell-specific RC-circuit and not with a filter-only model. However, after stimulation with triton X-100 there was a substantial change in model probability, indicating that there is no relevant intact monolayer anymore.

As a future prospect, this algorithm could also be used for more complex experimental setups of co-cultures like of endothelial and smooth muscle cells.

Author Contributions: Conceptualization, F.Z., T.N. and P.D.; methodology, F.Z. and P.D.; software, F.Z. and P.D.; validation, F.Z. and P.D.; formal analysis, F.Z.; investigation, F.Z., A.D. and F.V.H.; resources, F.Z., A.D., F.V.H., T.N. and P.D.; data curation, F.Z. and P.D.; writing—original draft preparation, F.Z.; writing—review and editing, T.N. and P.D.; visualization, F.Z.; supervision, T.N. and P.D.; and project administration, P.D. The manuscript contains parts of the doctoral thesis of F.Z. All authors have read and agreed to the published version of the manuscript.

Funding: This research received no external funding.

Institutional Review Board Statement: The study was conducted in accordance with the Declaration of Helsinki and approved by the Ethics Committee of the Medical Faculty Carl Gustav Carus of TU Dresden (permission EK 203112005).

Data Availability Statement: The raw data of this study are available on reasonable request from the corresponding author. Examples of Python programs for the application of nested sampling to Bayesian inference are found in the zenodo archive https://zenodo.org/record/6325568 (accessed on 25 July 2023).

Acknowledgments: The authors are grateful to the Center for Information Services and High Performance Computing (Zentrum für Informationsdienste und Hochleistungsrechnen (ZIH)) at TU Dresden for providing its facilities for high-throughput calculations.

Conflicts of Interest: The authors declare no conflict of interest.

Abbreviations

The following abbreviations are used in this manuscript:

CPE	constant phase element
TER	transendothelial resistance
R	resistance
C	capacitance
HUVEC	human umbilical vein endothelial cells
thr	thrombin
dd_{H_2O}	deionized water

References

1. Thom, V.T.; Wendel, M.; Deussen, A. Regulation of Ecto-5′-Nucleotidase by Docosahexaenoic Acid in Human Endothelial Cell *Cell. Physiol. Biochem.* **2013**, *32*, 355–366. [CrossRef] [PubMed]
2. Khedr, S.; Deussen, A.; Kopaliani, I.; Zatschler, B.; Martin, M. Effects of tryptophan-containing peptides on angiotensin-convertin enzyme activity and vessel tone ex vivo and in vivo. *Eur. J. Nutr.* **2018**, *57*, 907–915. [CrossRef] [PubMed]
3. Lum, H.; Aschner, J.L.; Phillips, P.G.; Fletcher, P.W.; Malik, A.B. Time course of thrombin-induced increase in endothelia permeability: Relationship to Ca2+i and inositol polyphosphates. *Am. J. Physiol. Lung Cell. Mol. Physiol.* **1992**, *263*, L219–L22 [CrossRef] [PubMed]
4. Rabiet, M.J.; Plantier, J.L.; Rival, Y.; Genoux, Y.; Lampugnani, M.G.; Dejana, E. Thrombin-Induced Increase in Endothelia Permeability Is Associated With Changes in Cell-to-Cell Junction Organization. *Arterioscler. Thromb. Vasc. Biol.* **1996**, *16*, 488–49 [CrossRef] [PubMed]
5. Ehret, R.; Baumann, W.; Brischwein, M.; Schwinde, A.; Stegbauer, K.; Wolf, B. Monitoring of cellular behavior by impedanc measurements on interdigitated electrode structures. *Biosens. Bioelectron.* **1997**, *12*, 29–41. [CrossRef] [PubMed]
6. Feroz, F.; Hobson, M.P. Multimodal nested sampling: An efficient and robust alternative to Markov Chain Monte Carlo method for astronomical data analyses. *Mon. Not. R. Astron. Soc.* **2008**, *384*, 449–463. [CrossRef]
7. von Toussaint, U. Bayesian inference in physics. *Rev. Mod. Phys.* **2011**, *83*, 943–999. [CrossRef]
8. Gregory, P. *Bayesian Logical Data Analysis for the Physical Sciences: A Comparative Approach with Mathematica® Support*; Cambridg University Press: Cambridge, UK, 2005. [CrossRef]
9. Handley, W.; Hobson, M.; Lasenby, A. POLYCHORD: Next-generation nested sampling. *Mon. Not. R. Astron. Soc.* **2015**, *45* 4384–4398. [CrossRef]
10. Feroz, F.; Hobson, M.P.; Bridges, M. MultiNest: An efficient and robust Bayesian inference tool for cosmology and particle physic *Mon. Not. R. Astron. Soc.* **2009**, *398*, 1601–1614. [CrossRef]
11. Buchner, J.; Georgakakis, A.; Nandra, K.; Hsu, L.; Rangel, C.; Brightman, M.; Merloni, A.; Salvato, M.; Donley, J.; Kocevski, D X-ray spectral modelling of the AGN obscuring region in the CDFS: Bayesian model selection and catalogue. *Astron. Astrophys* **2014**, *564*, A125. [CrossRef]
12. Wegener, J.; Seebach, J. Experimental tools to monitor the dynamics of endothelial barrier function: A survey of in vitr approaches. *Cell Tissue Res.* **2014**, *355*, 485–514. [CrossRef] [PubMed]
13. Fricke, H. The Electric Capacity of Suspensions of Red Corpuscles of a Dog. *Phys. Rev.* **1925**, *26*, 682–687. [CrossRef]
14. Wegener, J.; Sieber, M.; Galla, H.J. Impedance analysis of epithelial and endothelial cell monolayers cultured on gold surfaces. *J Biochem. Biophys. Methods* **1996**, *32*, 151–170. [CrossRef] [PubMed]
15. Reindl, A.; Bier, M. Impedance spectroscopy of ions at liquid-liquid interfaces. *Phys. Rev. E* **2013**, *88*, 052312. [CrossRef] [PubMed]

physical sciences forum

MDPI

·oceeding Paper

Bayesian Data Analysis Method for an Experiment to ·.'easure the Gravitational Acceleration of Antihydrogen †

Danielle Hodgkinson *, Joel Fajans and Jonathan S. Wurtele

Department of Physics, University of California at Berkeley, Berkeley, CA 94720, USA; joel@physics.berkeley.edu (J.F.); wurtele@berkeley.edu (J.S.W.)
* Correspondence: danielle.louise.hodgkinson@cern.ch
† Presented at the 42nd International Workshop on Bayesian Inference and Maximum Entropy Methods in Science and Engineering, Garching, Germany, 3–7 July 2023.

Abstract: The ALPHA-g experiment at CERN intends to observe the effect of gravity on antihydrogen. In ALPHA-g, antihydrogen is confined to a magnetic trap with an axis aligned parallel to the Earth's gravitational field. An imposed difference in the magnetic field of the confining coils above and below the trapping region, known as a bias, can be delicately adjusted to compensate for the gravitational potential experienced by the trapped anti-atoms. With the bias maintained, the magnetic fields of the coils can be ramped down slowly compared to the anti-atom motion; this releases the antihydrogen and leads to annihilations on the walls of the apparatus, which are detected by a position-sensitive detector. If the bias cancels out the gravitational potential, antihydrogen will escape the trap upwards or downwards with equal probability. Determining the downward (or upward) escape probability, p, from observed annihilations is non-trivial because the annihilation detection efficiency may be up–down asymmetric; some small fraction of antihydrogen escaping downwards may be detected in the upper region (and vice versa) meaning that the precise number of trapped antihydrogen atoms is unknown. In addition, cosmic rays passing through the apparatus lead to a background annihilation rate, which may also be up–down asymmetric. We present a Bayesian method to determine p by assuming annihilations detected in the upper and lower regions are independently Poisson distributed, with the Poisson mean expressed in terms of experimental quantities. We solve for the posterior p using the Markov chain Monte Carlo integration package, Stan. Further, we present a method to determine the gravitational acceleration of antihydrogen, a_g, by modifying the analysis described above to include simulation results. In the modified analysis, p is replaced by the simulated probability of downward escape, which is a function of a_g.

Keywords: Markov chain Monte Carlo; Stan; antimatter; CERN

1. Introduction

The universe appears to consist almost entirely of matter with a paucity of antimatter. Known as the baryon asymmetry problem, this disparity remains one of the large outstanding questions in physics. Experiments that investigate the standard model's predictions of the fundamental properties of antimatter have the potential to illuminate this open question. The ALPHA experiment at CERN traps and probes the simplest antimatter atom—antihydrogen—to enable comparisons to the well-known properties of hydrogen. ALPHA's new apparatus, known as ALPHA-g, intends to measure the effect of gravity on trapped antihydrogen [1–3]. According to Einstein's weak equivalence principle, antihydrogen accelerates at the same rate as hydrogen in the gravitational field of the Earth; a difference in the acceleration rates would be an extraordinary observation and could lead to a new way of understanding the excess of matter in our universe.

In ALPHA, antihydrogen is created by merging antiproton and positron plasmas. A schematic of the ALPHA experiment is shown in Figure 1. Antiprotons, from CERN's antiproton decelerator, enter the experiment from the left and arrive at the antiproton catching trap. Positrons are generated from radioactive beta decay in a Surko-type accumulator [4]. These charged plasmas are first confined by Penning–Malmberg traps; the electric field of hollow cylindrical electrodes provides axial confinement and an axial magnetic field provides radial confinement. The plasmas then travel along beamlines that are guided around the corner into the ALPHA-g atom trap by magnetic fields where they are caught in another set of Penning–Malmberg traps.

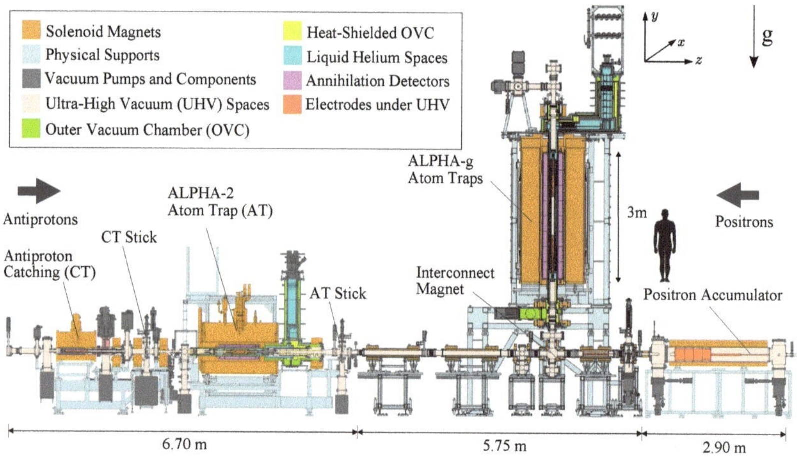

Figure 1. The ALPHA apparatus. Figure adapted from [5].

Charge-neutral antihydrogen [6] cannot be trapped with electric fields. A magnetic field configuration with a minimum magnitude at the trap center can trap antihydrogen via the interaction of its magnetic moment with the magnetic field, $\mathbf{B}$, which results in a trapping potential, $U_T = -\mu_B(|\mathbf{B}| - |\mathbf{B}_{\text{min}}|)$, where μ_B is the Bohr magneton and $\mathbf{B}_{\text{min}}$ is the minimum magnetic field in the trap. This configuration prevents antihydrogen from annihilating on the trap wall. The magnetic field magnitude, $|\mathbf{B}|$, which increases as a function of the radius within the trapping region, is generated by an octupole magnet and axially spaced coils. The coils confine antihydrogen axially. This magnetic 'bowl' potential is superimposed onto the Penning–Malmberg trapping fields prior to merging antiproton and positron plasmas. The merging technique forms antihydrogen in a thermal distribution with a temperature of $\sim$30 K. The depth of the magnetic bowl is only approximately 0.5 K; hence, most antihydrogen escapes quickly from the trap and only the fraction with the lowest energy remains confined.

The ALPHA-g superconducting magnets are shown in Figure 2. ALPHA designed the 'up–down measurement' section to measure the sign of the gravitational force on antihydrogen in the Earth's gravitational field. Initially, both the long octupole and lower short octupole are energized to provide radial antihydrogen confinement and the coils above and below the trapping regions. The 'lower' and 'upper' coils, are energized to provide axial confinement.

The experiment relies on the principle that trapped antihydrogen experiences a gravitational potential as well as a magnetic confining potential in ALPHA-g. Since the force of gravity is (in theory) parallel to the trap axis, the total potential for a trapped antihydrogen atom is

$$U(x,y,z) = -\mu_B\Big[|\mathbf{B}(x,y,z)| - |\mathbf{B}_{\text{min}}(x,y,z)|\Big] + m a_g z, \quad (1)$$

where m is the antihydrogen mass and a_g is the parameter of interest: the antihydrogen gravitational acceleration. When the magnetic fields provided by the lower and upper coils are equal, the gravitational potential results in an up–down asymmetry in the total potential, which is shown as 'no compensation' in Figure 3. Note that Figure 3 assumes antihydrogen behaves identically to hydrogen in the gravitational field of the Earth ($a_g = g$, where g is the hydrogen gravitational acceleration).

The magnetic fields created by the lower and upper coils can be deliberately unbalanced by supplying slightly different currents. This difference in the magnetic field magnitude, a bias, is defined as

$$\phi = \frac{\mu_B\left[|\boldsymbol{B}(x=0, y=0, z_u)| - |\boldsymbol{B}(x=0, y=0, z_l|\right]}{m(z_u - z_l)}, \quad (2)$$

where z_u and z_l are the axial centers of the upper and lower coils, respectively. The bias compensates for the gravitational potential over the length of the trap and can be delicately adjusted and varied over different experimental trials. Some biases will overcompensate for the gravitational potential and others will undercompensate, as shown in Figure 3. When the gravitational force is perfectly compensated, the total potential is up–down symmetric.

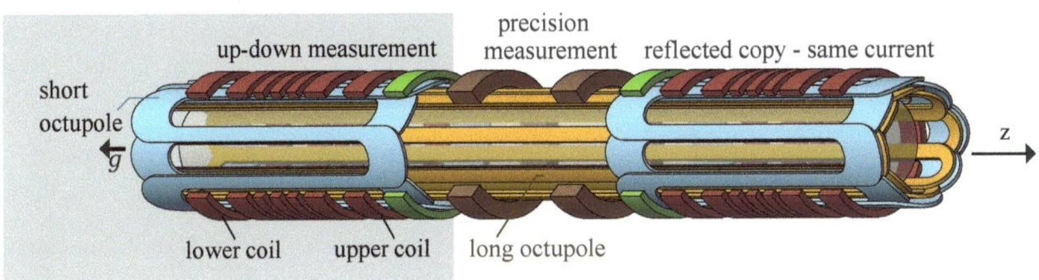

Figure 2. ALPHA-g superconducting magnets. The diagram is rotated such that the gravitational force points to the left. For the experiment described in the text, antihydrogen is confined in the 'up–down measurement' region with magnetic potential produced initially by energising the long octupole (yellow), short octupole (blue) and the lower and upper coils (red). Figure courtesy of Chukman So.

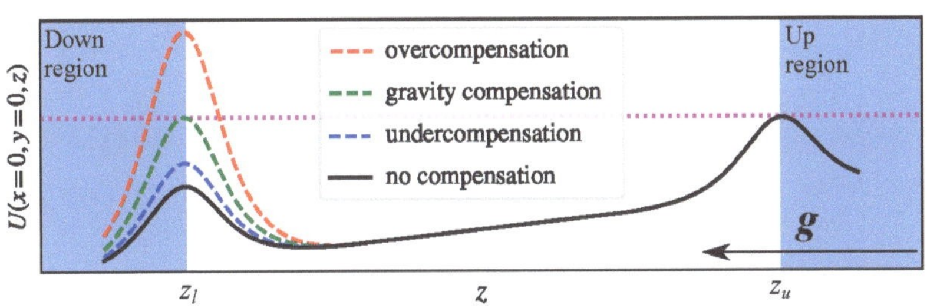

Figure 3. An exaggerated diagram of the on-axis magnetic potential during an up–down measurement. If the magnetic field produced by the lower and upper coils is equal (black solid curve), the gravitational potential leads to the (exaggerated) up/down asymmetric potential shown. By varying the relative magnetic field produced by the coils, the gravitational potential can be compensated. The red, blue and green dashed lines show varying degrees of compensation. The regions in which detected antihydrogen is considered to have escaped downwards or upwards are highlighted in blue. The centers of the lower and upper coils are marked as z_l and z_u, respectively.

To make a measurement of the gravitational potential, we must perform a controlled ramp down of the magnetic fields, during which we monitor the direction of antihydrogen escapes. As the current in both the lower and upper coils is ramped down slowly compared to the antihydrogen motion, the bias is held constant. If the total potential is not perfectly symmetric during coil ramp down, antihydrogen will preferentially escape either upwards or downwards. When antihydrogen escapes the trap, it will annihilate with the internal trap structures; these annihilations are reconstructed by a time and position-sensitive detector.

During an experimental trial, annihilations in the up and down regions (defined in Figure 3) are counted.

When performing these trials at different biases, we count the annihilations in the up and down regions to find a balance point between the bias and the gravitational potential. An equal probability to escape downwards (as opposed to upwards) indicates correct compensation of the gravitational potential. When this is achieved, you can equate a measurement of the bias with the gravitational potential.

We use standard ALPHA simulation techniques [7–9] and detailed Biot–Savart models of the ALPHA-g magnetic fields to predict the probability, p, of an individual antihydrogen atom escaping the trap downwards. At a set bias, ϕ, and a set a_g, $p \approx n_d/(n_d + n_u)$, where n_d and n_u are the simulated counts in the up and down regions, respectively. Since the full simulation results are still being analysed, Figure 4 shows estimations of p as a function of bias for $a_g = g$ ('normal gravity', green) and $a_g = -g$ ('repulsive gravity', blue), called an 'escape curve'. Note that when the imposed bias cancels the gravitational potential we have that $p = 0.5$, and the antihydrogen has equal probability of upward and downward escape. The goal of the up–down measurement is to determine the sign of the gravitational acceleration of antihydrogen by comparing an experimental escape curve to the simulation results. Determining the experimental escape curve from the data is challenging due to the following:

1. There is an experimentally limited amount of trapped antihydrogen.
2. The efficiency of detection in the up and down regions may be asymmetric.
3. Some small fraction of antihydrogen escaping downwards may be detected in the up region (and vice versa).
4. The precise number of total trapped anti-atoms is unknown experimentally.
5. Cosmic rays passing through the apparatus lead to a background annihilation rate, which may be up–down asymmetric.

In this work, we present a Bayesian data analysis method that considers the effects above and calculates the posterior p from the experimental data for a given ϕ in order to enable the experimental data to be compared to the simulation results. In addition, we present a modification of this Bayesian technique to enable a measured value to be assigned to a_g by including simulated escape curves in a likelihood analysis.

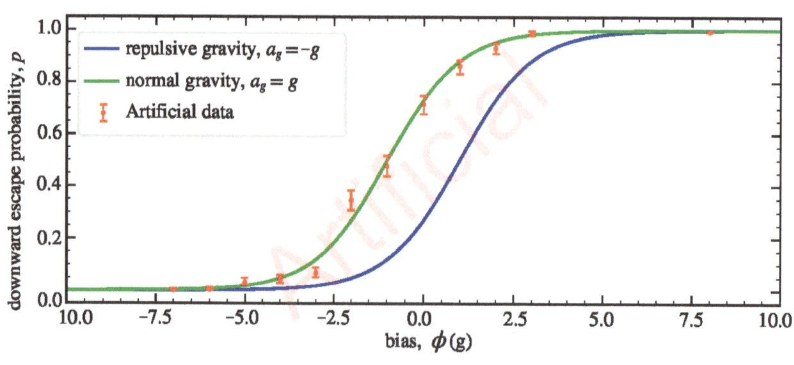

Figure 4. Estimation of simulated downward escape probability, p, as a function of bias, ϕ, for simulations assuming normal gravity ($a_g = g$, green) and repulsive gravity ($a_g = -g$, blue). Artificial data was generated using binomial sampling from the estimation of the normal gravity simulation (green line). The posterior p for the artificial data was determined using the Bayesian method described in the text. Red dots are the maximum a posteriori probability (MAP) estimate of the posterior p and the error bars are 68 % credible intervals.

2. Materials and Methods

2.1. Determining the Experimental Escape Curve

For each experimental bias, labelled with index i, our goal is to determine the posterior p_i. Our observed experimental data will be the up and down counts, $n_{u,i}$ and $n_{d,i}$, respectively. Let antihydrogen escaping the trap in the (correct) upwards (subscript u) direction be detected with efficiency η_{uu} and in the (incorrect) down (d) category with efficiency η_{du}. Similarly, antihydrogen escaping in the d direction will be detected in the correct category with efficiency η_{dd} and in the incorrect category with efficiency η_{ud}. Therefore, the detector efficiency has four components: $\boldsymbol{\eta} = (\eta_{uu}, \eta_{dd}, \eta_{du}, \eta_{ud})$.

Antihydrogen is formed in the up–down measurement trap with radial confinement provided by both the short and long octupole magnets. This maximizes the depth of the magnetic trapping potential, optimising the trapping rate. However, the superconducting windings of the long octupole (the end turns that connect the eight poles) are asymmetric about the trapping region (see Figure 2). To avoid an off-axis asymmetry in the trap magnetic fields that may mimic a bias, the long octupole is ramped down prior to the up–down measurement. This ramp down is slow compared to the anti-atom motion, meaning the antihydrogen will lose energy adiabatically [8] during the process. As a result, more antihydrogen can be trapped than by using the short octupole alone. The number of annihilations observed in the long octupole ramp down, $n_{l,i}$, provides an additional piece of information. Assuming antihydrogen is always formed in the same energy distribution in the initial trap, it provides a calibration to the total number of anti-atoms escaping during the coil ramp down. Since antihydrogen is expected to be formed in a distribution with a temperature ($\sim$30 K) that is large compared to the trap depth ($\approx$0.5 K), the energy distribution of the trapped antihydrogen is expected to be largely insensitive to fluctuations in antiproton and positron plasma conditions.

Cosmic rays passing through the apparatus can, to some extent, be distinguished from antihydrogen annihilations using reconstruction algorithms [10]. However, a small background rate of cosmic ray annihilations remains. We define this rate as having three components, $\boldsymbol{\rho} = (\rho_u, \rho_d, \rho_l)$. Components ρ_u and ρ_d are the mean cosmic ray annihilation rates in the up and down regions, respectively (see Figure 2), and ρ_l is the mean rate in the detection region for the long octupole ramp down, which spans the down region, the up region and the space in between. The mean cosmic ray annihilation rates can be measured experimentally when there are no particles trapped in the apparatus.

We assume the experimental counts ($n_{u,i}$, $n_{d,i}$ and $n_{l,i}$) are independently Poisson distributed such that the conditional probability is defined as

$$\mathrm{P}(n_{u,i}, n_{d,i,}, n_{l,i} | p_i, \boldsymbol{\rho}, \boldsymbol{\eta}, \lambda_i, \beta) = \frac{(\mu_{u,i})^{n_{u,i}}}{n_{u,i}!} e^{-\mu_{u,i}} \frac{(\mu_{d,i})^{n_{d,i}}}{n_{d,i}!} e^{-\mu_{d,i}} \frac{(\mu_{l,i})^{n_{l,i}}}{n_{l,i}!} e^{-\mu_{l,i}}, \quad (3)$$

where the Poisson means, $\mu_{u,i}$, $\mu_{d,i}$ and $\mu_{l,i}$, can be expressed in terms of p_i and the other experimental parameters,

$$\mu_{l,i} = \lambda_i + \rho_l \tau_{l,i}, \quad (4)$$

$$\mu_{u,i} = \Big[\eta_{\mathrm{uu}}\{1 - p_i\} + \eta_{\mathrm{ud}} p_i\Big] \beta \lambda_i + \rho_{\mathrm{u}} \tau_{m,i}, \quad (5)$$

$$\mu_{d,i} = \Big[\eta_{\mathrm{du}}\{1 - p_i\} + \eta_{\mathrm{dd}} p_i\Big] \beta \lambda_i + \rho_{\mathrm{d}} \tau_{m,i}, \quad (6)$$

where $\tau_{l,i}$ and $\tau_{m,i}$ represent the total duration of the long octupole and coil ramp down, respectively, and can be measured experimentally with high enough accuracy to neglect measurement uncertainty. The value λ_i is the mean number of antihydrogen atoms that are detected during the long octupole ramp down. Note that we have assumed a constant of proportionality, β, between λ_i and the mean number of antihydrogen atoms escaping during the coil ramp down that is shared between all biases.

We perform up–down measurements at $|\phi| \gg g$ to calibrate the detector efficiency. For $|\phi| \gg g$, almost all antihydrogen is expected to escape the trap downwards or upwards, depending on the sign of ϕ. Even at these extreme biases, a small fraction of anti-atoms are expected to escape in the non-preferential direction, which is likely due to a transfer of transverse to axial anti-atom energy [9] during the coil ramp down. Since the simulation predicts that some small fraction of antihydrogen will escape in the non-preferential direction, we include simulated results at these calibration biases in the likelihood analysis. The probability to observe $N_{u,j}$ and $N_{d,j}$ simulated counts in the up and down categories at a calibration bias is

$$P(N_{u,j}, N_{d,j}|p_j) = \frac{(\mu_{u,j})^{N_{u,j}}}{N_{u,j}!}e^{-\mu_{u,j}}\frac{(\mu_{d,j})^{N_{d,j}}}{N_{j,c}!}e^{-\mu_{d,j}}, \tag{7}$$

with Poisson means given by

$$\mu_{u,j} = (N_{u,j} + N_{d,j})(1 - p_j) \tag{8}$$

and

$$\mu_{d,j} = (N_{u,j} + N_{d,j})p_j, \tag{9}$$

where p_j is the probability of downward escape at the calibration bias.

The likelihood for a full set of experimental data at a total of K biases, of which a subset K_c are calibration biases, is

$$P(\overbrace{\{n_{u,0}, n_{d,0}, n_{l,0}\} \dots \{n_{u,K}, n_{d,K}, n_{l,K}\}, \{N_{u,0}, N_{d,0}\} \dots \{N_{u,K_c}, N_{d,K_c}\}}^{\mathcal{O}} | p_0 \dots p_K, \boldsymbol{\rho}, \boldsymbol{\eta}, \lambda_0 \dots \lambda_K, \beta) = \prod_{i=0}^{K} P(n_{u,i}, n_{d,i,}, n_{l,i}|p_i, \boldsymbol{\rho}, \boldsymbol{\eta}, \lambda_i, \beta) \prod_{j=0}^{K_c} P(N_{u,j}, N_{d,j}|p_j), \tag{10}$$

where the expressions for the conditional probabilities on the right hand side are given by Equations (3) and (7). For simplicity, we denote the set of observed experimental data at K biases and observed simulation counts at K_c biases using the symbol $\mathcal{O}$. Bayes theorem states

$$P(p_i \dots p_K, \boldsymbol{\rho}, \boldsymbol{\eta}, \lambda_i \dots \lambda_K, \beta|\mathcal{O}) \propto P(\mathcal{O}|p_i \dots p_K, \boldsymbol{\rho}, \boldsymbol{\eta}, \lambda_i \dots \lambda_K, \beta)P(p_i \dots p_K, \boldsymbol{\rho}, \boldsymbol{\eta}, \lambda_i \dots \lambda_K, \beta), \tag{11}$$

and the posterior distribution of p_i can be found by marginalising the nuisance parameters,

$$P(p_i|\mathcal{O}) \propto \int d\beta \iint_{j=0\dots K} d\lambda_j \int d\boldsymbol{\eta} \int d\boldsymbol{\rho} \iint_{j=0\dots K, j\neq i} dp_j P(\mathcal{O}|p_0 \dots p_K, \boldsymbol{\rho}, \boldsymbol{\eta}, \lambda_0 \dots \lambda_K, \beta)P(p_0 \dots p_K, \boldsymbol{\rho}, \boldsymbol{\eta}, \lambda_0 \dots \lambda_K, \beta), \tag{12}$$

where the integrals labelled with index j represent j-dimensional integration. We assume a prior for each parameter that is independent of all other parameters such that $P(p_0 \dots p_K, \boldsymbol{\rho}, \boldsymbol{\eta}, \lambda_0 \dots \lambda_K, \beta) = P(p_0)P(p_1)\dots P(\beta)$. Each p_i is assigned a conservative prior that is uniform between 0 and 1. The priors on the cosmic background annihilation rates ($\boldsymbol{\rho}$) will be informed by detector counting experiments when the particle traps are empty. The detector efficiencies ($\boldsymbol{\eta}$) are, in general, assigned uniform priors between 0 and 1, but it was necessary to pin η_{uu} to a fixed value to avoid a numerical convergence issue; this did not impact the posterior p_i. Broad uniform priors are assigned to the λ_i and β parameters. To solve for $P(p_i|\mathcal{O})$ at each bias, we use the Markov chain Monte Carlo integration package, Stan [11], which uses no-U-turn sampling [12] (NUTS). Stan libraries are available in many programming languages; we use the Python library, referred to as PyStan.

In Figure 4 we show the result for an experimental escape curve, generated using artificial data. For each of the discrete bias values that make up the artificial dataset, $p(\phi)$ is determined from the normal gravity ($a_g = g$) curve. Then, 200 binomial samples are generated in the up or down categories with probability of downward escape $p(\phi)$. With current antihydrogen accumulation rates in ALPHA-g, we anticipate approximately three days of data collection per bias to trap this amount of antihydrogen. The sampling provides a set of artificial up (n_u) and down (n_d) counts at the discrete biases. We also assume $n_l = n_u + n_d$, asserting a 1:1 relationship between the long octupole and coil ramp-down counts, which is approximately the ratio predicted by simulation. We plot the maximum a posteriori probability (MAP) estimate and 68 % credible intervals of $P(p_i|\mathcal{O})$. The artificial data curve is clearly discernible from the repulsive gravity curve with reasonable assumed experimental statistics. We have neglected the uncertainty associated with the accuracy of the measurement of the bias, which will form a horizontal error bar on the real experimental data.

2.2. Determining the Antihydrogen Gravitational Acceleration

Here we present a modified version of the likelihood analysis, in which we include simulated escape curves at a range of different assumed gravitational accelerations, a_g. Simulation results indicate that changing the assumed a_g is equivalent to translating the simulated escape curve along the bias axis. For example, the normal gravity ($a_g = g$) simulated escape curve is indistinguishable from a translation of the repulsive gravity ($a_g = -g$) curve by −2 g along the bias axis. As before, we assume counts in the up and down regions and in the long octupole ramp down, which are independently Poisson distributed, and hence the likelihood of the experimental counts is given by Equation (3) with modified Poisson means defined as

$$\mu'_{l,i} = \lambda_i + \rho_l \tau_{l,i}, \tag{13}$$

$$\mu'_{u,i} = \Big[\eta_{\mathrm{uu}}\{1 - p_{\mathrm{sim}}(\phi_i + a_g)\} + \eta_{\mathrm{ud}} p_{\mathrm{sim}}(\phi_i + a_g)\Big]\beta\lambda_i + \rho_{\mathrm{u}}\tau_{m,i}, \tag{14}$$

$$\mu'_{d,i} = \Big[\eta_{\mathrm{du}}\{1 - p_{\mathrm{sim}}(\phi_i + a_g)\} + \eta_{\mathrm{dd}} p_{\mathrm{sim}}(\phi_i + a_g)\Big]\beta\lambda_i + \rho_{\mathrm{d}}\tau_{m,i}, \tag{15}$$

where the experimental downward escape probability, p_i (see Equations (4)–(6)), has been replaced by a function, p_{sim}, which represents the simulated downward escape probability and is evaluated at the bias, ϕ_i, plus an offset, a_g. Monte Carlo simulations model the antihydrogen dynamics at discrete biases and discrete assumed gravitational accelerations. For each assumed gravitational acceleration, the simulated escape curve (containing simulations at a set of discrete biases) is translated along the bias axis to line up with the no-gravity simulation. For example, an escape curve assuming normal gravity ($a_g = g$) is translated by $-g$ along the bias axis. Finite simulation statistics lead to a spread of the simulated points about a central value, which is included by assigning a Gaussian prior to p_{sim}. The value p_{sim} is determined as a continuous function of bias using spline interpolation. Since a_g represents the offset from an assumption of no gravity it is the measured antihydrogen gravitational acceleration (the parameter of interest).

The posterior a_g can be found by applying Bayes' theorem and integrating over the nuisance parameters, similar to Equation (12). We assign a uniform prior to a_g that is significantly broader than the likely range, which can be determined by inspection of Figure 4. Again, we use a Stan [11] model to perform this calculation. Figure 5 shows the posterior probability distribution for a_g using the artificial data described in Section 2.1. The result is an antihydrogen gravitational acceleration of $a_g = 1.0^{+0.2}_{-0.1}\,g$ to one decimal place, trivially confirming that the artificial data were sampled from the escape curve with $a_g = g$.

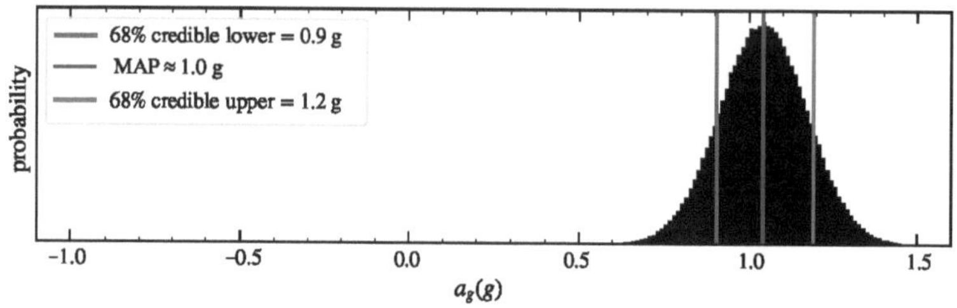

Figure 5. Posterior probability density of the gravitational acceleration of antihydrogen, a_g, for an artificial dataset. Blue and red/green vertical lines mark the maximum a posteriori probability (MAP) estimate and the 68 % credible intervals, respectively.

3. Discussion

In the ALPHA-g up–down measurement, antihydrogen will be confined by a magnetic field to enable the direction of its gravitational interaction with the Earth to be measured. An imposed difference in the magnetic field above and below the trapping region can compensate for the gravitational potential. Experimental trials involve slowly removing the magnetic confining potential with a maintained bias and counting the number of antihydrogen annihilations that are detected above and below the trapping region.

We presented a Bayesian data analysis method to determine the probability of an antihydrogen atom escaping the trap downwards, p_i, at a given bias during the experiment. Counts, detected in the regions above and below the trapping region, are assumed to be independently Poisson distributed. The method accounts for asymmetries in the efficiency by which antihydrogen annihilations are detected in the upper and lower regions of ALPHA-g, the small fraction of antihydrogen escaping downwards that may be detected in the upper region (and vice versa), the unknown number of total trapped anti-atoms and the (potentially up–down asymmetric) rates of cosmic ray annihilations. A demonstration of the method on an artificial dataset, with experimentally feasible statistics, predicts that we will be able to distinguish repulsive gravity from normal gravity. An alternative approach to this analysis [13] yields similar results.

Simulations indicate that changing the assumed value of the antihydrogen gravitational acceleration, a_g, is equivalent to translating the escape curve along the bias axis. Accordingly, we have developed a Bayesian analysis of the experimental value of a_g from the data by modifying the likelihood analysis. We applied the method to the artificial dataset, which trivially reproduces the assumed a_g. Again, the alternative approach of Ref. [13] yields similar results. The width of the posterior distribution reaffirms that $a_g = g$ and $a_g = -g$ will be distinguishable with a level of statistics that are feasibly obtainable. In this determination of a_g, we have neglected the uncertainty in the simulated magnetic fields and the unknown initial distribution of antihydrogen energies. For analysis of the real experimental data, these factors will provide an additional uncertainty in a_g that is not quantified here.

Author Contributions: Conceptualization, J.F.; methodology, D.H., J.F. and J.S.W.; software, D.H.; validation, N/A; formal analysis, D.H.; investigation, N/A; resources, J.F. and J.S.W.; data curation, N/A; writing—original draft preparation, D.H.; writing—review & editing, D.H., J.F. and J.S.W.; Visualization, D.H.; Supervision, J.F. and J.S.W.; project administration, J.F. and J.S.W.; funding acquisition, J.F. and J.S.W. All authors have read and agreed to the published version of the manuscript.

Funding: This work was supported by the DOE OFES and NSF-DOE Program in Basic Plasma Science.

Institutional Review Board Statement: Not applicable.

Informed Consent Statement: Not applicable.

Data Availability Statement: Data sharing is not applicable to this article.

Acknowledgments: We thank Andrew Charman (Department of Physics, University of California at Berkeley) for sharing his knowledge of Bayesian statistics and of the Stan programming language. We thank Karen Zukor for her edits to the language in this proceeding.

Conflicts of Interest: The authors declare no conflict of interest.

References

1. Zhmoginov, A.I.; Charman, A.E.; Shalloo, R.; Fajans, J.; Wurtele, J.S. Nonlinear dynamics of anti-hydrogen in magnetostatic traps: implications for gravitational measurements. *Class. Quantum Gravity* **2013**, *30*, 205014. [CrossRef]
2. Charman, A.E.; Amole, C.; Menary, S.; Capra, A.; Baquero-Ruiz, M.; Fajans, J.; Ashkezari, M.D.; Bertsche, W.; Butler, E.; Cesar, C.L.; et al. Description and first application of a new technique to measure the gravitational mass of antihydrogen. *Nat. Commun.* **2013**, *4*, 1785. [CrossRef]
3. Hangst, J.S. *Addendum to the ALPHA Proposal; The ALPHA-g Apparatus;* Report Number: CERN-SPSC-2016-031; SPSC-P-325-ADD-1; CERN: Geneva, Switzerland, 2016.
4. Surko, C.M.; Greaves, R.G.; Charlton, M. Stored positrons for antihydrogen production. *Hyperfine Interact.* **1997**, *109*, 181–188. [CrossRef]
5. Baker, C.J.; Bertsche, W.; Capra, A.; Cesar, C.L.; Charlton, M.; Christensen, A.J.; Collister, R.; Cridland Mathad, A.; Eriksson, S.; Evans, A.; et al. Design and performance of a novel low energy multispecies beamline for an antihydrogen experiment. *Phys. Rev. Accel. Beams* **2023**, *26*, 040101. [CrossRef]
6. Ahmadi, M.; Baquero-Ruiz, M.; Bertsche, W.; Butler, E.; Capra, A.; Carruth, C.; Cesar, C.L.; Charlton, M.; Charman, A.E.; Eriksson, S.; et al. An improved limit on the charge of antihydrogen from stochastic acceleration. *Nature* **2016**, *529*, 373–376. [CrossRef] [PubMed]
7. Amole, C.; Andresen, G.B.; Ashkezari, M.D.; Baquero-Ruiz, M.; Bertsche, W.; Butler, E.; Cesar, C.L.; Chapman, S.; Charlton, M.; Deller, A.; et al. Discriminating between antihydrogen and mirror-trapped antiprotons in a minimum-B trap. Antihydrogen and mirror-trapped antiproton discrimination: Discriminating between antihydrogen and mirror-trapped antiprotons in a minimum-B trap. *New J. Phys.* **2012**, *14*, 105010. [CrossRef]
8. Hodgkinson, D. On the Dynamics of Adiabatically Cooled Antihydrogen in an Octupole-Based Ioffe-Pritchard Magnetic Trap. Ph.D. Thesis, The University of Manchester, Wetzlar, Germany, 2022.
9. Zhong, M.; Fajans, J.; Zukor, A. Axial to transverse energy mixing dynamics in octupole-based magnetostatic antihydrogen traps. *New J. Phys.* **2018**, *20*, 053003. [CrossRef]
10. Hoecker, A.; Speckmayer, P.; Stelzer, J.; Therhaag, J.; von Toerne, E.; Voss, H.; Backes, M.; Carli, T.; Cohen, O.; Christov, A.; et al. TMVA—Toolkit for Multivariate Data Analysis. *arXiv* **2007**, arXiv:0703039.
11. Stan Development Team. Stan Modeling Language Users Guide and Reference Manual, 2.32. Available online: https://mc-stan.org (accessed on 12 July 2023).
12. Hoffman, M.D.; Gelman, A. The No-U-Turn Sampler: Adaptively Setting Path Lengths in Hamiltonian Monte Carlo. *J. Mach. Learn. Res.* **2014**, *15*, 1593–1623.
13. Urioni, M. (University of Brescia, Brescia, Italy); Stracka, S. (INFN Pisa, Pisa, Italy); Bonomi, G. (University of Brescia, Brescia, Italy). Unpublished work, 2023.

physical sciences forum

MDPI

Proceeding Paper

Learned Harmonic Mean Estimation of the Marginal Likelihood with Normalizing Flows †

Alicja Polanska [1,*], Matthew A. Price [1], Alessio Spurio Mancini [1] and Jason D. McEwen [1,2]

1 Mullard Space Science Laboratory, University College London, Dorking RH5 6NT, UK; m.price.17@ucl.ac.uk (M.A.P.); a.spuriomancini@ucl.ac.uk (A.S.M.); jason.mcewen@ucl.ac.uk (J.D.M.)
2 Alan Turing Institute, London NW1 2DB, UK
* Correspondence: alicja.polanska.22@ucl.ac.uk

† Presented at the 42nd International Workshop on Bayesian Inference and Maximum Entropy Methods in Science and Engineering, Garching, Germany, 3–7 July 2023.

Abstract: Computing the marginal likelihood (also called the Bayesian model evidence) is an important task in Bayesian model selection, providing a principled quantitative way to compare models. The learned harmonic mean estimator solves the exploding variance problem of the original harmonic mean estimation of the marginal likelihood. The learned harmonic mean estimator learns an importance sampling target distribution that approximates the optimal distribution. While the approximation need not be highly accurate, it is critical that the probability mass of the learned distribution is contained within the posterior in order to avoid the exploding variance problem. In previous work, a bespoke optimization problem is introduced when training models in order to ensure this property is satisfied. In the current article, we introduce the use of normalizing flows to represent the importance sampling target distribution. A flow-based model is trained on samples from the posterior by maximum likelihood estimation. Then, the probability density of the flow is concentrated by lowering the variance of the base distribution, i.e., by lowering its "temperature", ensuring that its probability mass is contained within the posterior. This approach avoids the need for a bespoke optimization problem and careful fine tuning of parameters, resulting in a more robust method. Moreover, the use of normalizing flows has the potential to scale to high dimensional settings. We present preliminary experiments demonstrating the effectiveness of the use of flows for the learned harmonic mean estimator. The harmonic code implementing the learned harmonic mean, which is publicly available, has been updated to now support normalizing flows.

Keywords: Bayesian model selection; harmonic mean estimator; normalizing flows

1. Introduction

Model selection is the task of evaluating which of the statistical models under consideration best describes observed data. In the Bayesian formalism, this involves computing the marginal likelihood (also called the Bayesian model evidence), which gives a way to a quantitative comparison of the suitability of models for a given problem. Model selection is relevant in a range of fields such as astronomy, biostatistics, economics, medical research, and many more. However, in practice, Bayesian model selection is often difficult, as computing the marginal likelihood requires evaluating a high-dimensional integral, which can be a challenging task. A number of methods to compute the marginal likelihood have been proposed (for reviews see [1,2]), such as nested sampling [3,4].

The learned harmonic mean estimator was proposed recently by some of the authors of the current article as an effective technique with which to compute the marginal likelihood [5]. The estimator requires only samples from the posterior and so is agnostic to the method used to generate samples, in contrast to nested sampling. Thus, the learned harmonic mean can be easily coupled with various MCMC sampling techniques or variational inference approaches. This property also allows the estimator to be adapted to address Bayesian model selection for simulation-based inference (SBI) [6], where an explicit likelihood is unavailable or infeasible. The learned harmonic mean estimator is implemented in the harmonic Python code (https://github.com/astro-informatics/harmonic, accessed on 30 June 2023).

In this work, we introduce the use of normalizing flows [7] for the learned harmonic mean estimator, which addresses some limitations of the models used previously. The use of normalizing flows eliminates the need for bespoke training, resulting in a more robust and scalable approach, which is now implemented in the harmonic code. We first review the learned harmonic mean estimator, before describing how normalizing flows may be used for the estimator and their main advantages in this context. We then present a number of experiments that demonstrate the effectiveness of the use of flows in the learned harmonic mean estimator.

2. The Harmonic Mean Estimator

Bayesian model selection requires the computation of the marginal likelihood given by

$$z = p(y \mid M) = \int \mathrm{d}\theta \, p(y \mid \theta, M) \, p(\theta \mid M) = \int \mathrm{d}\theta \mathcal{L}(\theta) \pi(\theta), \tag{1}$$

where y denotes observed data, θ the parameters of interest, and M the model under consideration. We adopt the shorthand notation for the likelihood of $\mathcal{L}(\theta) = p(y \mid \theta, M)$ and the prior of $\pi(\theta) = p(\theta \mid M)$.

The harmonic mean estimator was first proposed by [8], who showed that the marginal likelihood z can be estimated from the harmonic mean of the likelihood, given posterior samples. This follows by considering the expectation of the reciprocal of the likelihood with respect to the posterior distribution, leading to the following estimator of the reciprocal of the marginal likelihood $\rho = 1/z$:

$$\hat{\rho} = \frac{1}{N} \sum_{i=1}^{N} \frac{1}{\mathcal{L}(\theta_i)}, \quad \theta_i \sim p(\theta \mid y), \tag{2}$$

where N specifies the number of samples θ_i drawn from the posterior $p(\theta \mid y)$. The marginal likelihood can then be estimated from its reciprocal straightforwardly [5]. It was soon realized that the original harmonic mean estimator can fail catastrophically [9], as it can suffer from an exploding variance.

The estimator can also be interpreted as importance sampling. Consider the reciprocal marginal likelihood, which may be expressed in terms of the prior and posterior as:

$$\rho = \int \mathrm{d}\theta \, \frac{1}{\mathcal{L}(\theta)} \, p(\theta|y) = \int \mathrm{d}\theta \, \frac{1}{z} \, \frac{\pi(\theta)}{p(\theta \mid y)} \, p(\theta \mid y). \tag{3}$$

It is clear that the estimator has an importance sampling interpretation where the importance sampling target distribution is the prior $\pi(\theta)$, while the sampling density is the posterior $p(\theta|y)$, in contrast to typical importance sampling scenarios.

For importance sampling to be effective, one requires the sampling density to have fatter tails than the target distribution, i.e., to have greater probability mass in the tails of the distribution. Typically, the prior has fatter tails than the posterior since the posterior updates our initial understanding of the underlying parameters θ that are encoded in the prior, in the presence of new data y. For the harmonic mean estimator, the importance sampling density (the posterior) typically does not have fatter tails than the target (the prior) and so importance sampling is not effective. This explains why the original harmonic mean estimator can be problematic.

In [10], an arbitrary density $\varphi(\theta)$ is introduced to relate the reciprocal of the marginal likelihood to the likelihood through the following expectation:

$$\rho = \mathbb{E}_{p(\theta|y)}\left[\frac{\varphi(\theta)}{\mathcal{L}(\theta)\pi(\theta)}\right]. \tag{4}$$

The above expression motivates the estimator:

$$\hat{\rho} = \frac{1}{N}\sum_{i=1}^{N}\frac{\varphi(\theta_i)}{\mathcal{L}(\theta_i)\pi(\theta_i)}, \quad \theta_i \sim p(\theta|y). \tag{5}$$

The normalized density $\varphi(\theta)$ can be interpreted as an alternative importance sampling target distribution, hence we refer to this approach as the re-targeted harmonic mean estimator. Note that the original harmonic mean estimator is recovered for the target distribution $\varphi(\theta) = \pi(\theta)$.

The learned harmonic mean estimator is introduced in [5], where the target density $\varphi(\theta)$ is learned by machine learning techniques. It is shown in [5] that the optimal target distribution is the posterior. Since the target must be normalized, the normalized posterior is clearly not accessible since its normalizing constant is precisely the term of interest. The learned harmonic mean approximates the optimal target of the posterior with a learned model that is normalized. While the approximation need not be highly accurate, it is critical that the probability mass of the learned distribution is contained within the posterior in order to avoid the exploding variance problem. In [5], a bespoke optimization problem is introduced when training models in order to ensure this property is satisfied. Specifically, the model is fitted by minimizing the variance of the resulting estimator, while ensuring it is also unbiased, and with possible regularization. Such an approach requires a careful selection of an appropriate model and its hyperparameters for a problem at hand, determined by cross-validation. Furthermore, only simple classical machine learning models were considered in [5], which in many cases struggle to scale to high-dimensional settings.

3. Learning the Target Distribution Using Normalizing Flows

In this paper, we learn the target distribution of the learned harmonic mean estimator [5] using normalizing flows. Using normalizing flows renders the previous bespoke approach to training no longer necessary since it provides an elegant way to ensure that the probability mass of the learned distribution is contained within the posterior, thereby resulting in a learned harmonic mean estimator that is more flexible and robust. Furthermore, normalizing flows also offer the potential to scale to higher dimensional settings. We first introduce normalizing flows, before describing how they may be used for the learned harmonic mean estimator and their main advantages in this context.

3.1. Normalizing Flows

Normalizing flows are a class of probabilistic models that allow one to evaluate the density of and sample from a learned probability distribution (for a review, see [7]). They consist of a series of transformations that are applied to a simple base distribution. A vector θ of an unknown distribution $p(\theta)$ can be expressed through a transformation T of a vector z sampled from a base distribution $q(z)$:

$$\theta = T(z), \text{ where } z \sim q(z). \tag{6}$$

Typically, the base distribution is chosen so that its density can be evaluated simply and so that it can be sampled from easily. Often a Gaussian is used for the base distribution. The unknown distribution can then be recovered by the change of variables formula:

$$p(\theta) = q(z)|\det J_T(z)|^{-1}, \tag{7}$$

where $J_T(z)$ is the Jacobian corresponding to transformation T. In a flow-based model, T consists of a series of learned transformations that are each invertible and differentiable, so that the full transformation is also invertible and differentiable. This allows us to compose multiple simple transformations with learned parameters, into what is called a flow, obtaining a normalized approximation of the unknown distribution that we can sample from and evaluate. Careful attention is given to construction of the transformations such that the determinant of the Jacobian can be computed easily.

A relatively simple example of a normalizing flow is the real-valued non-volume preserving (real NVP) flow introduced in [11]. It consists of a series of bijective transformations given by affine coupling layers. Consider the D dimensional input z, split into elements up to and following d, respectively, $z_{1:d}$ and $z_{d+1:D}$, for $d < D$. Given input z, the output y of an affine couple layer is calculated by

$$y_{1:d} = z_{1:d}; \tag{8}$$

$$y_{d+1:D} = z_{d+1:D} \odot \exp(s(z_{1:d})) + t(z_{1:d}), \tag{9}$$

where $\odot$ denotes Hadamard (elementwise) multiplication. The scale s and translation t are typically represented by neural networks with learnable parameters that take as input $z_{1:d}$. This construction is easily invertible and ensures the Jacobian is a lower-triangular matrix, making its determinant efficient to calculate.

3.2. Concentrating the Probability Density for the Learned Harmonic Mean Estimator

Normalizing flows meet the core requirements of the learned target distribution of the learned harmonic mean estimator: namely, they provide a normalized probability distribution for which one can evaluate probability densities. In this work, we use them to introduce an elegant way to ensure the probability mass of the learned distribution is contained within the posterior. We thereby avoid the exploding variance issue of the original harmonic mean estimator and can evaluate the marginal likelihood accurately without the need for fine-tuning.

Reducing the variance of the base distribution, or equivalently lowering its "temperature" in a statistical mechanics perspective, clearly concentrates the probability density of the base distribution. This has the effect of also concentrating the probability density of the transformed distribution due to the continuity and differentiability of the flow. Consequently, once a flow is trained to approximate the posterior, by lowering the temperature of the base distribution (i.e., reducing its variance) we can concentrate the learned distribution to ensure its probability mass is contained within the posterior, as illustrated in Figure 1.

The learned distributions considered previously for the learned harmonic mean estimator [5] required the introduction of a bespoke optimization problem for training in order to ensure the learned target is contained within the posterior. This requires careful selection of an appropriate model and its hyperparameters, determined by cross-validation. The introduced normalizing flow approach renders bespoke training no longer necessary. Instead, we train a flow in the usual manner, based on maximum likelihood estimation, before concentrating its probability density. There is only one parameter to consider, the temperature $T \in (0, 1)$. Moreover, we expect a common value of $T \sim 0.9$ to be suitable for most problems. Once we have a flow with its probability density concentrated for $\varphi(\theta)$, the learned harmonic mean estimator can be computed in the usual manner [5]. Using normalizing flows with the learned harmonic mean thus provides a much more robust method. Furthermore, an added benefit of using flows is that we can draw samples from the flow distribution efficiently, in order to easily visualize the concentrated target distribution and compare it to the posterior.

In this preliminary work, we consider real NVP flows only, as described above, which are implemented in the harmonic code. In the future, we will consider more expressive and scalable flows. The use of normalizing flows for the learned distribution therefore has the potential to extend the learned harmonic mean estimator to problems with complex, high-dimensional posteriors.

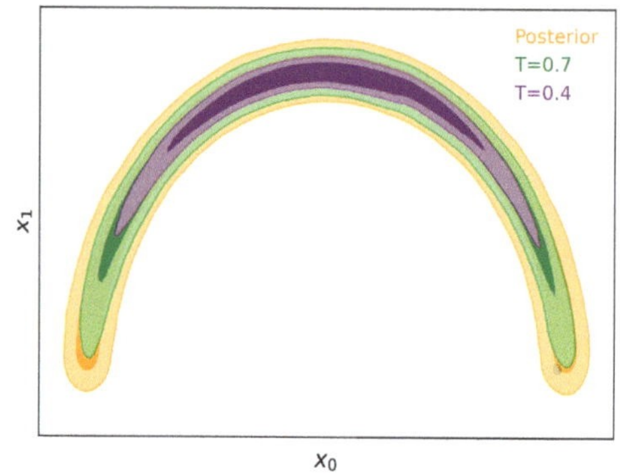

Figure 1. Diagram illustrating the concentration of the probability density of a normalizing flow. The flow is trained on samples from the posterior, giving us a normalized approximation of the posterior distribution. The temperature of the base distribution $T \in (0, 1)$ is reduced, which concentrates the probability density of the transformed distribution, ensuring that it is contained within the posterior. The concentrated flow can then be used as the target distribution for the learned harmonic mean estimator, avoiding the exploding variance issue of the original harmonic mean estimator.

4. Experiments

To validate the effectiveness of the method described in Section 3, we repeat a number of the numerical experiments carried out in [5] but using normalizing flows as the target distribution of the learned harmonic mean estimator. In the experiments that follow, we consider a real NVP flow where the scale and translation networks of the affine coupling layers are given by two-layer dense neural networks with a leaky ReLU in between. The scaling layers additionally include a proceeding softplus activation. We typically consider a flow with six coupling layers, where typically only the first two include scaling, and permute elements of the vector between coupling layers to ensure the flow transforms all elements. We consider a Gaussian base distribution with unit variance. We use the emcee package [12] to generate MCMC samples from the posterior. We then train the real NVP flow on half of the samples by maximum likelihood and calculate the marginal likelihood using the remaining samples by the learned harmonic mean estimator with the flow concentrated to temperature T. In all experiments, we consider an identical temperature of $T = 0.9$, which works well throughout, demonstrating that T does not require fine-tuning. We consider a relatively simple flow in this preliminary work and a small number of simple experiments. In future work, we will consider more expressible and scalable flows, and further experiments to thoroughly evaluate the robustness and scalability of the method.

4.1. Rosenbrock

A common benchmark problem to test methods that compute the marginal likelihood is a likelihood specified by the Rosenbrock function, which exhibits a narrow curving degeneracy. We consider the Rosenbrock likelihood in $d = 2$ dimensions and a simple uniform prior with $x_0 \in [-10, 10]$ and $x_1 \in [-5, 15]$. We sample the resulting posterior distribution, drawing 5000 samples for 200 chains, with burn-in of 2000 samples, yielding 3000 posterior samples per chain. Figure 2 shows a corner plot of the training samples from the posterior (red) and from the normalizing flow (blue) at temperature $T = 0.9$. It can be seen that the concentrated flow approximates the posterior well and has thinner tails, as required for the marginal likelihood estimate to be stable and accurate.

This process is repeated 100 times and the marginal likelihood is computed for each trial. Figure 3 shows a summary of the estimates across all the runs. The dashed red line in Figure 3a indicates the ground truth computed through numerical integration, which is tractable in two dimensions. It can be seen that the learned harmonic mean estimator using a real NVP flow provides an accurate and unbiased estimate of the marginal likelihood.

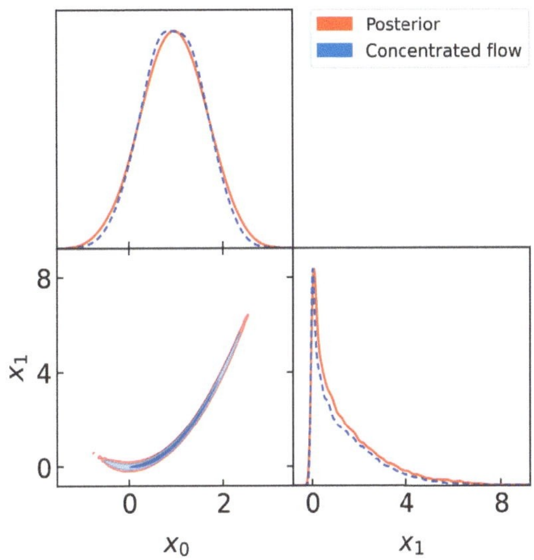

Figure 2. Corner plot of samples from the posterior (red) and real NVP flow with temperature $T = 0.9$ (blue) for the Rosenbrock benchmark problem. The target distribution given by the concentrated flow is contained within the posterior and has thinner tails, as required for the learned harmonic mean estimator.

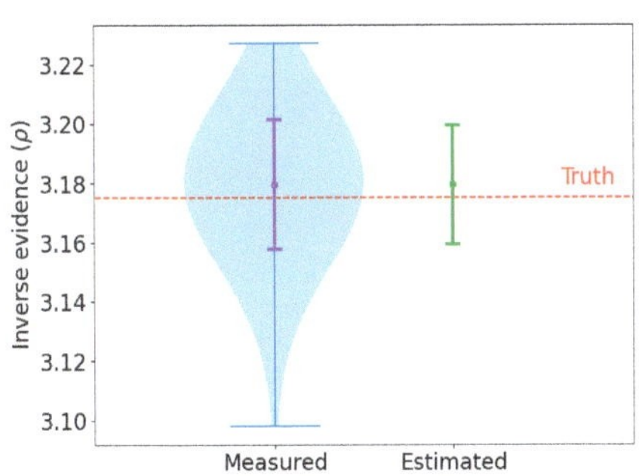

(**a**) Inverse evidence

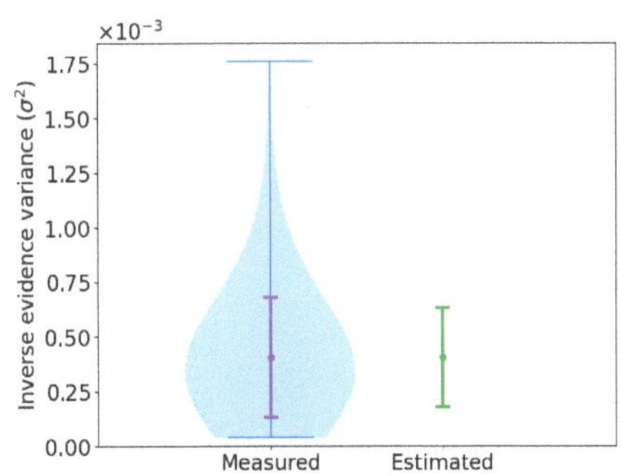

(**b**) Variance of inverse evidence

Figure 3. Marginal likelihood computed by the learned harmonic mean estimator with a concentrated flow for the Rosenbrock benchmark problem. One hundred experiments are repeated to recover empirical estimates of the statistics of the estimator. In panel (**a**), the distribution of marginal likelihood values are shown (measured) along with the estimate of the standard deviation computed by the error estimator (estimated). The ground truth is indicated by the red dashed line. In panel (**b**), the distribution of the variance estimator is shown (estimated) along with the standard deviation computed by the variance-of-variance estimator (estimated). The learned harmonic mean estimator and its error estimators are highly accurate.

4.2. Normal-Gamma

We consider the Normal-Gamma example, for which the marginal likelihood can be computed analytically [2,5]. It was found that the marginal likelihood values computed by the original harmonic mean estimator do not vary with a varying prior [2], highlighting this example as a pathological failure of the original harmonic mean estimator. We consider the same pathological example here and demonstrate that our learned harmonic mean estimator with normalizing flows is highly accurate (as is the learned harmonic mean with other models; [5]). We consider the Normal-Gamma model [5,13] with data $y_i \sim \mathrm{N}(\mu, \tau^{-1})$, for $i \in \{1, \ldots, n\}$, with mean μ and precision (inverse variance) τ. A normal prior is assumed for μ and a Gamma prior for τ:

$$\mu \sim \mathrm{N}\big(\mu_0, (\tau_0 \tau)^{-1}\big), \ \tau \sim \mathrm{Ga}(a_0, b_0), \tag{10}$$

with mean $\mu_0 = 0$, shape $a_0 = 10^{-3}$, and rate $b_0 = 10^{-3}$. The precision scale factor τ_0 is varied to observe the impact of changing prior on the computed marginal likelihood.

We draw 1500 samples for 200 chains, with burn-in of 500 samples, yielding 1000 posterior samples per chain. Figure 4 shows a corner plot of the training samples from the posterior for $\tau = 0.001$ (red) and from the normalizing flow (blue) at temperature $T = 0.9$. Again, it can be seen the concentrated learned target is close to the posterior but with thinner tails, as expected. We consider priors with $\tau \in \{10^{-4}, 10^{-3}, 10^{-2}, 10^{-1}, 1\}$. Figure 5 shows the relative accuracy of the marginal likelihood computed by the learned harmonic mean estimator using normalizing flows, that is, the ratio of the estimated marginal likelihood to the analytic ground truth. We additionally consider a concentrated flow with $T = 0.95$ to demonstrate that accuracy is not highly dependent on the temperature parameter. It can be seen that the estimate remains accurate and is indeed sensitive to the prior for both temperatures. The estimates for the flow with T closer to one have a slightly lower variance, as one would expect, since the broader target φ makes more efficient use of samples.

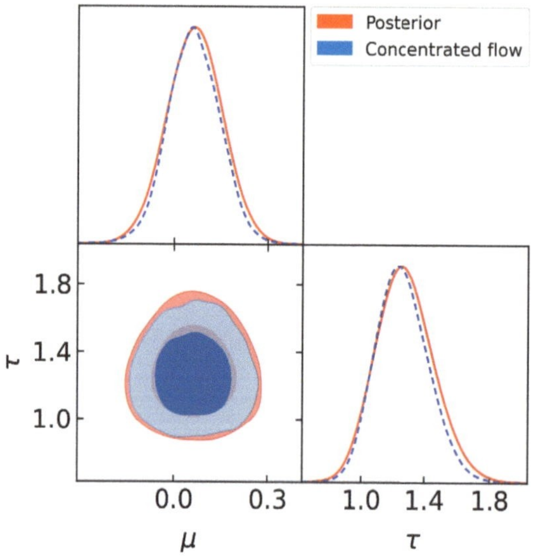

Figure 4. Corner plot of samples from the posterior (red) and real NVP flow trained on the posterior samples with temperature $T = 0.9$ (blue) for the Normal-Gamma example with $\tau = 0.001$. The target distribution given by the concentrated flow is contained within the posterior and has thinner tails, as required for the learned harmonic mean estimator.

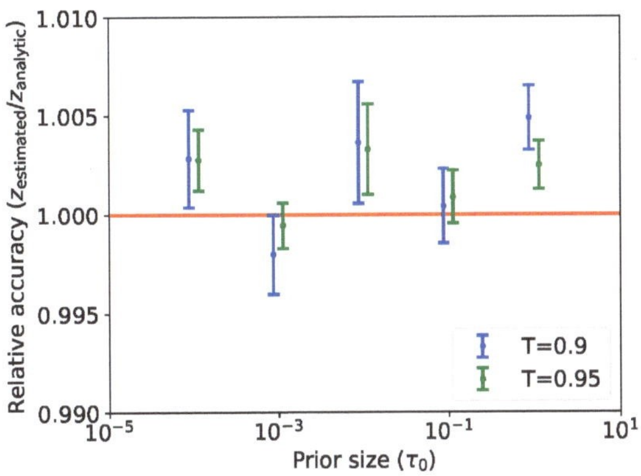

Figure 5. Ratio of marginal likelihood values computed by the learned harmonic mean estimator with a concentrated flow to those computed analytically for the Normal-Gamma problem. Error bars corresponding to the estimated standard deviation of the learned harmonic estimator are also shown. Notice that the marginal likelihood values computed by the learned harmonic mean estimator are highly accurate and are indeed sensitive to changes in the prior. Predictions made with flow at temperature $T = 0.9$ (blue) and $T = 0.95$ (green) are shown, which are slightly offset for ease of visualization, demonstrating that accuracy is not highly sensitive to the choice of T.

4.3. Logistic Regression Models: Pima Indian Example

We consider the comparison of two logistic regression models using the Pima Indians data, which is another common benchmark problem for comparing estimators of the marginal likelihood. The original harmonic mean estimator has been shown to fail catastrophically for this example [2]. The Pima Indians data [14], originally from the National Institute of Diabetes and Digestive and Kidney Diseases, were compiled from a study of indicators of diabetes in $n = 532$ Pima Indian women aged 21 or over. Seven primary predictors of diabetes were recorded, including: number of prior pregnancies (NP); plasma glucose concentration (PGC); diastolic blood pressure (BP); triceps skin fold thickness (TST); body mass index (BMI); diabetes pedigree function (DP); and age (AGE). The probability of diabetes p_i for person $i \in \{1, \ldots, n\}$ can be modelled by the logistic function. An independent multivariate Gaussian prior with precision $\tau = 0.01$ is assumed for parameters θ. Two different logistic regression models are compared, with different subsets of covariates:

$$\text{Model } M_1: \quad \text{covariates} = \{\text{NP, PGC, BMI, DP}\} \text{ (and bias)};$$
$$\text{Model } M_2: \quad \text{covariates} = \{\text{NP, PGC, BMI, DP, AGE}\} \text{ (and bias)}.$$

A reversible jump algorithm [15] is used by [2] to compute a benchmark Bayes factor BF_{12} of 13.96 ($\log \text{BF}_{12} = 2.6362$), which is treated as ground truth.

We draw 5000 samples from for 200 chains, with burn-in of 1000 samples, yielding 4000 posterior samples per chain. We train a flow consisting of six scaled layers followed by two unscaled ones. Figure 6 shows a corner plot of the training samples from the posterior (red) and from the normalizing flow (blue) at temperature $T = 0.9$. Again, it can be seen that the concentrated learned target is close to the posterior but with thinner tails, as expected. We compute the marginal likelihood for Model 1 and Model 2 using our learned harmonic mean estimator. The log evidence found for Model 1 and 2 is -257.2300 ± 0.0020 and -259.8602 ± 0.0031, respectively, resulting in the estimate $\log \text{BF}_{12} = 2.6302 \pm 0.0051$, which is in close agreement with the benchmark.

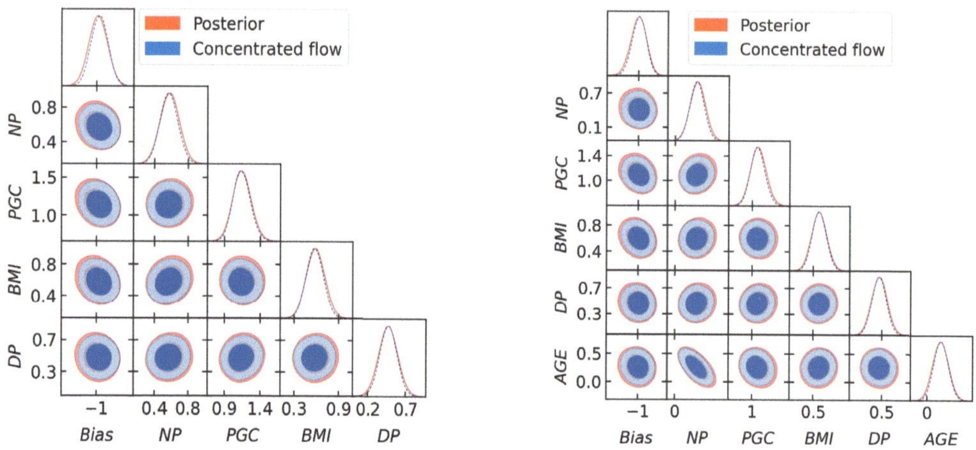

(**a**) Model 1 (**b**) Model 2

Figure 6. Corner plot of the samples from the posterior (red) and real NVP flow trained on the posterior samples with temperature $T = 0.9$ (blue) for the Pima Indian benchmark problem for $\tau = 0.01$. The target distribution given by the concentrated flow is contained within the posterior and has thinner tails, as required for the learned harmonic mean estimator.

5. Conclusions

In this work, we propose using normalizing flows for the learned harmonic mean estimator of the marginal likelihood. The flow may be fitted to posterior samples by the usual maximum likelihood estimation. Its probability density may then be concentrated by lowering the temperature of the base distribution, ensuring that the probability mass of the transformed distribution is contained within the posterior to avoid the exploding variance issue of the original harmonic mean estimator. The use of flows therefore results in a more robust learned harmonic mean estimator. We perform a number of experiments to compute the marginal likelihood with the proposed approach, using a real NVP flow, finding excellent agreement with ground truth values. In this preliminary work, we consider only simple real NVP flows and a simple set of experiments. In a follow-up article, we will consider more expressive and scalable flows to address problems with complex, high-dimensional posteriors. We will also perform a more extensive set of numerical experiments to thoroughly assess performance. This preliminary work nevertheless suggests that the learned harmonic mean estimator with normalizing flows provides an effective technique with which to compute the marginal likelihood for Bayesian model selection. Furthermore, it is applicable for any MCMC sampling technique or variational inference approach.

Author Contributions: Conceptualization, A.P., M.A.P. and J.D.M.; methodology, A.P., M.A.P. and J.D.M.; software, A.P., M.A.P. and J.D.M.; validation, A.P., M.A.P. and J.D.M.; investigation, A.P.; resources, J.D.M.; writing—original draft preparation, A.P. and J.D.M.; writing—review and editing, A.P., M.A.P., A.S.M. and J.D.M.; visualization, A.P.; supervision, M.A.P., A.S.M. and J.D.M.; funding acquisition, J.D.M. All authors have read and agreed to the published version of the manuscript.

Funding: A.P. is supported by the UCL Centre for Doctoral Training in Data Intensive Science (STFC grant number ST/W00674X/1). A.S.M. acknowledges support from the MSSL STFC Consolidated Grant ST/W001136/1. This research was also funded by EPSRC grant number EP/W007673/1.

Data Availability Statement: The harmonic code and experiments are available at https://github.com/astro-informatics/harmonic, accessed on 30 June 2023.

Conflicts of Interest: The authors declare no conflict of interest.

References

1. Clyde, M.; Berger, J.; Bullard, F.; Ford, E.; Jefferys, W.; Luo, R.; Paulo, R.; Loredo, T. Current challenges in Bayesian model choice. In *Statistical Challenges in Modern Astronomy IV*; Astronomical Society of the Pacific: San Francisco, CA, USA, 2007; Volume 371, p. 224.
2. Friel, N.; Wyse, J. Estimating the evidence—A review. *Stat. Neerl.* **2012**, *66*, 288–308. [CrossRef]
3. Skilling, J. Nested sampling for general Bayesian computation. *Bayesian Anal.* **2006**, *1*, 833–859. [CrossRef]
4. Ashton, G.; Bernstein, N.; Buchner, J.; Chen, X.; Csányi, G.; Fowlie, A.; Feroz, F.; Griffiths, M.; Handley, W.; Habeck, M.; et al. Nested sampling for physical scientists. *Nat. Rev. Methods Prim.* **2022**, *2*, 39. [CrossRef]
5. McEwen, J.D.; Wallis, C.G.; Price, M.A.; Spurio Mancini, A. Machine learning assisted Bayesian model comparison: Learnt harmonic mean estimator. *arXiv* **2023**, arXiv:2111.12720.
6. Spurio Mancini, A.; Docherty, M.M.; Price, M.A.; McEwen, J.D. Bayesian model comparison for simulation-based inference. *arXiv* **2022**, arXiv:2207.04037.
7. Papamakarios, G.; Nalisnick, E.; Rezende, D.J.; Mohamed, S.; Lakshminarayanan, B. Normalizing flows for probabilistic modeling and inference. *J. Mach. Learn. Res.* **2021**, *22*, 2617–2680.
8. Newton, M.A.; Raftery, A.E. Approximate Bayesian inference with the weighted likelihood bootstrap. *J. R. Stat. Soc. Ser. Stat. Methodol.* **1994**, *56*, 3–26. [CrossRef]
9. Neal, R.M. Contribution to the discussion of "Approximate Bayesian inference with the weighted likelihood bootstrap" by Newton MA, Raftery AE. *J. R. Stat. Soc. Ser. Stat. Methodol.* **1994**, *56*, 41–42.
10. Gelfand, A.E.; Dey, D.K. Bayesian model choice: Asymptotics and exact calculations. *J. R. Stat. Soc. Ser. Stat. Methodol.* **1994**, *56*, 501–514. [CrossRef]
11. Dinh, L.; Sohl-Dickstein, J.; Bengio, S. Density estimation using real nvp. *arXiv* **2016**, arXiv:1605.08803.
12. Foreman-Mackey, D.; Hogg, D.W.; Lang, D.; Goodman, J. emcee: The MCMC Hammer. *Publ. Astron. Soc. Pac.* **2013**, *125*, 306–312. [CrossRef]
13. Bernardo, J.; Smith, A. *Bayesian Theory*; John Wiley & Sons: New York, NY, USA, 1994.
14. Smith, J.W.; Everhart, J.E.; Dickson, W.; Knowler, W.C.; Johannes, R.S. Using the ADAP learning algorithm to forecast the onset of diabetes mellitus. In Proceedings of the Annual Symposium on Computer Application in Medical Care, Washington, DC, USA, 6–9 November 1988; American Medical Informatics Association: Bethesda, MD, USA, 1988; p. 261.
15. Green, P.J. Reversible jump Markov chain Monte Carlo computation and Bayesian model determination. *Biometrika* **1995**, *82*, 711–732. [CrossRef]

physical sciences forum

MDPI

Proceeding Paper

Quantification of Endothelial Cell Migration Dynamics Using Bayesian Data Analysis †

Anselm Hohlstamm, Andreas Deussen, Stephan Speier and Peter Dieterich *

Institut für Physiologie, Medizinische Fakultät Carl Gustav Carus, TU Dresden, Fetscherstraße 74, 01307 Dresden, Germany; anselm.hohlstamm@tu-dresden.de (A.H.); andreas.deussen@tu-dresden.de (A.D.); stephan.speier@tu-dresden.de (S.S.)

* Correspondence: peter.dieterich@tu-dresden.de

† Presented at the 42nd International Workshop on Bayesian Inference and Maximum Entropy Methods in Science and Engineering, Garching, Germany, 3–7 July 2023.

Abstract: Endothelial cells keep a tight and adaptive inner cell layer in blood vessels. Thereby, the cells develop complex dynamics through integrating active individual and collective cell migration, cell-cell interactions as well as interactions with external stimuli. It is the aim of this study to quantify and model these underlying dynamics. Therefore, we seeded and stained human umbilical vein endothelial cells (HUVECs) and recorded their positions every 10 min for 48 h via live-cell imaging. After image segmentation and tracking of several 10.000 cells, we applied Bayesian data analysis to models assessing the experimentally obtained cell trajectories. By analyzing the mean squared velocities, we found a dependence on the local cell density. Based on this connection, we developed a model, which approximates the time-dependent frequency of cell divisions. Furthermore, we determined two different phases of velocity deceleration, which are influenced by the emergence of correlated cell movements and time-dependent aging in this non-stationary system. By integrating the findings of correlation functions, we will be able to develop a comprehensive model to improve the understanding of endothelial cell migration in the future.

Keywords: cell migration; endothelium; Bayesian analysis

1. Introduction

Coordinated cell migration in the human body is essential throughout the whole life. Cells need to be in the correct positions during the development of an embryo or for the formation of organs, as well as for keeping them functioning by means of regeneration and repair [1,2]. In addition, migration plays a key part in the immune response of the organism [3,4]. However, the negative aspects of cell migration can be observed in pathological processes, for example, during cancer invasion [2]. Endothelial cells form in many vessels a barrier between the bloodstream and the surrounding tissues. Here, migration helps in stabilizing this barrier as well as repairing it if needed. Furthermore, a coordinated movement of endothelial cells is fundamental during the formation of new vessels [1]. On the pathophysiological side, there are many diseases like atherosclerosis, edema formation or respiratory distress syndrome, which arise from the dysfunction of the aforementioned barrier [1].

When modeling cell migration, several biological properties and environmental conditions have to be accounted for. For instance, cells are sensitive to chemotactic stimuli [4] but also to mechanical cues like the stiffness of the substrate, which influences mean cell speeds [5,6]. In collective settings, the regulation of adhesion molecules mediates the connection to neighboring cells as well as the adhesion to the extracellular matrix [1]. While endothelial cells can migrate individually under certain conditions, they also move in groups or sheets [1]. In a 2D tissue culture, human umbilical vein endothelial cells (HUVECs) also show collective movement patterns. Thereby, cells not only move around while staying connected but also influence each other in their behavior [7]. This generates the complex dynamics of active individual and collective cell migration.

Here, we analyze the dynamics of confluent endothelial cell migration and quantify changes in the mean squared velocity over time. We also determine the effect of the average cell density and its changes on mean cell velocities within the monolayer by coupling them in a combined mathematical model.

2. Materials and Methods

HUVECs were isolated from umbilical cords according to the previously described protocol [8]. The cells were then collected from T25 flasks, counted and filled into a microscope slide, which was placed in the incubator. After 20 h, and 4 h before the start of the experiment was scheduled, the cell nuclei were stained with Hoechst 33342 (final concentration 50 ng/mL).

The live-cell imaging microscope Zeiss Axio Observer.Z1 / 7 enabled us to observe the cells at 37 °C and 5% carbon dioxide for 48 h. By taking several micrographs of a large area shortly after one another and stitching them together, we gained pictures of a connected cell layer, comprising an area of more than 42 mm^2. We acquired phase contrast and fluorescence images of the cells every 10 min to capture the signals of the stained cell nuclei. For this study, we performed ten experiments with the cells from five different umbilical cords.

To analyze the micrographs with up to 50.000 cell signals per frame, we developed an automated image segmentation software. Using the programming language Python, we were able to extract the positions of the labeled cell nuclei, which were then connected for each cell to a trajectory as a function of time using the Crocker-Gier algorithm [9]. Final trajectories (Figure 1) of at least nine time steps were taken into account while the distance between consecutive positions must not exceed 10-12 μm, depending on the cell density. Cell velocities were calculated as the difference quotient of two consecutive cell positions.

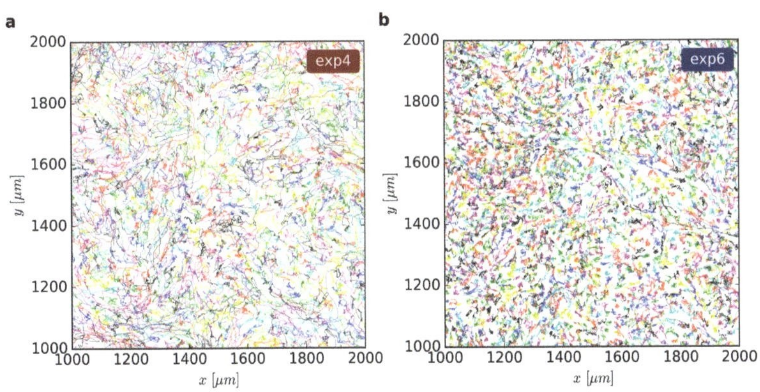

Figure 1. In a section (1 mm^2) of the cell layer, all cell trajectories are shown for the experiments 4 (**a**) and 6 (**b**). The trajectories are labeled sequentially with one out of seven colors in the order of their first appearance and the attributed cell number. The overall cell density in (**a**) is much lower than in (**b**). This leads to different dynamics and pattern formation.

The parameters and evidence of the models were estimated with the help of Bayesian data analysis (for details see [4]). A numerical analysis was performed with the MultiNest-sampling algorithm [10,11] in its Python implementation [12]. Uniform priors were applied for parameters N_0, β, t_0, A, A_1 and A_2. For all remaining parameters (α_0, α_1, σ, τ, τ_1, τ_2, A_0, c and two parameters adjusting the experimental error), Jeffreys' priors were used. The agreement of data and model was assessed using a Gaussian likelihood function. The model means and scatterings were obtained by averaging over the resulting parameter samplings.

3. Results

To obtain a first overview of the temporal development of cell movements, we calculated the mean squared velocity for each time step as the average of all cells that could be tracked for the entire observation time (Figure 2).

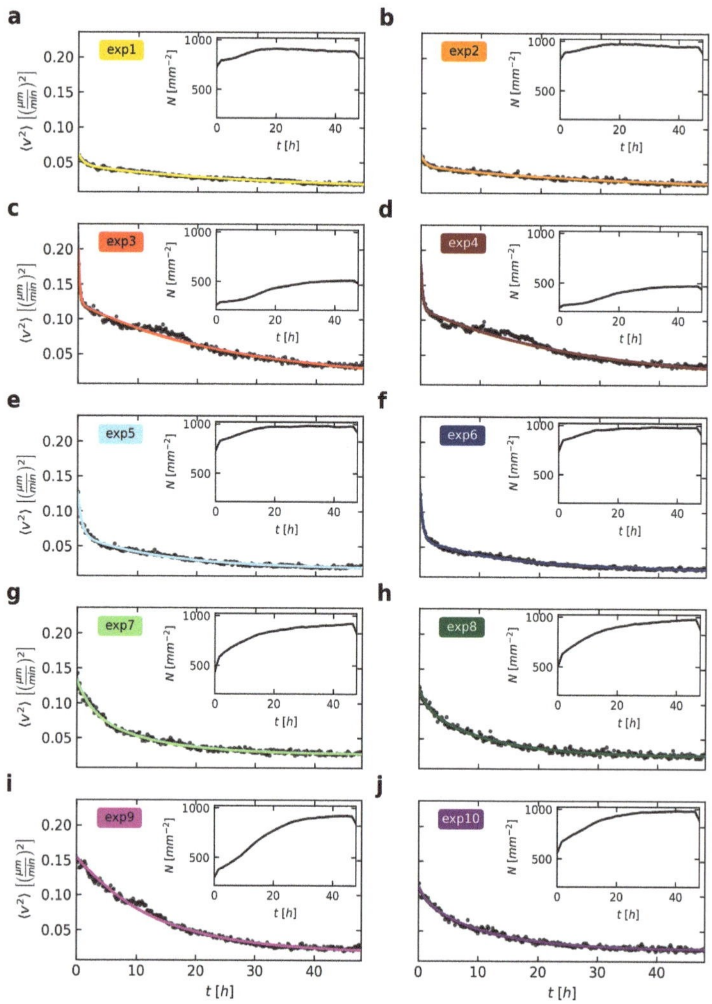

Figure 2. The mean squared velocities and cell densities (inset) are shown for each time step in all ten experiments (black dots, **a**–**j**). Additionally, the mean squared velocities are fitted with two exponential functions (Equation (1), solid colored line). In certain time intervals of the experiments (especially in **c**,**d**,**i**), the fit transiently deviates from the course of the data points. The quite small uncertainties (two standard deviations) of the model fit are marked as a gray area (barely visible). The calculation of the velocities is based on all cells that could be tracked for the entire duration of the experiment (48 h). The values for the cell densities consider all cells with the exception of the boundary area. The cell numbers in the first and last nine time steps are artificially lower due to the defined minimum trajectory length during cell tracking. Experiments with cells from the same umbilical cord are shown in the same row.

The cells decrease their mean squared velocity over time (Figure 2). However, this decline in velocities varies strongly between the ten experiments. We also noted different cell numbers at the beginning of the imaging as well as differing growth in the observed cell layers (Figure 2, insets). With the cells moving around and proliferating vividly, the cell arrangement usually gets more homogenous and macroscopically ordered over time. The resulting overall higher cell densities confine cell movements and therefore decrease mean squared velocities. To highlight this inverse relation, we chose the same axis limits for the velocities and densities for all experiments in Figure 2. If cell densities are high, then the corresponding velocities are, in comparison with the other experiments, quite low, and vice versa (e.g., Figure 2 a as opposed to c).

To develop a model describing the decline in mean cell velocities, the effect of several simultaneously ongoing biological processes should be quantified separately. First, we found that the cells rapidly slowed down at the start of the experiment. This is followed by a second, slower decline before the mean squared velocity seems to converge to a certain limit greater than zero for long times depending on the cell density. In some experiments (Figure 2c,d) the two declines seem to happen sequentially, whereas there are other examples (Figure 2h) that show a much smoother transition. We speculate that these two processes actually occur simultaneously with differing parameters since this seems to be more likely from a biological view. Hence, the little plateau, as observed especially in experiments 3 and 4 (Figure 2c,d) after 10 h, could be caused by a third phenomenon. Based on this consideration, we described the general decline using two exponential functions, with parameters $A_{1,2}$ and $\tau_{1,2}$ and a basal squared velocity A_0.

$$\langle v^2(t)\rangle = A_0 + A_1 e^{-\frac{t}{\tau_1}} + A_2 e^{-\frac{t}{\tau_2}}. \tag{1}$$

As shown in Figure 2, this fit, obtained using Bayesian analysis, well captures the effects of the two processes that are slowing down the cells with the time scales of τ_1 and τ_2, ranging between ~16 and ~142 min and between ~655 and ~1780 min, respectively. These declines could be caused by the continuous cell aging and strengthening of cell-cell contacts. However, there are still data points that deviate from the model of Equation (1) during time ranges mainly within the first 20 h. Up to now, we did not include cell division activities in our model, which should be related to the velocities of cells: On the one hand, cells can move around more freely in areas with low density. They are less prone to collide with each other or have more space to change their directions. On the other hand, temporal phases of very active cell proliferation seem to be marked by a transiently constant or only very slowly decreasing cell velocity (see e.g., Figure 2c,d,i). Although this effect is short-lived, it shows that the separation of dividing cells can lead to accelerations in the system. By contrast, these cell divisions can also cause an overall deceleration due to an increasing confinement in densely packed cell clusters. That is why the mean squared velocities in experiments 3 and 4 show a transient plateau (Figure 2c,d) in their decline or why τ_2, as a means to describe the second, slower decline, might differ in these experiments.

In order to quantify the connection between mean cell velocity and cell density, we calculated the relative changes in cell numbers for each time step (Figure 3).

The cells usually go through one phase of more intense cell division, which builds up at first and reaches a maximum before it subsides again. However, the onset, duration and strength of this phase differ strongly between the experiments. Occasionally, the maximum of cell division lies probably even before the start of the experimental observation. We can describe the relative increase $\alpha(t)$ in the cell numbers N per time t with

$$\alpha(t) = \frac{1}{N}\frac{dN}{dt} = \alpha_0 + \alpha_1 e^{-\left|\frac{t-t_0}{\sigma}\right|^{\beta}}. \tag{2}$$

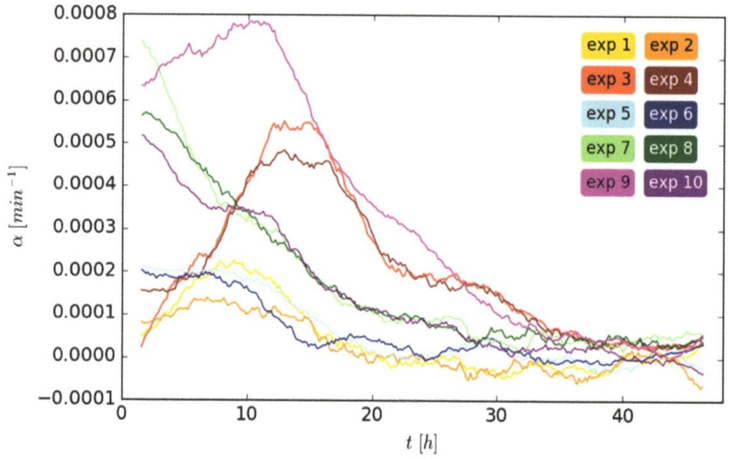

Figure 3. The changes in cell numbers *α(t)* (Equation (2)) are calculated for each time step. A uniform filter with a size of 300 min was applied to smoothen the data. Due to the defined minimum length of cell trajectories during image analysis, the first and last nine time steps were discarded.

Here, α_0 describes the basal cell division, α_1 the strength, σ the duration and t_0 the time point of the maximum of the cell division phase. The exponent β changes the shape of the curve with $\beta = 2$ being the special case of a Gaussian function.

Since changes in cell density can modify the velocities of cells, we couple the cell division activity of Equation (2) linearly with factor c to the mean squared velocity of Equation (1). This temporarily raises the mean squared velocity during strong increases in cell numbers for $c > 0$.

$$\langle v^2(t)\rangle = A_0 + A_1 e^{-\frac{t}{\tau_1}} + A_2 e^{-\frac{t}{\tau_2}} + c\left(\alpha_0 + \alpha_1 e^{-|\frac{t-t_0}{\sigma}|^{\beta}}\right). \quad (3)$$

Depending on the time point of the maximum cell division rate, the cell density can be calculated by integrating Equation (2), leading to the following result:

$$N(t) = \begin{cases} N_0 e^{\frac{\alpha_1\sigma}{\beta}\{\gamma(\frac{1}{\beta},\, (\frac{t-t_0}{\sigma})^{\beta}) - \gamma(\frac{1}{\beta},\, (-\frac{t_0}{\sigma})^{\beta})\} + \alpha_0 t} & if\ t_0 < 0\,, \\ N_0 e^{-\frac{\alpha_1\sigma}{\beta}\{\gamma(\frac{1}{\beta},\, (\frac{t_0-t}{\sigma})^{\beta}) - \gamma(\frac{1}{\beta},\, (\frac{t_0}{\sigma})^{\beta})\} + \alpha_0 t} & if\ t_0 > 0\ and\ t < t_0\,, \\ N_0 e^{\frac{\alpha_1\sigma}{\beta}\{\gamma(\frac{1}{\beta},\, (\frac{t_0}{\sigma})^{\beta}) + \gamma(\frac{1}{\beta},\, (\frac{t-t_0}{\sigma})^{\beta})\} + \alpha_0 t} & if\ t_0 > 0\ and\ t > t_0\,. \end{cases} \quad (4)$$

Here, the lower incomplete gamma function is used according to Equation (5)

$$\gamma(a,x) = \int_0^x t^{a-1} e^{-t}\, dt \quad (5)$$

and N_0 denotes the initial cell number at $t = 0$. The obtained parameters of the simultaneous fit, calculated individually for each experiment, indicate that the trend of the velocities consists of a faster and a slower decline (Figure 4). After about one to three hours (elapsed time equals two τ_1), the cells have slowed down considerably (Figure 4a). After that time, the mean squared velocity decreases only slowly, with τ_2 usually being greater than 10 h (Figure 4b). The exception from these findings is the ninth experiment, where the two declines are merged and therefore difficult to separate. The first two experiments show a particularly slow decline since the cell density had always been high and the cells from this umbilical cord showed less activity. At large, the obtained values for the parameters are roughly of the same magnitude compared to the counterparts of the isolated fit (Equation (1) and Figure 2). This indicates that there are at least two time-dependent phases, with the cell division causing the time-independent alterations of the overall dynamics. This third influence is especially strong in experiments 3, 4 and 9, as indicated by parameter c (Figure 4c), where the mean squared velocities were also the highest. The first experiment shows slightly more cell division than the second while the mean squared velocities are almost the same (Figure 2a,b). Interestingly, the optimal parameters for the combined fit include a higher c and subsequently higher τ_1 and τ_2 (Figure 4) instead of these three parameters just being smaller. This might emphasize the importance of considering cell division in the model.

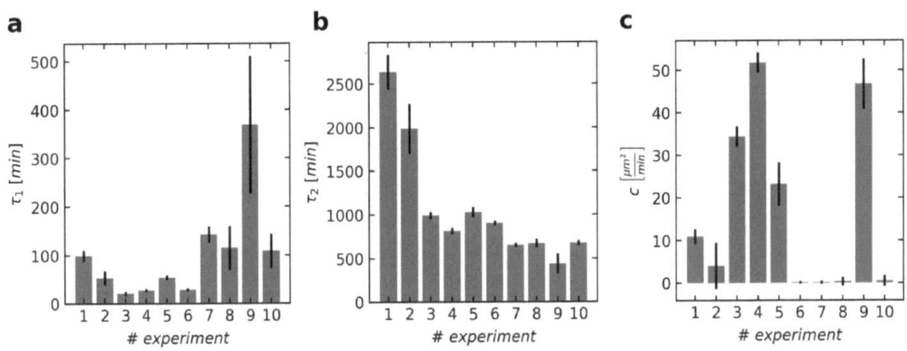

Figure 4. The mean and standard deviation for the estimated parameters τ_1 (**a**), τ_2 (**b**) and c (**c**) are shown for each experiment. Here, τ_1 and τ_2 describe the two time scales with which the exponential functions are decreasing. Parameter c depicts the coupling strength between the rate of cell division and the mean squared velocity.

Based on the estimated parameters, we plotted the fits for the mean squared velocity and density (Figure 5).

A comparison by eye shows that the model is suitable to describe the data of the cell densities as long as the cell numbers do not decrease (Figure 5a,b). Furthermore, the fit of the mean squared velocities was improved by the coupling to cell division activities, especially in experiments 3, 4 and 9. To verify that indeed two time scales are necessary to describe the overall decline, we compared the resulting Bayesian evidences with two alternative models. Model 1 (12 parameters) uses one exponential, Model 2 (13 parameters) uses one stretched exponential function in addition to the cell division term (Table 1). So far, Model 3 (14 parameters) was applied (Figure 5), which uses, in comparison with Model 1, an additional exponential function (2 in total).

Table 1. Three different models (including one exponential, one stretched exponential or two exponential functions) were used for the fit of the mean squared velocity.

Model	Formula	
1	$\langle v^2(t)\rangle = A_0 + Ae^{-\frac{t}{\tau}} + c\left(\alpha_0 + \alpha_1 e^{-\left\vert\frac{t-t_0}{\sigma}\right\vert^{\beta}}\right)$	(6)
2	$\langle v^2(t)\rangle = A_0 + Ae^{-\left\vert\frac{t}{\tau}\right\vert^{l}} + c\left(\alpha_0 + \alpha_1 e^{-\left\vert\frac{t-t_0}{\sigma}\right\vert^{\beta}}\right)$	(7)
3	$\langle v^2(t)\rangle = A_0 + A_1 e^{-\frac{t}{\tau_1}} + A_2 e^{-\frac{t}{\tau_2}} + c\left(\alpha_0 + \alpha_1 e^{-\left\vert\frac{t-t_0}{\sigma}\right\vert^{\beta}}\right)$	(8)

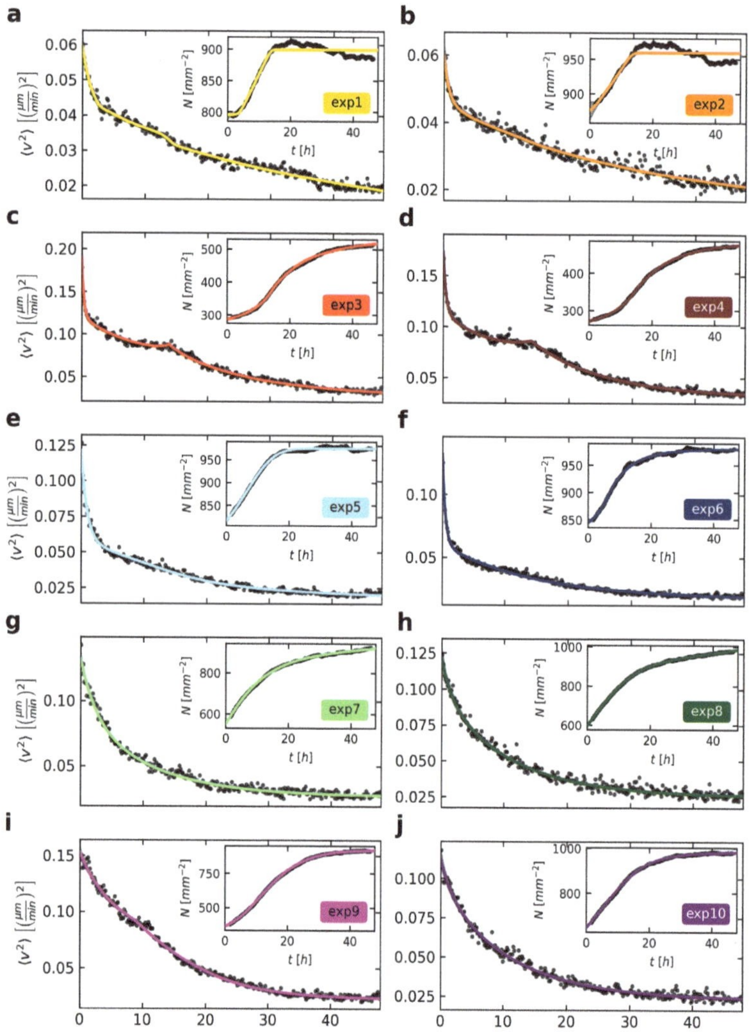

Figure 5. The fits to the mean squared velocity (**a**–**j**), as shown in Figure 2, were improved using Equation (3). The shown data of the cell density (inset) was fitted simultaneously for each experiment according to Equation (4). The quite small uncertainties (two standard deviations) of the model fit are represented as a gray area (almost only visible in the inset of (**b**)). Note that the ranges of the y-axes are optimally adjusted to each data set in contrast to Figure 2.

The comparison of the evidences of the applied models shows a clear preference for Model 3 for nine out of ten experiments (Figure 6). This endorses the theory that there is usually a decline consisting of two parts with different temporal dynamics. Only the ninth experiment could also be described with just one exponential function. These cells showed the strongest and longest-lasting cell division phase out of all experiments, which led to a more than two-fold increase in cell density. This influence was incorporated into the fit for the velocity via a high coupling parameter c (Figure 4c). Thus, the effect of the proliferation-independent cell movements is less recognizable, especially because the processes occur largely simultaneously. This might have also caused the small difference between τ_1 and τ_2 and its higher error (Figure 4a,b).

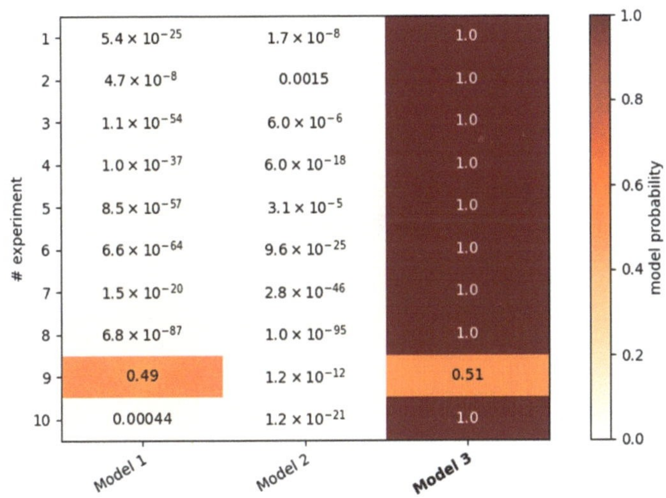

Figure 6. Three models (Table 1) were used for the simultaneous fit of the mean squared velocity and density. The evidence, obtained by Bayesian analysis, is used to compare the accuracy of the different models. The model probability is shown color-coded and in numbers for each experiment.

4. Discussion

As long as the cells are motile and their numbers are not decreasing, Model 3 is able to fit the cell density and mean squared velocity for all experiments despite their interindividual differences. Bayesian analysis was used to quantify the parameters of the suspected underlying biological processes. It is noteworthy that the parameter estimation succeeded in finding suitable results for the parameters included in both fits, promoting the idea that not only the cell density but also the velocities are positively influenced by cell divisions. It should be noted that the number of cell divisions was derived from the changes in cell density. The actual number of cell divisions could be higher, since we did not quantify the number of cell deaths. The comparison of three different models showed that the best approach consists of the sum of two exponentially decaying functions. We considered in all models the relatively small but accumulating amount of increased velocities generated by cell division. Although the two new cells shift initially only a few micrometers apart from each other in opposite directions, the influences of several cell divisions do not even each other out. Nevertheless, it is taken into account that strong cell proliferation increases cell density and eventually diminishes cell velocities. The basic velocity level could be connected to the absolute value of the cell numbers. Furthermore, the effect of aging or increasing cell localization on the dynamics might be fitted in a better way than with just an exponential function. Here, the necessity to use two of these functions for the velocity already shows that there are many more biophysical mechanisms influencing cell movements. In the future, we will use and extend this approach to assess other aspects of cell migration.

Author Contributions: Conceptualization, A.H., A.D. and P.D.; methodology, A.H. and P.D.; software, A.H. and P.D.; validation, A.H. and P.D.; formal analysis, A.H.; investigation, A.H. and P.D.; resources, A.H., A.D., S.S. and P.D.; data curation, A.H. and P.D.; writing—original draft preparation, A.H.; writing—review and editing, A.D., S.S. and P.D.; visualization, A.H.; supervision, A.D. and P.D.; project administration, P.D. All authors have read and agreed to the published version of the manuscript.

Funding: This research received no external funding.

Institutional Review Board Statement: The study was conducted in accordance with the Declaration of Helsinki, and approved by the Ethics Committee at TU Dresden (EK 203112005).

Informed Consent Statement: Not applicable.

Data Availability Statement: Data sharing is not applicable to this article.

Conflicts of Interest: The authors declare no conflict of interest.

References

1. Michaelis, U.R. Mechanisms of endothelial cell migration. *Cell. Mol. Life Sci.* **2014**, *71*, 4131–4148. [CrossRef] [PubMed]
2. Friedl, P.; Gilmour, D. Collective cell migration in morphogenesis, regeneration and cancer. *Nat. Rev. Mol. Cell Biol.* **2009**, *1* 445–457. [CrossRef] [PubMed]
3. Lämmermann, T.; Germain, R.N. The multiple faces of leukocyte interstitial migration. *Semin. Immunopathol.* **2014**, *36*, 227–25 [CrossRef] [PubMed]
4. Dieterich, P.; Lindemann, O.; Moskopp, M.L.; Tauzin, S.; Huttenlocher, A.; Klages, R.; Chechkin, A.; Schwab, A. Anomalou diffusion and asymmetric tempering memory in neutrophil chemotaxis. *PLoS Comput. Biol.* **2022**, *18*, e1010089. [CrossRef] [PubMed]
5. Lo, C.M.; Wang, H.B.; Dembo, M.; Wang, Y.L. Cell movement is guided by the rigidity of the substrate. *Biophys. J.* **2000**, *7* 144–152. [CrossRef] [PubMed]
6. Peyton, S.R.; Putnam, A.J. Extracellular matrix rigidity governs smooth muscle cell motility in a biphasic fashion. *J. Cell Physio* **2005**, *204*, 198–209. [CrossRef] [PubMed]
7. Rørth, P. Collective cell migration. *Annu. Rev. Cell Dev. Biol.* **2009**, *25*, 407–429. [CrossRef] [PubMed]
8. Khedr, S.; Deussen, A.; Kopaliani, I.; Zatschler, B.; Martin, M. Effects of tryptophan-containing peptides on angiotensin-convertin enzyme activity and vessel tone ex vivo and in vivo. *Eur. J. Nutr.* **2018**, *57*, 907–915. [CrossRef] [PubMed]
9. Crocker, J.; Grier, D. Methods of Digital Video Microscopy for Colloidal Studies. *J. Colloid Interface Sci.* **1996**, *179*, 298–31 [CrossRef]
10. Skilling, J. Nested sampling for general Bayesian computation. *Bayesian Anal.* **2006**, *1*, 833–859. [CrossRef]
11. Feroz, F.; Hobson, M.P.; Bridges, M. MultiNest: An efficient and robust Bayesian inference tool for cosmology and particle physic *Mon. Not. R. Astron. Soc.* **2009**, *398*, 1601–1614. [CrossRef]
12. Buchner, J.; Georgakakis, A.; Nandra, K.; Hsu, L.; Rangel, C.; Brightman, M.; Merloni, A.; Salvato, M.; Donley, J.L.; Kocevski, D X-ray spectral modelling of the AGN obscuring region in the CDFS: Bayesian model selection and catalogue. *Astron. Astrophys* **2014**, *564*, A125. [CrossRef]

physical sciences forum

MDPI

Proceeding Paper

Variational Bayesian Approximation (VBA) with Exponential Families and Covariance Estimation †

Seyedeh Azadeh Fallah Mortezanejad [1] and Ali Mohammad-Djafari [2,3,*]

1 School of Automotive and Traffic Engineering, Jiangsu University, Zhenjiang 212013, China; azadehfallah131@gmail.com
2 International Science Consulting and Training (ISCT), 91440 Bures sur Yvette, France
3 Shanfeng Company, Shaoxing 312352, China
* Correspondence: djafari@ieee.org
† Presented at the 42nd International Workshop on Bayesian Inference and Maximum Entropy Methods in Science and Engineering, Garching, Germany, 3–7 July 2023.

Abstract: Variational Bayesian Approximation (VBA) is a fast technique for approximating Bayesian computation. The main idea is to assess the joint posterior distribution of all the unknown variables with a simple expression. Mean–Field Variational Bayesian Approximation (MFVBA) is a particular case developed for large–scale problems where the approximated probability law is separable in all variables. A well–known drawback of MFVBA is that it tends to underestimate the variances in the variables, even though it estimates the means well. It can lead to poor inference results. We can obtain a fixed point algorithm to evaluate the means in exponential families for the approximating distribution. However, this does not solve the problem of underestimating the variances. In this paper, we propose a modified method of VBA with exponential families to first estimate the posterior mean and then improve the estimation of the posterior covariance. We demonstrate the performance of the procedure with an example.

Keywords: Variational Bayesian Approximation (VBA); Kullback–Leibler divergence; Mean–Field Variational Bayesian Approximation (MFVBA); Linear Response Variational Bayes (LRVB)

1. Introduction

Bayesian computation uses Bayes' theorem to revise the probabilities of unknowns based on prior knowledge and data. When the likelihood expressions and prior are available, we can get the expression of the posterior law. However, this expression needs the denominator integral computation of the Bayes formula, called evidence. Approximate Bayesian Computation (ABC) are a class of algorithms used to perform Bayesian computation without exactly computing the evidence. The first class of these methods generate samples from the posterior law. We can mention all of the Monte Carlo sampling methods, such as slice sampling, nested sampling, and Markov chain Monte Carlo (MCMC) methods. The second category of these methods includes those that directly compute the means and variances, such as the Variational Bayesian Approximation (VBA) methods.

MCMC is a class of algorithms for sampling from a probability distribution and estimating the posterior distribution. Metropolis et al. (1953) [1] introduced the first MCMC algorithm called the Metropolis algorithm. Hastings (1970) [2] brought it closer to applied statistics, generalizing the Metropolis algorithm for asymmetric proposal distributions.

VBA methods were introduced by Jordan et al. (1998) [3]. They were motivated by the demand to improve the Bayesian inference for large and complex models intractable by conventional methods such as MCMC. VBA is a fast and robust scheme to perform Bayesian computations in high–dimensional problems. The main idea is to approximate the high–complexity posterior probability law with a simple and lower–complexity probability law using the Kullback–Leibler divergence as the approximation criterion.

In many cases, the VBA methods are faster than the MCMC methods. VBA can release two types of posterior approximations. The first is the free–form product of distributions such as the conjugated exponential family known as the Mean–Field Approximation (MFA). The second one is the fixed–form for posterior distributions like multivariate Gaussian with a proper parametrization for the model, Sarkka and Nummenmaa (2009) [4]. MFA is a general approach in the graphical models, particularly for the hierarchical Bayesian models.

The foundation of VBA is minimizing the Kullback–Leibler divergence (KL) between the approximated probability distribution and the exact posterior distribution. Rohde and Wand (2016) [5] worked on semi–parametric Mean–Field Variational Bayesian Approximation (MFVBA) as a combination of the KL measure and MFA. A significant drawback of MFVBA is that it underestimates the variable uncertainties and is uninformed concerning the variable covariance of models. Giordano et al. (2015, 2018) [6,7] rectified the MFVBA and named Linear Response Variational Bayes (LRVB) as being able to provide proper uncertainties.

MFVBA is a valuable approximation method for estimating the posterior mean in the Bayesian framework. While using an exponential family for approximation, we get a fixed point algorithm for reckoning the means. In this paper, we show how to use VBA with exponential families to approximate the posterior means and simultaneously evaluate the posterior covariance more precisely. Via an example, we also compare the MFVBA and MCMC posterior distributions.

This paper's organization is as follows: In Section 2, we demonstrate the main idea of MFVBA with a particular case in exponential family (EF) distributions. In Section 3, we modify the covariance matrix resulting from the MFVBA to get a better approximation. In Section 4, we give details of the covariance matrix expression in EF. We show that the covariance computation needs high–dimensional matrix inversion, which is very costly in high–dimensional problems. In Section 5, we provide an example of how to compute the posterior distribution and compare the computational costs of the proposed method and the MCMC. In Section 6, we present the main conclusion on the paper content.

2. Mean–Field Variational Bayesian Approximation (MFVBA)

VBA approximates posterior distributions when the exact expression of the posterior is too complex or costly. MFA simplifies the VBA by assuming that the final distribution is factorized into independent margins, which means that each unknown parameter has its distribution independent of the others.

Suppose that p is approximated by q^* and $\boldsymbol{\theta}$ is the unknown vector of parameters. We define $\boldsymbol{V} = \text{Cov}_{q^*}(\boldsymbol{\theta})$ and $\boldsymbol{\Sigma} = \text{Cov}_p(\boldsymbol{\theta})$. We know that $\boldsymbol{V} < \boldsymbol{\Sigma}$. $\boldsymbol{V}$ is a poor estimate (underestimate) of $\boldsymbol{\Sigma}$ even if $\boldsymbol{m}^* = \text{E}_{q^*}\{\boldsymbol{\theta}\} = \text{E}_p\{\boldsymbol{\theta}\}$. The question is how to find a better estimate for $\boldsymbol{\Sigma}$. One solution for $\boldsymbol{\Sigma}$ in the case of the exponential family, proposed by Giordano et al. (2015) [6], is $\boldsymbol{\Sigma} = (\boldsymbol{I} - \boldsymbol{V}\boldsymbol{H})^{-1}\boldsymbol{V}$ via the LRVB method, where $\boldsymbol{I}$ and $\boldsymbol{H}$ are an identity matrix and Hessian matrix of the log probability $\ln p$, respectively. To show this, let us go through the details step by step.

2.1. Main Idea

- General Bayesian framework: Given the likelihood $p(\mathcal{D}|\boldsymbol{\theta})$ and the prior $p(\boldsymbol{\theta})$, the expression of the posterior $p(\boldsymbol{\theta}|\mathcal{D})$ is given by:

$$p(\boldsymbol{\theta}|\mathcal{D}) = \frac{p(\mathcal{D}|\boldsymbol{\theta})p(\boldsymbol{\theta})}{p(\mathcal{D})}. \tag{1}$$

- MFVBA: Approximate $p(\boldsymbol{\theta}|x)$ by a separable $q(\boldsymbol{\theta})$:

$$q(\boldsymbol{\theta}) = \prod_j q_j(\theta_j), \tag{2}$$

by minimizing:

$$\mathrm{KL}[q:p] = \int q \ln \frac{q}{p} = \mathrm{E}_q\{\ln q\} - \mathrm{E}_q\{\ln p\}. \tag{3}$$

- Noting by $L = \mathrm{E}_q\{\ln p\}$ the expected posterior likelihood and by $S = -\mathrm{E}_q\{\ln q\}$ the entropy of q, we can also write:

$$q^* = \arg\min_q\{\mathrm{KL}[q:p]\} = \arg\max_q\{E = S + L\}. \tag{4}$$

- E is the evidence lower bound.

2.2. VBA and Exponential Family

- If q is chosen to be in an exponential family:

$$q(\boldsymbol{\theta}|\boldsymbol{\eta}) = \exp\left[\boldsymbol{\eta}'\boldsymbol{\theta} - A(\boldsymbol{\eta})\right], \tag{5}$$

then it is entirely characterized by its mean $\boldsymbol{m} = \mathrm{E}_q\{\boldsymbol{\theta}\}$ and so is:

$$q^*(\boldsymbol{\theta}|\boldsymbol{\eta}^*) = \exp\left[\boldsymbol{\eta}^{*\prime}\boldsymbol{\theta} - A(\boldsymbol{\eta}^*)\right], \tag{6}$$

then the same characterization by the mean $\boldsymbol{m}^* = \mathrm{E}_{q^*}\{\boldsymbol{\theta}\}$.
- We can then define the objective E as a function of $\boldsymbol{m}$, and the first–order condition of the optimality is:

$$\frac{\partial E}{\partial \boldsymbol{m}}\Big|_{\boldsymbol{m}=\boldsymbol{m}^*} = 0. \tag{7}$$

2.3. MFVBA, Exponential Family (EF) and Fixed Point Algorithm

- VBA+exponential family with $\mathrm{E}_q\{\boldsymbol{\theta}\}$:

$$\frac{\partial E}{\partial \boldsymbol{m}}\Big|_{\boldsymbol{m}=\boldsymbol{m}^*} = 0 \rightarrow \frac{\partial E}{\partial \boldsymbol{m}}\Big|_{\boldsymbol{m}=\boldsymbol{m}^*} + \boldsymbol{m} = \boldsymbol{m} \rightarrow M(\boldsymbol{m}) = \boldsymbol{m}. \tag{8}$$

- Iterating on this fixed point algorithm:

$$M(\boldsymbol{m}) = \boldsymbol{m}^{(k-1)} \quad \text{with} \quad M(\boldsymbol{m}) := \frac{\partial E}{\partial \boldsymbol{m}} + \boldsymbol{m}, \tag{9}$$

converges to $\boldsymbol{m}^* = \mathrm{E}_{q^*}\{\boldsymbol{\theta}\}$.

3. VBA, EF and Covariance Estimation

- Noting from $\boldsymbol{V} = \mathrm{Cov}_{q^*}(\boldsymbol{\theta})$ and $\boldsymbol{\Sigma} = \mathrm{Cov}_p(\boldsymbol{\theta})$, we know that $\boldsymbol{V} < \boldsymbol{\Sigma}$.
- $\boldsymbol{V}$ is a poor estimate (underestimate) of $\boldsymbol{\Sigma}$ even if $\boldsymbol{m}^* = \mathrm{E}_{q^*}\{\boldsymbol{\theta}\} = \mathrm{E}_p\{\boldsymbol{\theta}\}$.
- The question is now: How to find a better estimate for $\boldsymbol{\Sigma}$.
- One solution proposed by Giordano et al. (2015) [6] is the LRVB.
The main idea is to perturb $p(\boldsymbol{\theta}|x)$ by some exponential family to obtain $p_{\boldsymbol{t}}(\boldsymbol{\theta}|x)$:

$$p_{\boldsymbol{t}}(\boldsymbol{\theta}|x) = p(\boldsymbol{\theta}|x)\exp\left[\boldsymbol{t}'\boldsymbol{\theta} + C(\boldsymbol{t})\right] \tag{10}$$

such that:

$$\ln p_{\boldsymbol{t}}(\boldsymbol{\theta}|x) = \ln p(\boldsymbol{\theta}|x) + \boldsymbol{t}'\boldsymbol{\theta} + C(\boldsymbol{t}) \tag{11}$$

- $C(\boldsymbol{t})$ is the cumulate generating function of $p_{\boldsymbol{t}}(\boldsymbol{\theta}|x)$, so

$$\boldsymbol{\Sigma} = \mathrm{Cov}_p(\boldsymbol{\theta}) = \frac{\partial^2 C(\boldsymbol{t})}{\partial \boldsymbol{t}'\partial \boldsymbol{t}}\Big|_{\boldsymbol{t}=\boldsymbol{0}} = \frac{\partial \mathrm{E}_{p_{\boldsymbol{t}}}\{\boldsymbol{\theta}\}}{\partial \boldsymbol{t}'}\Big|_{\boldsymbol{t}=\boldsymbol{0}} = \frac{\partial \boldsymbol{m}_{\boldsymbol{t}}^*}{\partial \boldsymbol{t}'}\Big|_{\boldsymbol{t}=\boldsymbol{0}} \tag{12}$$

4. Linear Perturbation Method for Better Covariance Estimation

To summarize this method:

- Perturb $p(\boldsymbol{\theta}|x)$ by an exponential family term to obtain:

$$p_{\boldsymbol{t}}(\boldsymbol{\theta}|x) = p(\boldsymbol{\theta}|x)\exp\left[\boldsymbol{t}'\boldsymbol{\theta} + C(\boldsymbol{t})\right].$$

Thus, we have:

$$\begin{aligned} -\mathrm{KL}[q:p_{\boldsymbol{t}}] &= -\mathrm{KL}[q:p] + \boldsymbol{t}'\boldsymbol{m} \longrightarrow E_{\boldsymbol{t}} = E + \boldsymbol{t}'\boldsymbol{m}, \\ \frac{\partial E}{\partial \boldsymbol{m}} &= 0 \longrightarrow \boldsymbol{m} = \mathrm{E}_q\{\boldsymbol{m}\}, \quad M(\boldsymbol{m}) = \boldsymbol{m}, \end{aligned}$$

and finally:

$$E_{\boldsymbol{t}} = E + \boldsymbol{t}'\boldsymbol{m} \longrightarrow M_{\boldsymbol{t}}(\boldsymbol{m}) = M(\boldsymbol{m}) + \boldsymbol{t}. \tag{13}$$

- Replacing these relations, we have:

$$\frac{\partial \boldsymbol{m}_{\boldsymbol{t}}^*}{\partial \boldsymbol{t}'} = \left.\frac{\partial M_{\boldsymbol{t}}}{\partial \boldsymbol{m}'}\right|_{\boldsymbol{m}=\boldsymbol{m}_{\boldsymbol{t}}^*} \frac{\partial \boldsymbol{m}}{\partial \boldsymbol{t}} + \frac{\partial M_{\boldsymbol{t}}}{\partial \boldsymbol{t}} = \left.\frac{\partial M_{\boldsymbol{t}}}{\partial \boldsymbol{m}'}\right|_{\boldsymbol{m}=\boldsymbol{m}_{\boldsymbol{t}}^*} \frac{\partial \boldsymbol{m}}{\partial \boldsymbol{t}} + \boldsymbol{I}.$$

We obtain:

$$\boldsymbol{\Sigma} = \frac{\partial \boldsymbol{m}_{\boldsymbol{t}}^*}{\partial \boldsymbol{t}'} = \frac{\partial M}{\partial \boldsymbol{m}}\boldsymbol{\Sigma} + \boldsymbol{I} = \left(\frac{\partial^2 E}{\partial \boldsymbol{m}\partial \boldsymbol{m}'} + \boldsymbol{I}\right) + \boldsymbol{I} \longrightarrow \boldsymbol{\Sigma} = -\left(\frac{\partial^2 E}{\partial \boldsymbol{m}\partial \boldsymbol{m}'}\right),$$

and finally:

$$E = L + S \longrightarrow \boldsymbol{\Sigma} = -\left(\frac{\partial^2 L}{\partial \boldsymbol{m}\partial \boldsymbol{m}'} + \frac{\partial^2 S}{\partial \boldsymbol{m}\partial \boldsymbol{m}'}\right). \tag{14}$$

- Noting that the entropy of an exponential family is:

$$S = \boldsymbol{\eta}'\boldsymbol{m} + A(\boldsymbol{\eta}) \longrightarrow \frac{\partial S}{\partial \boldsymbol{m}} = -\boldsymbol{\eta} \longrightarrow \frac{\partial^2 S}{\partial \boldsymbol{m}\partial \boldsymbol{m}'} = -\frac{\partial \boldsymbol{\eta}}{\partial \boldsymbol{m}} = \boldsymbol{V}^{-1}.$$

We get:

$$\boldsymbol{\Sigma} = -\left(\frac{\partial^2 L}{\partial \boldsymbol{m}\partial \boldsymbol{m}'} + \frac{\partial^2 S}{\partial \boldsymbol{m}\partial \boldsymbol{m}'}\right)^{-1} = -(\boldsymbol{H} - \boldsymbol{V}^{-1})^{-1}.$$

- Finally, using a matrix inversion lemma, we hold:

$$\boldsymbol{\Sigma} = (\boldsymbol{I} - \boldsymbol{V}\boldsymbol{H})^{-1}\boldsymbol{V}. \tag{15}$$

4.1. Dealing with Nuisance Parameters

- If we are only interested in some of the variables $\boldsymbol{\alpha}$, i.e., $\boldsymbol{\theta} = (\boldsymbol{\alpha}, \boldsymbol{z})$ where $\boldsymbol{z}$ are nuisance parameters that we are not interested in, then we can write:

$$\boldsymbol{\Sigma} = \begin{bmatrix} \boldsymbol{\Sigma}_\alpha & \boldsymbol{\Sigma}_{\alpha z} \\ \boldsymbol{\Sigma}_{z\alpha} & \boldsymbol{\Sigma}_z \end{bmatrix}.$$

- Looking for the partition of $\boldsymbol{\theta} = (\boldsymbol{\alpha}, \boldsymbol{z})$, we obtain:

$$\boldsymbol{\Sigma}_\alpha = (\boldsymbol{I}_\alpha - \boldsymbol{V}_\alpha\boldsymbol{H}_\alpha - \boldsymbol{V}_\alpha\boldsymbol{H}_{\alpha z}(\boldsymbol{I}_z - \boldsymbol{V}_z\boldsymbol{H}_z)^{-1}\boldsymbol{V}_z\boldsymbol{H}_{z\alpha})^{-1}\boldsymbol{V}_\alpha.$$

4.2. The Whole Algorithm

- Having chosen the prior $p(\boldsymbol{\theta})$ and the likelihood $p(\mathcal{D}|\boldsymbol{\theta})$, find the expression of the posterior law:

$$p(\boldsymbol{\theta}|\mathcal{D}) \propto p(\mathcal{D}|\boldsymbol{\theta})p(\boldsymbol{\theta}).$$

- Choose an exponential family $q(\boldsymbol{\theta})$ and find the expressions of:

$$L = \mathrm{E}_q\{\ln p\}, \quad S = \mathrm{E}_q\{\ln q\}, \quad \text{and} \quad E = L + S,$$

as a function of the expectation parameters $\boldsymbol{m} = \mathrm{E}_q\{\boldsymbol{\theta}\}$.
- Find the expression of $\boldsymbol{M} = \nabla_{\boldsymbol{m}} E + \boldsymbol{m}$ and write the following fix point update $\boldsymbol{M}(\boldsymbol{m}) = \boldsymbol{m}^{(k-1)}$ until convergence.
- When $\boldsymbol{m}^{\star}$ is obtained, then compute:

$$\boldsymbol{V} = \frac{\partial^2 S}{\partial \boldsymbol{m} \partial \boldsymbol{m}'}|_{\boldsymbol{m}^{\star}} \quad \text{and} \quad \boldsymbol{H} = \frac{\partial^2 L}{\partial \boldsymbol{m} \partial \boldsymbol{m}'}|_{\boldsymbol{m}^{\star}}.$$

- Compute $\boldsymbol{\Sigma} = (\boldsymbol{I} - \boldsymbol{V}\boldsymbol{H})^{-1}\boldsymbol{V}$.

5. Numerical Experimentations of Normal–Poisson Distribution

The Normal–Poisson example is a non–conjugacy generalized mix model. The observed data are from a Poisson distribution named y_n and a design vector x_n, for $n = 1, \ldots, N$. The considered model is in Figure 1:

$$y_n | z_n \sim \mathcal{P}(\exp(z_n)), \tag{16}$$

where

$$z_n | \beta, \tau \sim \mathcal{N}\left(\beta x_n, \tau^{-1}\right), \quad \beta \sim \mathcal{N}\left(0, \sigma_\beta^2\right), \quad and\, \tau \sim \Gamma(\alpha_\tau, \beta_\tau).$$

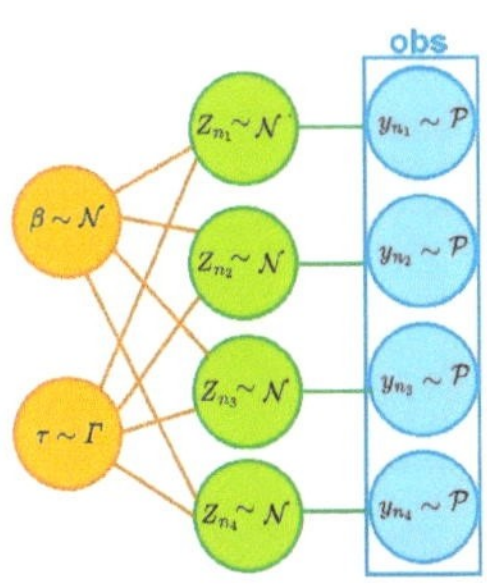

Figure 1. The Normal–Poisson hierarchical model pattern when $N = 4$ with sample size 100.

In the first step, we approximate the posterior joint distribution of β, τ and $z = (z_1, \cdots, z_n)$ for $n = 1, \ldots, N$ via MFVBA:

$$q(\beta, \tau, z) = q_{N+1}(\beta)\, q_{N+2}(\tau) \prod_{n=1}^{N} q_n(z_n).$$

We need to write $L = \ln p(\beta, \tau, z, y)$:

$$\ln p(\beta, \tau, z, y) = -\frac{\beta^2}{2\sigma_\beta^2} + (\alpha_\tau + \frac{N}{2} - 1)\ln(\tau) - \beta_\tau \tau - \frac{\tau}{2}\sum_{n=1}^{N} z_n^2 - \frac{\tau\beta^2}{2}\sum_{n=1}^{N} x_n^2 \tag{17}$$
$$+ \tau\beta \sum_{n=1}^{N} z_n x_n + \sum_{n=1}^{N} z_n y_n - \sum_{n=1}^{N} \exp(z_n) - \sum_{n=1}^{N} \ln(y_n!) + C,$$

where C is a constant and does not depend on unknown variables. We start with computing the distributions of z_n for $n = 1, \ldots, N$. For the simplicity of notations, we assume that n is a natural number in the interval $[1, N]$, and $q_{N+2\backslash n}$ means all $N + 2$ distribution items except the nth one, and $\langle\cdot\rangle_q$ means the expectation over density q. Thus, $q_n(z_n)$, the density of z_n, is equivalent to the following:

$$\langle \ln p(\beta,\tau,z,y)\rangle_{q_{N+2\backslash n}} \propto -\frac{\langle\tau\rangle_{q_{N+2}}}{2} z_n^2 + (\langle\tau\rangle_{q_{N+2}}\langle\beta\rangle_{q_{N+1}} x_n + y_n) z_n - \exp(z_n).$$

Since this expression does not follow a specific distribution because of the existence of $\exp(z_n)$, we omit this part according to its expectation, which is a function of $\langle z_n\rangle_{q_n}$ and $\langle z_n^2\rangle_{q_n}$. We can calculate the expectations using the Normal moment–generating distribution function. The logical reason is that its expectation depends only on $\langle z_n\rangle_{q_n}$ and $\langle z_n^2\rangle_{q_n}$. Moreover, we have $\langle\tau\rangle_{q_{N+2}} = \frac{\alpha_\tau}{\beta_\tau}$ and $\langle\beta\rangle_{q_{N+1}} = 0$. Thus,

$$\langle \ln p(\beta,\tau,z,y)\rangle_{q_{N+2\backslash n}} \propto -\frac{\alpha_\tau}{2\beta_\tau}\left(z_n^2 - \frac{2\beta_\tau y_n}{\alpha_\tau} z_n\right),$$

which means a Normal distribution for z_n, $\mathcal{N}(\frac{\beta_\tau \bar{y}_n}{\alpha_\tau}, \frac{\beta_\tau}{\alpha_\tau})$ for $n = 1,\ldots,N$. The process is quite similar for $q_{N+2}(\tau)$ and $q_{N+1}(\beta)$. The corresponding equivalents of distributions for q_{N+2} and q_{N+1} are shown below, respectively:

$$\begin{aligned}\langle \ln p(\beta,\tau,z,y)\rangle_{q_{1,\ldots,N+1}} \propto\ & (\alpha_\tau + \frac{N}{2} - 1)\ln(\tau) \\ & - \frac{\tau}{2}\left(2\beta_\tau + \sum_{n=1}^{N}(\langle z_n^2\rangle_{q_n} + \langle\beta^2\rangle_{q_{N+1}} x_n^2 - 2\langle\beta\rangle_{q_{N+1}} x_n \langle z_n\rangle_{q_n})\right),\end{aligned}$$

and

$$\langle \ln p(\beta,\tau,z,y)\rangle_{q_{1,\ldots,N,N+2}} \propto -\frac{\beta^2}{2\sigma_\beta^2} - \frac{\langle\tau\rangle_{q_{N+2}}\beta^2}{2}\sum_{n=1}^{N} x_n^2 + \langle\tau\rangle_{q_{N+2}}\beta \sum_{n=1}^{N}\langle z_n\rangle_{q_n} x_n.$$

Thus, we have:

$$\tau \sim \Gamma\left(\alpha_\tau + \frac{N}{2},\ \frac{1}{2}\left(2\beta_\tau + \frac{\beta_\tau}{\alpha_\tau}N + \frac{\beta_\tau^2}{\alpha_\tau^2}\sum_{n=1}^{N}\bar{y}_n^2 + \sigma_\beta^2\sum_{n=1}^{N} x_n^2\right)\right),$$

and

$$\beta \sim \mathcal{N}\left(\frac{\sigma_\beta^2(2\alpha_\tau + N)\frac{\beta_\tau}{\alpha_\tau}\sum_{n=1}^{N}\bar{y}_n x_n}{2\beta_\tau + \frac{\beta_\tau}{\alpha_\tau}N + \frac{\beta_\tau^2}{\alpha_\tau^2}\sum_{n=1}^{N}\bar{y}_n^2 + \sigma_\beta^2(2\alpha_\tau + N + 1)\sum_{n=1}^{N} x_n^2},\right.$$
$$\left.\frac{\sigma_\beta^2\left(2\beta_\tau + \frac{\beta_\tau}{\alpha_\tau}N + \frac{\beta_\tau^2}{\alpha_\tau^2}\sum_{n=1}^{N}\bar{y}_n^2 + \sigma_\beta^2\sum_{n=1}^{N} x_n^2\right)}{2\beta_\tau + \frac{\beta_\tau}{\alpha_\tau}N + \frac{\beta_\tau^2}{\alpha_\tau^2}\sum_{n=1}^{N}\bar{y}_n^2 + \sigma_\beta^2(2\alpha_\tau + N + 1)\sum_{n=1}^{N} x_n^2}\right).$$

In the second step, we calculate the variance diagonal matrix $\boldsymbol{V}$:

$$\boldsymbol{V} = \begin{bmatrix} \Sigma_\beta & 0 & 0 \\ 0 & \Sigma_\tau & 0 \\ 0 & 0 & \boldsymbol{\Sigma_z} \end{bmatrix},$$

where

$$\Sigma_\beta = \frac{\sigma_\beta^2\left(2\beta_\tau + \frac{\beta_\tau}{\alpha_\tau}N + \frac{\beta_\tau^2}{\alpha_\tau^2}\sum_{n=1}^N \bar{y}_n^2 + \sigma_\beta^2\sum_{n=1}^N x_n^2\right)}{2\beta_\tau + \frac{\beta_\tau}{\alpha_\tau}N + \frac{\beta_\tau^2}{\alpha_\tau^2}\sum_{n=1}^N \bar{y}_n^2 + \sigma_\beta^2(2\alpha_\tau + N + 1)\sum_{n=1}^N x_n^2},$$

$$\Sigma_\tau = \frac{2(2\alpha_\tau + N)}{\left(2\beta_\tau + \frac{\beta_\tau}{\alpha_\tau}N + \frac{\beta_\tau^2}{\alpha_\tau^2}\sum_{n=1}^N \bar{y}_n^2 + \sigma_\beta^2\sum_{n=1}^N x_n^2\right)^2},$$

and

$$\Sigma_{z_n} = \frac{\beta_\tau}{\alpha_\tau}, \quad n = 1, \dots, N.$$

Thus, $\boldsymbol{\Sigma}_{\boldsymbol{z}} = diag(\Sigma_{z_1}, \cdots, \Sigma_{z_N})$. The third step is to compute the Hessian matrix $\boldsymbol{H}$:

$$\boldsymbol{H} = \begin{bmatrix} \frac{\partial^2 L}{\partial \beta^2} & \frac{\partial^2 L}{\partial \beta \partial \tau} & \frac{\partial^2 L}{\partial \beta \partial \boldsymbol{z}} \\ \frac{\partial^2 L}{\partial \tau \partial \beta} & \frac{\partial^2 L}{\partial \tau^2} & \frac{\partial^2 L}{\partial \tau \partial \boldsymbol{z}} \\ \frac{\partial^2 L}{\partial \boldsymbol{z} \partial \beta} & \frac{\partial^2 L}{\partial \boldsymbol{z} \partial \tau} & \frac{\partial^2 L}{\partial \boldsymbol{z}^2} \end{bmatrix}.$$

Since $\boldsymbol{H}$ is symmetric in our case study, we need to calculate the main diagonal and the upper triangle. This covariance matrix depends on the values of $\boldsymbol{m}$:

$$\boldsymbol{m} = \left(E_q(\beta), E_q(\beta^2), E_q(\tau), E_q(\ln(\tau)), E_q(z_1), E_q(z_1^2), \dots, E_q(z_N), E_q(z_N^2)\right)^T.$$

Thus, the diagonal of $\boldsymbol{H}$ as a function of $\boldsymbol{m}$ is shown below:

$$\begin{aligned} \frac{\partial^2 L}{\partial \beta^2} &= -\frac{1}{\sigma_\beta^2} - E_q(\tau)\sum_{n=1}^N x_n^2, \\ \frac{\partial^2 L}{\partial \tau^2} &= -(\alpha_\tau + \frac{N}{2} - 1)E_q\left(\frac{1}{\tau^2}\right), \\ \frac{\partial^2 L}{\partial z_n^2} &= -E_q(\tau) - E_q(\exp(z_n)), \quad n = 1, \dots, N, \end{aligned}$$

where L is defined in Equation (17). The upper triangle is:

$$\begin{aligned} \frac{\partial^2 L}{\partial \beta \partial \tau} &= -E_q(\beta)\sum_{n=1}^N x_n^2 + \sum_{n=1}^N E_q(z_n)x_n, \\ \frac{\partial^2 L}{\partial \beta \partial z_n} &= E_q(\tau)x_n, \quad n = 1, \dots, N, \\ \frac{\partial^2 L}{\partial \tau \partial z_n} &= -E_q(z_n) + E_q(\beta)x_n, \quad n = 1, \dots, N. \end{aligned}$$

The last step is to compute $\boldsymbol{\Sigma}$ using (15). For numerical experimentation, we generate data from the Normal–Poisson model with these descriptions:

$$\sigma_\beta^2 = 10, \ \alpha_\tau = 1, \ \beta_\tau = 1, \ N = 4.$$

We generate x_n for $n = 1, \dots, N$ from the Normal distribution $\mathcal{N}(2, 1)$. The number of the reputation for vector y_n is 100, so y_n is a matrix $N \times 100$. The only data we use are the observation matrix y_n and vector x_n. The covariance matrix $\boldsymbol{V}$ from MFVBA is a diagonal matrix:

$$\boldsymbol{V} = \begin{bmatrix} 1.48927 & 0 & \mathbf{0}^T \\ 0 & 0.00071 & \mathbf{0}^T \\ \mathbf{0} & \mathbf{0} & \boldsymbol{I} \end{bmatrix},$$

where $\mathbf{0}$ and $\boldsymbol{I}$ are the zero vector with N dimensions and the identity matrix with $N \times N$ dimensions, respectively. The Hessian matrix $\boldsymbol{H}$ and matrix $\boldsymbol{I} - \boldsymbol{V}\boldsymbol{H}$ are:

$$\boldsymbol{H} = \begin{bmatrix} -0.67147 & 0.09617 & 0.05557 & 0.03031 & 0.13360 & 0.07026 \\ 0.09617 & -1639.9471 & -0.21738 & -0.02130 & 0.02651 & 0.06653 \\ 0.05557 & -0.21738 & -2.20665 & 0 & 0 & 0 \\ 0.03031 & -0.02130 & 0 & -1.78014 & 0 & 0 \\ 0.13360 & 0.02651 & 0 & 0 & -1.86901 & 0 \\ 0.07026 & 0.06653 & 0 & 0 & 0 & -1.69561 \end{bmatrix},$$

and

$$\boldsymbol{I} - \boldsymbol{V}\boldsymbol{H} = \begin{bmatrix} 2 & -0.14322 & -0.08276 & -0.04514 & -0.19897 & -0.10464 \\ -6.8318 \times 10^{-5} & 2.16501 & 0.00015 & 1.5130 \times 10^{-5} & -1.8832 \times 10^{-5} & -4.7264 \times 10^{-5} \\ -0.05557 & 0.21738 & 3.20665 & 0 & 0 & 0 \\ -0.03031 & 0.02130 & 0 & 2.78014 & 0 & 0 \\ -0.13360 & -0.02651 & 0 & 0 & 2.86901 & 0 \\ -0.07026 & -0.06653 & 0 & 0 & 0 & 2.69561 \end{bmatrix}.$$

The covariance matrix via LRVBA is:

$$\boldsymbol{\Sigma} = \begin{bmatrix} 0.74986 & 2.3409 \times 10^{-5} & 0.01299 & 0.00818 & 0.03492 & 0.01955 \\ 2.3408 \times 10^{-5} & 0.00033 & -2.1838 \times 10^{-5} & -2.2586 \times 10^{-6} & 4.1220 \times 10^{-6} & 8.7090 \times 10^{-6} \\ 0.01299 & -2.1838 \times 10^{-5} & 0.31208 & 0.00014 & 0.00060 & 0.00034 \\ 0.00817 & -2.2586 \times 10^{-6} & 0.00014 & 0.35978 & 0.00038 & 0.00021 \\ 0.03492 & 4.1220 \times 10^{-6} & 0.00060 & 0.00038 & 0.35018 & 0.00091 \\ 0.01955 & 8.7090 \times 10^{-6} & 0.00034 & 0.00021 & 0.00091 & 0.37148 \end{bmatrix}. \quad (18)$$

The sparsity patterns for the $\boldsymbol{V}, \boldsymbol{H}$ and $\boldsymbol{I} - \boldsymbol{V}\boldsymbol{H}$ matrixes are shown in Figure 2.

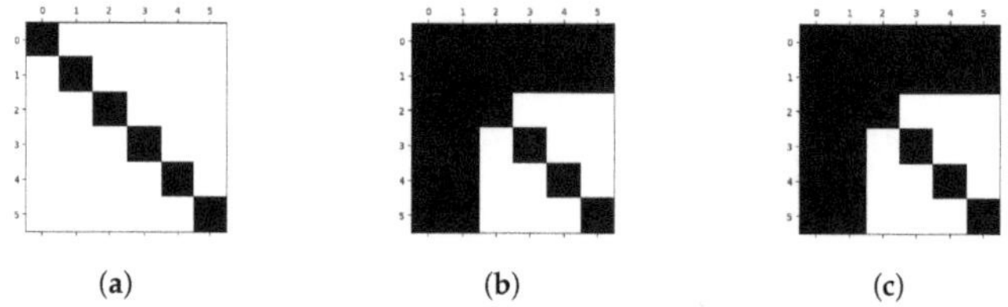

Figure 2. Sparsity patterns for the matrices using model (16). (**a**) VBA covariance matrix $\boldsymbol{V}$; (**b**) Hessian matrix $\boldsymbol{H}$; (**c**) $\boldsymbol{I} - \boldsymbol{V}\boldsymbol{H}$ matrix.

In addition, we simulate the posterior distribution via MCMC. This simulation has five chains and one divergence. The estimated marginal distribution for each unknown variable is shown in Figures 3 and 4 via MFVBA and MCMC, respectively. More details are in Table 1. The joint distribution of unknown variables is approximated via MFVBA and MCMC. The MFVBA distribution is the marginal density factorial of the unknown variables. Therefore, the variables are independent in MFVBA, and the covariance matrix is diagonal. LRVBA is a correction method on the covariance matrix. Thus, the matrix is no longer diagonal as presented in (18). Also, we add more details about the MCMC results in Table 1. The Monte Carlo Standard Error (MCSE) is a chain-accurate measure and provides a quantitative suggestion of how big the estimation noise is. The noises are small based on the mean and sd of the MCSE. The ess–bulk column delivers an evaluated bulk influential sample size using rank–normalised draws. The ess–tail column creates an approximated tail practical sample size by computing the effective minimum sample sizes between the 5% and 95% quantiles. The last column is the r–hat convergence diagnostic, which compares the between and within–chain estimation for model variables and other univariate quantities.

Table 1. Summary of the margin densities for unknown variables.

Unknown	MFVBA Method		LRVBA Method	MCMC Method								
	mean	sd	sd	mean	sd	hdi–3%	hdi–97%	MCSE–mean	MCSE–sd	ess–bulk	ess–tail	r–hat
β	0.044	1.220	0.866	−1.616	1.769	−5.175	1.620	0.148	0.194	186.0	43.0	1.11
τ	0.047	0.027	0.018	0.361	0.188	0.057	0.685	0.024	0.017	38.0	51.0	1.11
z_0	0.270	1	0.559	−1.316	0.177	−1.664	−1.028	0.008	0.006	468.0	316.0	1.02
z_1	0.050	1	0.600	−3.051	0.410	−3.783	−2.252	0.021	0.015	382.0	372.0	1.00
z_2	0.100	1	0.592	−2.375	0.332	−3.029	−1.790	0.015	0.011	476.0	415.0	1.02
z_3	0	1	0.610	−7.334	4.641	−14.366	−3.311	0.868	0.620	45.0	55.0	1.13

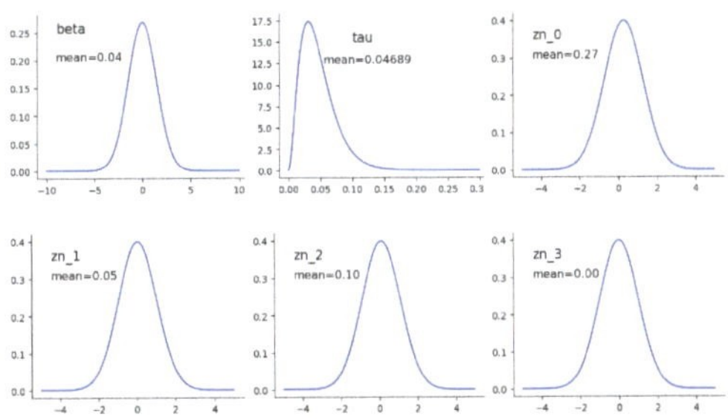

Figure 3. The final estimations of margin densities for each variable via MFVBA method.

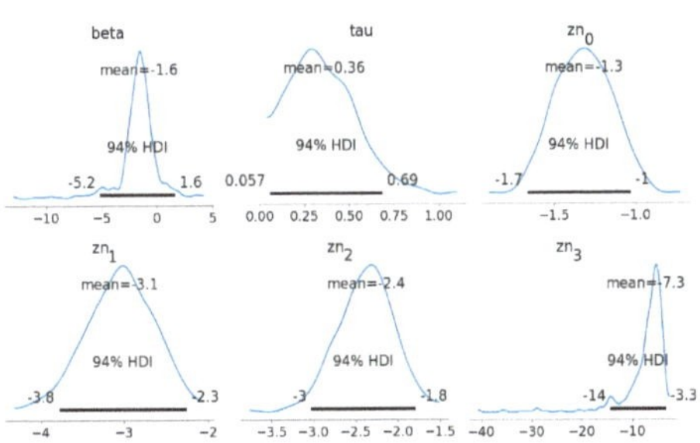

Figure 4. The final margins via MCMC method.

6. Conclusions

Bayesian computations are used to infer unknown parameters from data. ABC methods are applied to perform approximate computation, particularly in high–dimensional problems. Two classes of approximate computational methods are MCMC and VBA. MCMC generates samples from the probability function to approximate the posterior distribution. It generates a sequence of samples that converges to the target distribution. The VBA approximates the expression of the joint posterior distribution of unknown parameters in a complex model using a simple expression. VBA has some advantages over MCMC, such as being faster, more scalable and versatile. It also guarantees convergence to a local minimum of KL divergence.

VBA's drawback is that it underestimates the covariance matrix of the unknowns. For this aim, we use LRVB to approximate the covariance more precisely. However, the LRVB computations still need the inversion of some matrixes, which is costly when the unknown parameters' dimensions are high. We work on a numerical example to show the whole process of VBA and LRVB as well as the performance and ability of the proposed method and compare it with the classical MCMC results. As a result of the numerical example, the VBA implementation was much shorter than the MCMC, and we had an explicit form of joint distribution with a modified covariance matrix with the LRVB aid.

Author Contributions: Conceptualization, S.A.F.M. and A.M.-D.; methodology, S.A.F.M. and A.M.-D.; software, S.A.F.M.; validation, A.M.-D.; formal analysis, S.A.F.M. and A.M.-D.; investigation, S.A.F.M. and A.M.-D.; resources, S.A.F.M. and A.M.-D.; data curation, S.A.F.M. and A.M.-D.; writing—original draft preparation, S.A.F.M. and A.M.-D.; writing—review and editing, S.A.F.M. and A.M.-D.; visualization, S.A.F.M. and A.M.-D.; supervision, A.M.-D.; project administration, A.M.-D. All authors have read and agreed to the published version of the manuscript.

Funding: This research received no external funding.

Institutional Review Board Statement: Not applicable.

Informed Consent Statement: Not applicable.

Data Availability Statement: Data sharing is not applicable to this article.

Conflicts of Interest: The author Ali Mohammad-Djafari is Scientific Research Leader of Shanfeng Company. The authors declare no conflict of interest.

References

1. Metropolis, N.; Rosenbluth, A.W.; Rosenbluth, M.N.; Teller, A.H.; Teller, E. Equation of state calculations by fast computing machines. *J. Chem. Phys.* **1953**, *21*, 1087–1092. [CrossRef]
2. Hastings, W.K. Monte Carlo sampling methods using Markov chains and their applications. *Biometrika* **1970**, *57*, 97–109. [CrossRef]
3. Jordan, M.I.; Ghahramani, Z.; Jaakkola, T.S.; Saul, L.K. An introduction to variational methods for graphical models. In *Learning in Graphical Models*; Springer: Dordrecht, The Netherlands, 1998; pp. 105–161.
4. Sarkka, S.; Nummenmaa, A. Recursive noise adaptive Kalman filtering by variational Bayesian approximations. *IEEE Trans. Autom. Control* **2009**, *54*, 596–600. [CrossRef]
5. Rohde, D.; Wand, M.P. Semiparametric mean field variational Bayes: General principles and numerical issues. *J. Mach. Learn. Res.* **2016**, *17*, 5975–6021.
6. Giordano, R.J.; Broderick, T.; Jordan, M.I. Linear response methods for accurate covariance estimates from mean field variational Bayes. *Adv. Neural Inf. Process. Syst.* **2015**, *28*, 1441–1449.
7. Giordano, R.; Broderick, T.; Jordan, M.I. Covariances, robustness and variational bayes. *J. Mach. Learn. Res.* **2018**, *19*, 1–49.

physical sciences forum

MDPI

Proceeding Paper

Proximal Nested Sampling with Data-Driven Priors for Physical Scientists †

Jason D. McEwen [1,2,*], Tobías I. Liaudat [1,3], Matthew A. Price [1], Xiaohao Cai [4] and Marcelo Pereyra [5]

1 Mullard Space Science Laboratory, University College London (UCL), Dorking RH5 6NT, UK; tobiasliaudat@gmail.com (T.I.L.); m.price.17@alumni.ucl.ac.uk (M.A.P.)
2 Alan Turing Institute, London NW1 2DB, UK
3 Department of Computer Science, University College London (UCL), London WC1E 6BT, UK
4 School of Electronics and Computer Science, University of Southampton, Southampton SO17 1BJ, UK; x.cai@soton.ac.uk
5 School of Mathematical and Computer Sciences, Heriot-Watt University, Edinburgh EH14 4AS, UK; m.pereyra@hw.ac.uk
* Correspondence: jason.mcewen@ucl.ac.uk
† Presented at the 42nd International Workshop on Bayesian Inference and Maximum Entropy Methods in Science and Engineering, Garching, Germany, 3–7 July 2023.

Abstract: Proximal nested sampling was introduced recently to open up Bayesian model selection for high-dimensional problems such as computational imaging. The framework is suitable for models with a log-convex likelihood, which are ubiquitous in the imaging sciences. The purpose of this article is two-fold. First, we review proximal nested sampling in a pedagogical manner in an attempt to elucidate the framework for physical scientists. Second, we show how proximal nested sampling can be extended in an empirical Bayes setting to support data-driven priors, such as deep neural networks learned from training data.

Keywords: Bayesian model selection; nested sampling; proximal calculus

1. Introduction

In many of the sciences, not only is one interested in estimating the parameters of an underlying model, but deciding which model is best among a number of alternatives is of critical scientific interest. Bayesian model comparison provides a principled approach to model selection [1] that has found widespread application in the sciences [2].

Bayesian model comparison requires computation of the model evidence:

$$\mathcal{Z} = p(y \mid M) = \int \mathrm{d}x\, p(y \mid x, M) p(x \mid M) = \int \mathrm{d}x\, \mathcal{L}(x)\, \pi(x), \tag{1}$$

also called the marginal likelihood, where $y \in \mathbb{R}^m$ denotes data, $x \in \mathbb{R}^n$ parameters of interest, and M the model under consideration. We adopt the shorthand notation for the likelihood of $\mathcal{L}(x) = p(y \mid x, M)$ and prior of $\pi(x) = p(x \mid M)$. Evaluating the multi-dimensional integral of the model evidence is computationally challenging, particularly in high dimensions. While a number of highly successful approaches to computing the model evidence have been developed, such as nested sampling (e.g., [2–8]) and the learned harmonic mean estimator [9–11], previous approaches do not scale to the very high-dimensional settings of computational imaging, which is our driving motivation.

The proximal nested sampling framework was introduced recently by a number of authors of the current article in order to open up Bayesian model selection for high-dimensional imaging problems [12]. Proximal nested sampling is suitable for models for which the likelihood is log-convex, which are ubiquitous in the imaging sciences. By restricting the class of models considered, it is possible to exploit structure of the problem to enable computation in very high-dimensional settings of $\mathcal{O}(10^6)$ and beyond.

Proximal nested sampling draws heavily on convex analysis and proximal calculus. In this article we present a pedagogical review of proximal nested sampling, sacrificing some mathematical rigor in an attempt to provide greater accessibility. We also provide a concise review of convexity and proximal calculus to introduce the background underpinning the framework. We assume the reader is familiar with nested sampling, and hence we avoid repeating an introduction to nested sampling and instead refer the reader to other sources that provide excellent descriptions [2,3,8]. Finally, for the first time, we show in an empirical Bayes setting how proximal nested sampling can be extended to support data-driven priors, such as deep neural networks learned from training data.

2. Convexity and Proximal Calculus

We present a concise review of convexity and proximal calculus to introduce the background underpinning proximal nested sampling to make it more accessible.

2.1. Convexity

Proximal nested sampling draws on convexity, key concepts of which are illustrated in Figure 1. A set C is convex if for any $x_1, x_2 \in C$ and $\alpha \in (0,1)$ we have $\alpha x_1 + (1-\alpha)x_2 \in C$. The epigraph of a function $f : \mathbb{R}^n \to \mathbb{R}$ is defined by $\mathrm{epi}(f) = \{(x,\gamma) \in \mathbb{R}^n \times \mathbb{R} \,|\, f(x) \leq \gamma\}$. The function f is convex if and only if its epigraph is convex. A convex function is lower semicontinuous if its epigraph is closed (i.e., includes its boundary).

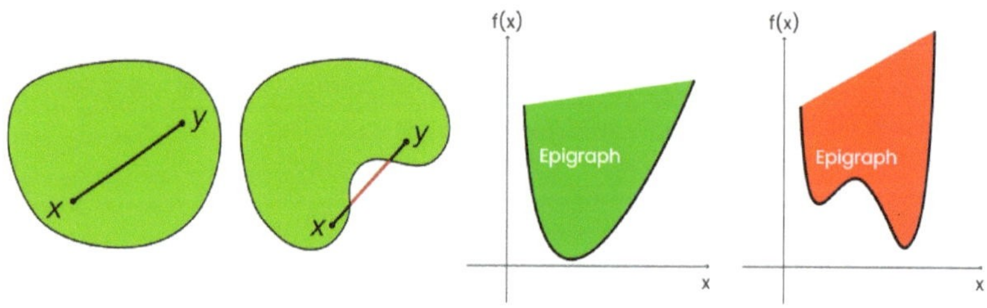

(**a**) Convex set (**b**) Non-convex set (**c**) Convex function (**d**) Non-convex function

Figure 1. Proximal nested sampling considers likelihoods that are log-convex and lower semicontinuous. A lower semicontinuous convex function has a convex and closed epigraph.

2.2. Proximity Operator

Proximal nested sampling leverages proximal calculus [13,14], a key component of which is the proximity operator, or prox. The proximity operator of the function f with parameter λ is defined by

$$\mathrm{prox}_f^\lambda(x) = \arg\min_u \left[f(u) + \|u - x\|^2/2\lambda\right]. \quad (2)$$

The proximity operator maps a point x towards the minimum of f while remaining in the proximity of the original point. The parameter λ controls how close the mapped point remains to x. An illustration is given in Figure 2.

The proximity operator can be considered as a generalisation of the projection onto a convex set. Indeed, the projection operator can be expressed as a prox by

$$\Pi_C(x) = \arg\min_u \left[\chi_C(u) + \|u - x\|^2/2\right], \quad (3)$$

with function f given by the characteristic function $\chi_C(x) = \infty$ if $x \notin C$ and zero otherwise.

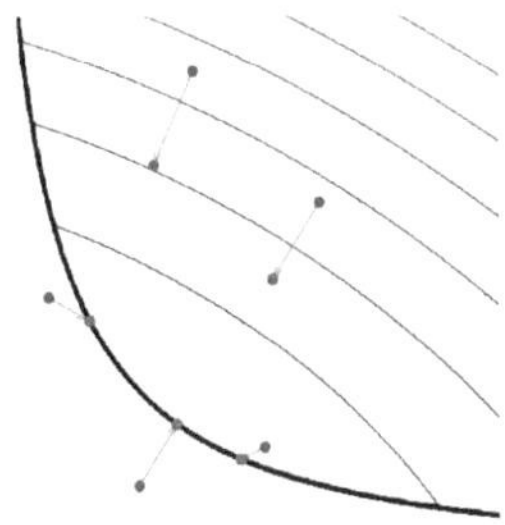

Figure 2. Illustration of the proximity operator (reproduced from [14]). The proximal operator maps the blue points to red points (i.e., from base to head of arrows). The thick black line defines the domain boundary, while the thin black lines define level sets (iso-contours) of f. The proximity operator maps points towards the minimum of f while remaining in the proximity of the original point.

2.3. Moreau–Yosida Regularisation

The final component required in the development of proximal nested sampling is Moreau–Yosida regularisation (e.g., [14]). The Moreau–Yosida envelop of a convex function $f : \mathbb{R}^n \to \mathbb{R}$ is given by the infimal convolution:

$$f^\lambda(x) = \inf_{u \in \mathbb{R}^N} f(u) + \frac{\|u - x\|^2}{2\lambda}. \tag{4}$$

The Moreau–Yoshida envelope of a function can be interpreted as taking its convex conjugate, adding regularisation, before taking the conjugate again [14]. Consequently, it provides a smooth regularised approximation of f, which is very useful to enable the use of gradient-based computational algorithms (e.g., [15]).

The Moreau–Yosida envelope exhibits the following properties. First, λ controls the degree of regularisation with $f^\lambda(x) \to f(x)$ as $\lambda \to 0$. Second, the gradient of the Moreau–Yosida envelope of f can be computed through its prox by $\nabla f^\lambda(x) = (x - \text{prox}_f^\lambda(x))/\lambda$.

3. Proximal Nested Sampling

The challenge of nested sampling in high-dimensional settings is to sample from the prior distribution subject to a hard likelihood constraint [2,3,8]. Proximal nested sampling addresses this challenge for the case of log-convex likelihoods, which are widespread in computational imaging problems. In this section we review the proximal nested sampling framework [12] in a pedagogical manner, sacrificing some mathematical rigor in an attempt to improve readability and accessibility.

3.1. Constrained Sampling Formulation

Consider a prior and likelihood $\pi(x) \propto \exp(-f(x))$ and $\mathcal{L}(x) \propto \exp(-g(x))$, where the log-likehood $g = -\log \mathcal{L}$ is a convex lower semicontinuous function. The log-prior $f = -\log \pi$ need only be differentiable or convex (it need not be convex if it is differentiable).

We consider sampling from the prior $\pi(x)$, such that $\mathcal{L}(x) > L^*$ for some likelihood value $L^* \geq 0$. Let $\iota_{L^*}(x)$ and $\chi_{L^*}(x)$ be the indicator function and characteristic function corresponding to this constraint, respectively, defined as

$$\iota_{L^*}(x) = \begin{cases} 1, & \mathcal{L}(x) > L^*, \\ 0, & \text{otherwise,} \end{cases} \quad \text{and} \quad \chi_{L^*}(x) = \begin{cases} 0, & \mathcal{L}(x) > L^*, \\ +\infty, & \text{otherwise.} \end{cases} \tag{5}$$

Since log is monotonic, $\mathcal{L}(x) > L^*$ is equivalent to $g(x) < \tau$ for $\tau = -\log L^*$. Explicitly define the convex set of the likelihood constraint by $\mathcal{B}_\tau = \{x \mid g(x) < \tau\}$. Then, $\chi_{L^*}(x)$ is equivalent to $\chi_{\mathcal{B}_\tau}(x)$, where $\chi_{\mathcal{B}_\tau}(x) = \infty$ if $x \notin \mathcal{B}_\tau$ and zero otherwise.

Let $\pi_{L^*}(x) = \pi(x)\iota_{L^*}(x)$ represent the prior distribution with the hard likelihood constraint $\mathcal{L}(x) > L^*$. Since $\iota_{L^*}(x) = \exp(-\chi_{L^*}(x))$, then we have

$$-\log \pi_{L^*}(x) = -\log \pi(x) + \chi_{\mathcal{B}_\tau}(x). \quad (6)$$

To sample from the constrained prior we require sampling techniques that firstly can scale to high-dimensional settings and that secondly can support the convex constraint $\chi_{\mathcal{B}_\tau}(x)$.

3.2. Langevin MCMC Sampling

Langevin Markov chain Monte Carlo (MCMC) sampling has been demonstrated to be highly effective at sampling in high-dimensional settings by exploiting gradient information [15,16]. The Langevin stochastic differential equation associated with distribution $p(x)$ is a stochastic process defined by

$$\mathrm{d}x(t) = \frac{1}{2}\nabla \log p(x(t))\mathrm{d}t + \mathrm{d}w(t), \quad (7)$$

where $w(t)$ is Brownian motion. This process converges to $p(x)$ as time t increases and is therefore useful for generating samples from $p(x)$. In practice, we compute a discrete-time approximation of $x(t)$ by the conventional Euler–Maruyama discretisation:

$$x^{(k+1)} = x^{(k)} + \frac{\delta}{2}\nabla \log p(x^{(k)}) + \sqrt{\delta}w^{(k+1)}, \quad (8)$$

where $w^{(k)}$ is a sequence of standard Gaussian random variables and δ is a step size.

Equation (8) provides a strategy for sampling in high-dimensions. However, notice that the updates rely on the score of the target distribution $\nabla \log p(\cdot)$. Nominally the target distribution must therefore be differentiable, which is not the case for our target of interest given by Equation (6). The prior may or may not be differentiable but the likelihood constraint certainly is not. Proximal versions of Langevin sampling have been developed to address the setting where the distribution is log-convex but not necessarily differentiable [15,16]. We follow a similar approach.

3.3. Proximal Nested Sampling Framework

The proximal nested sampling framework follows by taking the constrained sampling formulation of Equation (6), adopting Langevin MCMC sampling of Equation (8), and applying Moreau–Yosida regularisation of Equation (4) to the convex constraint $\chi_{\mathcal{B}_\tau}(x)$ to yield a differentiable target. This strategy yields (see [12]) the update equation:

$$x^{(k+1)} = x^{(k)} + \frac{\delta}{2}\nabla \log \pi(x^{(k)}) - \frac{\delta}{2\lambda}[x^{(k)} - \mathrm{prox}_{\chi_{\mathcal{B}_\tau}}(x^{(k)})] + \sqrt{\delta}w^{(k+1)}, \quad (9)$$

where δ is the step size and λ is the Moreau–Yosida regularisation parameter.

Further intuition regarding proximal nested sampling can be gained by examining the term $v^{(k)} = -[x^{(k)} - \mathrm{prox}_{\chi_{\mathcal{B}_\tau}}(x^{(k)})]$, together with Figure 3. The vector $v^{(k)}$ points from the sample $x^{(k)}$ to its projection onto the likelihood constraint. If the sample $x^{(k)}$ is already in the likelihood-restricted prior support $\mathcal{B}_\tau$, i.e., $x \in \mathcal{B}_\tau$, the term $v^{(k)}$ disappears and the Markov chain iteration simply involves the standard Langevin MCMC update. In contrast, if $x^{(k)}$ is not in $\mathcal{B}_\tau$, i.e., $x \notin \mathcal{B}_\tau$, then a step is taken in the direction $v^{(k)}$, which acts to move the next iteration of the Markov chain in the direction of the projection of $x^{(k)}$ onto the convex set $\mathcal{B}_\tau$. This term therefore acts to push the Markov chain back into the constraint set $\mathcal{B}_\tau$ if it wanders outside of it. (Note that proximal nested sampling has some similarity with Galilean [17] and constrained Hamiltonian [18] nested sampling. In these approaches, Makov chains are also considered and if the Markov chain steps outside of the likelihood-constraint then it is reflected by an approximation of the shape of the boundary.)

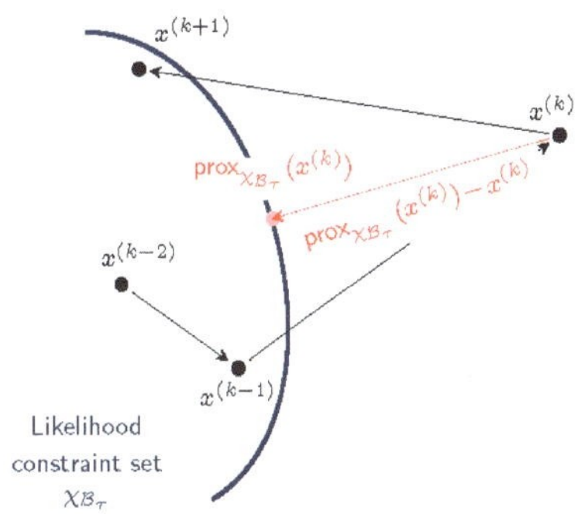

Figure 3. Diagram illustrating proximal nested sampling. If a sample $x^{(k)}$ outside of the likelihood constraint is considered, then proximal nested sampling introduces a term in the direction of the projection of $x^{(k)}$ onto the convex set defining the likelihood constraint, thereby acting to push the Markov chain back into the constraint set $\mathcal{B}_\tau$ if it wanders outside of it. A subsequent Metropolis–Hastings step can be introduced to enforce strict adherence to the convex likelihood constraint.

We have so far assumed that the (log) prior is differentiable (see Equation (9)). This may not be the case, as is typical for sparsity-promoting priors (e.g., $-\log \pi(x) = \|\Psi^\dagger x\|_1 + \text{const.}$ for some wavelet dictionary Ψ). Then, we make a Moreau–Yosida approximation of the log-prior, yielding the update equation:

$$x^{(k+1)} = x^{(k)} - \frac{\delta}{2\lambda}\left[x^{(k)} - \text{prox}^{\lambda}_{-\log\pi}(x^{(k)})\right] - \frac{\delta}{2\lambda}\left[x^{(k)} - \text{prox}_{\chi_{\mathcal{B}_\tau}}(x^{(k)})\right] + \sqrt{\delta} w^{(k+1)}. \quad (10)$$

For notational simplicity here we have adopted the same regularisation parameter λ for each Moreau–Yosida approximation.

With the current formulation we are not guaranteed to recover samples from the prior subject to the hard likelihood constraint due to the approximation introduced in the Moreu–Yosida regularisation and due to the approximation in discretising the underlying Langevin stochastic differential equation. We therefore introduce a Metropolis–Hastings correction step to eliminate the bias introduced by these approximations and ensure convergence to the required target density (see [12] for further details).

Finally, we adopt this strategy for sampling from the constrained prior in the standard nested sampling strategy to recover the proximal nested sampling framework. The algorithm can be initalised with samples from the prior as described by the update equations above but with the likelihood term removed, i.e., with $[x^{(k)} - \text{prox}_{\chi_{\mathcal{B}_\tau}}(x^{(k)})] \to 0$.

3.4. Explicit Forms of Proximal Nested Sampling

While we have discussed the general framework for proximal nested sampling, we have yet to address the issue of computing the proximity operators involved. As Equation (2) demonstrates, computing proximity operators involves solving an optimisation problem. Only in certain cases are closed form solutions available [13]. Explicit forms of proximal nested sampling must therefore be considered for the problem at hand.

We focus on a common high-dimensional inverse imaging problem where we acquire noisy observations, $y = \Phi x + n$, of an underlying image x via some measurement model Φ, in the presence of Gaussian noise n (without loss of generality we consider independent and identically distributed noise here). We consider a Gaussian negative likelihood, $-\log \mathcal{L}(x) = \|y - \Phi x\|_2^2 / 2\sigma^2 + \text{const.}$, and a sparsity-promoting prior, $-\log \pi(x) = \mu \|\Psi^\dagger x\|_1 + \text{const.}$, for some wavelet dictionary Ψ. The prox of the prior can be computed in closed-form by [13]

$$\text{prox}^{\lambda}_{-\log\pi}(x) = x + \Psi\left(\text{soft}_{\lambda\mu}(\Psi^\dagger x') - \Psi^\dagger x\right), \quad (11)$$

where $\mathrm{soft}_\lambda(\cdot)$ is the soft thresholding function with threshold λ (recall μ is the scale of the sparsity-promoting prior, i.e., the regularisation parameter, defined above). However, the prox of the likelihood is not so straightforward. The prox for the likelihood can be recast as a saddle-point problem that can be solved iteratively by a primal dual method initialised by the current sample position (see [12] for further details):

1. $z^{(i+1)} = z^{(i)} + \delta_1 \Phi \bar{x}^{(i)} - \mathrm{prox}_{\chi_{\mathcal{B}'_{\tau'}}}(z^{(i)} + \delta_1 \Phi \bar{x}^{(i)})$,

$$\text{where } \mathrm{prox}_{\chi_{\mathcal{B}'_{\tau'}}}(z) = \mathrm{proj}_{\mathcal{B}'_{\tau'}}(z) = \begin{cases} z, & \text{if } z \in \mathcal{B}'_{\tau'}, \\ \frac{z-y}{\|z-y\|_2}\sqrt{2\tau\sigma^2} + y, & \text{otherwise;} \end{cases}$$

2. $x^{(i+1)} = (x' + x^{(i)} - \delta_2 \Phi^\dagger z^{(i+1)})/2\,;$
3. $\bar{x}^{(i+1)} = x^{(i+1)} + \delta_3(x^{(i+1)} - x^{(i)})\,.$

Combining these algorithms to efficiently compute prox operators with the proximal nested sampling framework, we can compute the model evidence to perform Bayesian model comparison in high-dimensional settings. We can also obtain posterior distributions with the usual weighted samples from the dead points of nested sampling. This allows one to recover, for example, point estimates such as the posterior mean image.

4. Deep Data-Driven Priors

While hand-crafted priors, such as wavelet-based sparsity promoting priors, are common in computational imaging, they provide only limited expressivity. If example images are available an empirical Bayes approach with data-driven priors can be taken, where the prior is learned from training data. Since proximal nested sampling requires only the log-likelihood to be convex, complex data-driven priors, such as represented by deep neural networks, can be integrated into the framework. Through Tweedie's formula we describe how proximal nested sampling can be adapted to support data-driven priors, opening up Bayesian model selection for data-driven approaches. We take a similar approach to [19], where data-driven priors are integrated into Langevin MCMC sampling strategies, although in that work model selection is not considered.

4.1. Tweedie's Formula and Data-Driven Priors

Tweedie's formula is a remarkable result in Bayesian estimation credited to personal correspondence with Maurice Kenneth Tweedie [20]. Tweedie's formula has gained renewed interest in recent years [19,21–23] due to its connection to score matching [24–26] and denoising diffusion models [27,28], which, as of this writing, provide state-of-the-art performance in deep generative modelling.

Tweedie's result follows by considering the following scenario. Consider x sampled from a prior distribution $q(\cdot)$ and noisy observations $z \sim \mathcal{N}(x, \sigma^2 I)$. Tweedie's formula gives the posterior expectation of x given z as

$$E(x|z) = z + \sigma^2 \nabla \log p(z), \tag{12}$$

where $p(z)$ is the marginal distribution of z (for further details see, e.g., [21]). The critical advantage of Tweedie's formula is that it does not require knowledge of the underlying distribution $q(\cdot)$ but rather only the marginalised distribution of the observation. Equation (12) can be interpreted as a denoising strategy to estimate x from noisy observations z. Moreover, Tweedie's formula can also be used to relate a denoiser (potentially a trained deep neural network) to the score $\nabla \log p(z)$.

In a data-driven setting, where the underlying prior is implicitly specified by training data (which are considered to be samples from the prior), there is no guarantee that the underlying prior, and therefore the posterior, is well-suited for gradient-based Bayesian computation such as Langevin sampling; e.g., it may not be differentiable. Therefore, we consider a regularised version of the prior defined by Gaussian smoothing:

$$p_\epsilon(x) = (2\pi\epsilon)^{-n/2} \int \mathrm{d}x' \exp(| x - x'\|_2^2/(2\epsilon))q(x'). \tag{13}$$

This regularisation can also be viewed as adding a small amount of regularising Gaussian noise. We can therefore leverage Tweedie's formula to relate the regularised prior distribution $p_\epsilon(x)$ to a denoiser D_ϵ trained to recover x from noisy observations $x_\epsilon \sim \mathcal{N}(x, \epsilon I)$; i.e., the score of the regualised prior can be related to the denoiser by

$$\nabla \log p_\epsilon(x) = \epsilon^{-1}(D_\epsilon(x) - x). \tag{14}$$

Denoisers are commonly integrated in proximal optimisation algorithms in replacement of proximity operators, giving rise to so-called plug-and-play (PnP) methods [29,30] and, more recently, also into Bayesian computational algorithms [19]. Typically, denoisers are represented by deep neural networks, which can be trained by injecting a small amount of noise in training data and learning to denoise the corrupted data. While a noise level ϵ needs to be chosen, as discussed above, this is considered a regularisation of the prior and so the denoiser need not be trained on the noise level of a problem at hand. In this manner, the same denoiser can be used for multiple subsequent problems (hence the PnP name). The learned score of the regularised prior inherits the same properties as the denoiser, such as smoothness, and hence the denoiser should be considered carefully. Well-behaved denoisers have been considered already in PnP methods (in order to provide convergence guarantees) and a popular approach for imaging problems is the DnCNN model [30], which is based on a deep convolutional neural network and that is (Lipschitz) continuous.

4.2. *Proximal Nested Sampling with Data-Driven Priors*

By Tweedie's formula, the standard proximal nested sampling update of Equation (9) can be revised to integrate a learned denoiser, yielding

$$x^{(k+1)} = x^{(k)} - \frac{\alpha\delta}{2\epsilon}[x - D_\epsilon(x^{(k)})] - \frac{\delta}{2\lambda}[x^{(k)} - \mathrm{prox}_{\chi_{\mathcal{B}_\tau}}(x^{(k)})] + \sqrt{\delta}w^{(k+1)}, \tag{15}$$

where we have included a regularisation parameter α that allows us to balance the influence of the prior and the data fidelity terms [19]. We typically consider a deep convolutional neural network based on the DnCNN model [30] since it is (Lipschitz) continuous and has been demonstrated to perform very well in PnP settings [19,30]. Again, this sampling strategy can then be integrated into the standard nested sampling framework.

We can therefore support data-driven priors in the proximal nested sampling framework by integrating a deep denoiser that learns to denoise training data, using Tweedie's formula to relate this to the score of a regularised data-driven prior.

5. Numerical Experiments

The new methodology presented allows us to perform Bayesian model comparison between a data-driven and hand-crafted prior (validation of proximal nested sampling in a setting where the ground truth can be computed directly has been performed already [12]). We consider a simple radio interferometric imaging reconstruction problem as an illustration. We assume the same observational model as Section 3.4, with white Gaussian noise giving a signal-to-noise ratio (SNR) of 15 dB. The measurement operator Φ is a masked Fourier transform as a simple model of a radio interferometric telescope. The mask is built by randomly selecting 50% of the Fourier coefficients. A Gaussian likelihood is used in both models. For the hand-crafted prior, we consider a sparsity-promoting prior using a Daubechis 6 wavelet dictionary. We base the data-driven prior on a DnCNN [30] model trained on galaxy images extracted from the IllustrisTNG simulations [31]. We also consider an IllustrisTNG galaxy simulation, not used in training, as the ground-truth test image.

We generate samples following the proximal nested sampling strategies of Equations (10) and (15) for the hand-crafted and data-driven priors, respectively. Posterior inferences (e.g., posterior mean image) and the model evidence can then be computed from nested sampling samples in the usual manner. The step size δ is set to 10^{-7}, the Moreau–Yosida regularisation parameter λ to 5×10^{-7}, and the regularisation strength of wavelet-based model μ to 5×10^{4}. We consider noise level $\epsilon \simeq 8.34$ and set the regularisation parameter α of the data-driven prior to 3.5×10^{-7}. For the nested sampling methods, the number of live and dead samples is set to 10^2 and 2.5×10^3, respectively. For the Langevin sampling, we use a thinning factor of 20 and set the number of burn-in iterations to 10^2.

Results are presented in Figure 4. The data-driven prior results in a superior reconstruction with an improvement in SNR of 1.2 dB, although it may be difficult to tell simply from visual inspection of the recovered images. Computing the SNR of the reconstructed images requires knowledge of the ground truth, which clearly is not accessible in realistic settings involving real observational data. The Bayesian model evidence, computed by proximal nested sampling, proves a way to compare the hand-crafted and data-driven models without requiring knowledge of the ground truth and is therefore applicable in realistic scenarios. We compute log evidences of -2.96×10^3 for the hand-crafted prior and -1.35×10^3 for the data-driven prior. Consequently, the data-driven model is preferred by the model evidence, which agrees with the SNR levels computed from the ground truth. These results are all as one might expect since learned data-driven priors are more expressive than hand-crafted priors and can better adapt to model high-dimensional images.

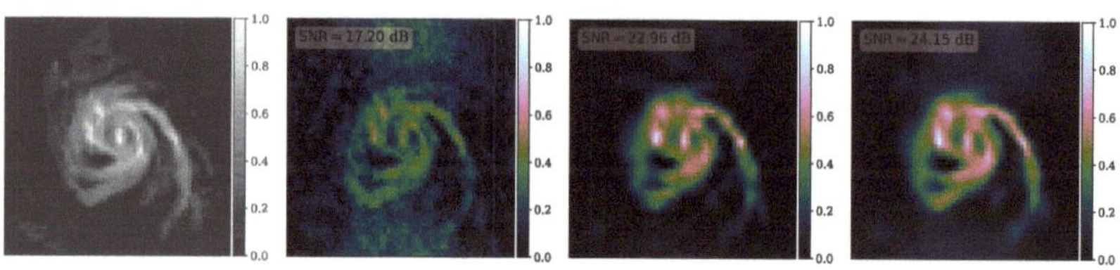

(**a**) Ground truth (**b**) Dirty (**c**) Hand-crafted prior (**d**) Data-driven prior

Figure 4. Results of radio interferometric imaging reconstruction problem. (**a**) Ground truth galaxy image. (**b**) Dirty reconstruction based on pseudo-inverting the measurement operator Φ. (**c**) Posterior mean reconstruction computed from proximal nested samples for the hand-crafted wavelet-sparsity prior. (**d**) Posterior mean reconstruction for the data-driven prior based on a deep neural network (DnCNN) trained on example images. Reconstruction SNR is shown on each image. The computed SNR levels demonstrate that the data-driven prior results in a superior reconstruction quality, although this may not be obvious from a visual assessment of the reconstructed images. Computing the reconstructed SNR requires knowledge of the ground truth, which is not available in realistic settings. The Bayesian model evidence proves a way to compare the hand-crafted and data-driven models without requiring knowledge of the ground truth. For this example, the Bayesian evidence correctly selects the data-driven prior as the best model.

6. Conclusions

Proximal nested sampling leverages proximal calculus to extend nested sampling to high-dimensional settings for problems involving log-convex likelihoods, which are ubiquitous in computational imaging. The purpose of this article is two-fold. First, we review proximal nested sampling in a pedagogical manner in an attempt to elucidate the framework for physical scientists. Second, we show how proximal nested sampling can be extended in an empirical Bayes setting to support data-driven priors, such as deep neural networks learned from training data. We show only preliminary results for learned proximal nested sampling and will present a more thorough study in a follow-up article.

Author Contributions: Conceptualization, J.D.M. and M.P.; methodology, J.D.M., X.C. and M.P.; software, T.I.L., M.A.P. and X.C.; validation, T.I.L., M.A.P. and X.C.; resources, J.D.M.; data curation, M.A.P.; writing—original draft preparation, J.D.M. and T.I.L.; writing—review and editing, J.D.M., T.I.L., M.A.P., X.C. and M.P.; supervision, J.D.M.; project administration, J.D.M.; funding acquisition, J.D.M., M.A.P. and M.P. All authors have read and agreed to the published version of the manuscript.

Funding: This research was funded by EPSRC grant number EP/W007673/1.

Institutional Review Board Statement: Not applicable.

Informed Consent Statement: Not applicable.

Data Availability Statement: The ProxNest code and experiments are available at https://github.com/astro-informatics/proxnest (accessed on 13 July 2023).

Conflicts of Interest: The authors declare no conflict of interest.

References

1. Robert, C.P. *The Bayesian Choice*; Springer: New York, NY, USA 2007.
2. Ashton, G.; Bernstein, N.; Buchner, J.; Chen, X.; Csányi, G.; Fowlie, A.; Feroz, F.; Griffiths, M.; Handley, W.; Habeck, M.; et al. Nested sampling for physical scientists. *Nat. Rev. Methods Prim.* **2022**, *2*, 39. [CrossRef]
3. Skilling, J. Nested sampling for general Bayesian computation. *Bayesian Anal.* **2006**, *1*, 833–859. [CrossRef]
4. Mukherjee, P.; Parkinson, D.; Liddle, A.R. A nested sampling algorithm for cosmological model selection. *Astrophys. J.* **2006**, *638*, L51–L54. [CrossRef]
5. Feroz, F.; Hobson, M.P. Multimodal nested sampling: an efficient and robust alternative to MCMC methods for astronomical data analysis. *Mon. Not. R. Astron. Soc. (MNRAS)* **2008**, *384*, 449–463. [CrossRef]
6. Feroz, F.; Hobson, M.P.; Bridges, M. MULTINEST: an efficient and robust Bayesian inference tool for cosmology and particle physics. *Mon. Not. R. Astron. Soc. (MNRAS)* **2009**, *398*, 1601–1614. [CrossRef]
7. Handley, W.J.; Hobson, M.P.; Lasenby, A.N. POLYCHORD: nested sampling for cosmology. *Mon. Not. R. Astron. Soc. Lett.* **2015**, *450*, L61–L65. [CrossRef]
8. Buchner, J. Nested sampling methods. *arXiv* **2021**, arXiv:2101.09675.
9. McEwen, J.D.; Wallis, C.G.R.; Price, M.A.; Docherty, M.M. Machine learning assisted Bayesian model comparison: the learnt harmonic mean estimator. *arXiv* **2022**, arXiv:2111.12720.
10. Spurio Mancini, A.; Docherty, M.M.; Price, M.A.; McEwen, J.D. Bayesian model comparison for simulation-based inference. *RAS Tech. Instrum.* **2023**, *2*, 710–722. [CrossRef]
11. Polanska, A.; Price, M.A.; Spurio Mancini, A.; McEwen, J.D. Learned harmonic mean estimation of the marginal likelihood with normalising flows. *Phys. Sci. Forum* **2023**, *9*, 10. [CrossRef]
12. Cai, X.; McEwen, J.D.; Pereyra, M. Proximal nested sampling for high-dimensional Bayesian model selection. *Stat. Comput.* **2022**, *32*, 87. [CrossRef]
13. Combettes, P.; Pesquet, J.C. *Proximal Splitting Methods in Signal Processing*; Springer: New York, NY, USA, 2011; pp. 185–212.
14. Parikh, N.; Boyd, S. Proximal algorithms. *Found. Trends Optim.* **2013**, *1*, 123–231.
15. Pereyra, M. Proximal Markov chain Monte Carlo algorithms. *Stat. Comput.* **2016**, *26*, 745–760. [CrossRef]
16. Durmus, A.; Moulines, E.; Pereyra, M. Efficient Bayesian computation by proximal Markov chain Monte Carlo: when Langevin meets Moreau. *SIAM J. Imaging Sci.* **2018**, *1*, 473–506. [CrossRef]
17. Skilling, J. Bayesian computation in big spaces-nested sampling and Galilean Monte Carlo. In Proceedings of the AIP Conference 31st American Institute of Physics, Zurich, Switzerland, 29 July–3 August 2012; Volume 1443, pp. 145–156.
18. Betancourt, M. Nested sampling with constrained hamiltonian monte carlo. *AIP Conf. Proc.* **2011**, *1305*, 165–172. [CrossRef]
19. Laumont, R.; Bortoli, V.D.; Almansa, A.; Delon, J.; Durmus, A.; Pereyra, M. Bayesian imaging using Plug & Play priors: when Langevin meets Tweedie. *SIAM J. Imaging Sci.* **2022**, *15*, 701–737.
20. Robbins, H. An Empirical Bayes Approach to Statistics. In *Proceedings of the Third Berkeley Symposium on Mathematical Statistics and Probability, Berkeley, CA, USA, December 1954 and July–August 1955*; University of California Press: Berkeley, CA, USA, 1956; Volume 3.1, pp. 157–163.
21. Efron, B. Tweedie's formula and selection bias. *J. Am. Stat. Assoc.* **2011**, *106*, 1602–1614. [CrossRef] [PubMed]
22. Kim, K.; Ye, J.C. Noise2score: tweedie's approach to self-supervised image denoising without clean images. *Adv. Neural Inf. Process. Syst.* **2021**, *34*, 864–874.
23. Chung, H.; Sim, B.; Ryu, D.; Ye, J.C. Improving diffusion models for inverse problems using manifold constraints. *arXiv* **2022**, arXiv:2206.00941.
24. Sohl-Dickstein, J.; Weiss, E.; Maheswaranathan, N.; Ganguli, S. Deep unsupervised learning using nonequilibrium thermodynamics. *PMLR* **2015**, *37*, 2256–2265.
25. Song, Y.; Ermon, S. Generative modeling by estimating gradients of the data distribution. In Proceedings of the 33rd Annual Conference on Neural Information Processing Systems (NeurIPS 2019), Vancouver, Canada, 8–14 December 2019.

26. Song, Y.; Ermon, S. Improved techniques for training score-based generative models. *Adv. Neural Inf. Process. Syst.* **202** *33*, 12438–12448.
27. Song, Y.; Sohl-Dickstein, J.; Kingma, D.P.; Kumar, A.; Ermon, S.; Poole, B. Score-based generative modeling through stochasti differential equations. *arXiv* **2020**, arXiv:2011.13456.
28. Rombach, R.; Blattmann, A.; Lorenz, D.; Esser, P.; Ommer, B. High-resolution image synthesis with latent diffusion models. I Proceedings of the IEEE/CVF Conference on Computer Vision and Pattern Recognition, New Orleans, LA, USA, 18–24 June 202 pp. 10684–10695.
29. Venkatakrishnan, S.V.; Bouman, C.A.; Wohlberg, B. Plug-and-play priors for model based reconstruction. In Proceedings of th 2013 IEEE Global Conference on Signal and Information Processing, IEEE, Austin, TX, USA, 3–5 December 2013; pp. 945–948.
30. Ryu, E.; Liu, J.; Wang, S.; Chen, X.; Wang, Z.; Yin, W. Plug-and-play methods provably converge with properly traine denoisers. In Proceedings of the International Conference on Machine Learning. PMLR, Long Beach, CA, USA, 9–15 June 201 pp. 5546–5557.
31. Nelson, D.; Springel, V.; Pillepich, A.; Rodriguez-Gomez, V.; Torrey, P.; Genel, S.; Vogelsberger, M.; Pakmor, R.; Marinacci, F Weinberger, .R.; et al. The IllustrisTNG Simulations: Public Data Release. *Comput. Astrophys. Cosmol.* **2019**, *6*, 2. [CrossRef]

physical sciences forum

MDPI

Proceeding Paper

Bayesian Inference and Deep Learning for Inverse Problems †

Ali Mohammad-Djafari [1,2,*], Ning Chu [2], Li Wang [3] and Liang Yu [4,5]

1 International Science Consulting and Training (ISCT), 91440 Bures sur Yvette, France
2 Zhejiang Shangfeng Special Blower Company, Shaoxing 312352, China; chuning1983@sina.com
3 School of Mathematics and Statistics, Central South University, Changsha 410017, China; li.wang.csu@csu.edu.cn
4 School of Civil Aviation, Northwestern Polytechnical University, Xi'an 710072, China
5 State Key Laboratory of Airliner Integration Technology and Flight Simulation, Shanghai 200126, China; liang.yu@sjtu.edu.cn
* Correspondence: djafari@ieee.org
† Presented at the 42nd International Workshop on Bayesian Inference and Maximum Entropy Methods in Science and Engineering, Garching, Germany, 3–7 July 2023.

Abstract: Inverse problems arise anywhere we have an indirect measurement. In general, they are ill-posed to obtain satisfactory solutions, which needs prior knowledge. Classically, different regularization methods and Bayesian inference-based methods have been proposed. As these methods need a great number of forward and backward computations, they become costly in computation, particularly when the forward or generative models are complex, and the evaluation of the likelihood becomes very costly. Using deep neural network surrogate models and approximate computation can become very helpful. However, in accounting for the uncertainties, we need first to understand Bayesian deep learning, and then we can see how we can use it for inverse problems. In this work, we focus on NN, DL, and, more specifically, the Bayesian DL particularly adapted for inverse problems. We first give details of Bayesian DL approximate computations with exponential families; then, we see how we can use them for inverse problems. We consider two cases: First, we consider the case where the forward operator is known and used as a physics constraint, and the second examines more general data-driven DL methods.

Keywords: bayesian inference; neural network; deep learning (DL); inverse problems; physics-based DL; infrared imaging

1. Introduction

Inverse problems arise almost everywhere in science and engineering. In fact, anytime and in any application when we have indirect measurements related to what we really want to measure through some mathematical relation, this is called the forward model. Then, we have to infer the desired unknown from the observed data using this forward model or a surrogate one. In general, many inverse problems are ill-posed, and many methods for finding well-posed solutions for them are mainly based either on regularization theory or Bayesian inference. We mention those in particular, which are based on the optimization of a criterion with two parts: a data model output matching criterion (likelihood part in the Bayesian) and a regularization term (prior model in the Bayesian). Different criteria for these two terms and a great number of standard and advanced optimization algorithms have been proposed and used with great success. When these two terms are distances, they can have a Bayesian maximum a posteriori (MAP) interpretation, where these two terms correspond, respectively, to the likelihood and prior probability models. The Bayesian approach gives more flexibility in choosing these terms via the likelihood and the prior probability distributions. This flexibility goes much further with the hierarchical models and appropriate hidden variables [1]. Also, the possibility of estimating the hyperparameters gives much more flexibility for semisupervised methods.

However, full Bayesian computations can become very heavy computationally. In particular, this occurs when the forward model is complex, and the evaluation of the likelihood has a high computational cost [2]. In those cases, using surrogate simpler models can become very helpful to reduce the computational costs, but then we have to account for the uncertainty quantification (UQ) of the obtained results [3]. Neural networks (NNs), with their diversity such as convolutional neural networks (CNNs), deep learning (DL), etc., have become tools for fast and low computational surrogate forward models.

Over the last decades, machine learning (ML) methods and algorithms have gained great success in many tasks, such as classification, clustering, segmentation, object detection, and many other areas. There are many different structures of neural networks (NNs), such as feed-forward, Convolutional NNs (CNNs), Deep NNs, etc. [4]. Using these methods directly for inverse problems, as intermediate preprocessing or as tools for performing fast approximate computation in different steps of regularization or Bayesian inference, has also been successful, but not as much as could be possible. Recently, physics-informed neural networks have gained great success in many inverse problems, thereby proposing interaction between the Bayesian formulation of forward models and optimization algorithms and ML-specific algorithms for intermediate hidden variables. These methods have become very helpful to obtain approximate practical solutions to inverse problems in real-world applications [5–11].

In this paper, first, in Section 2, some mathematical notations for dealing with NNs are given. In Section 3, a detailed presentation of the Bayesian inference and the approximate computation needed for BDL are given. Then, in Section 4, we consider a focus on the NN and DL methods for inverse problems. First, we present the same cases where we know the forward and its adjoint model. Then, we consider the case where we may not have this knowledge and want to propose directly data-driven DL methods [12,13].

2. Neural Networks, Deep Learning, and Bayesian DL

The main objective of the NN for a regression problem can be described as follows:

$$\mathbf{x}_i \rightarrow \boxed{\text{NN: } f(\mathbf{x}_i)} \rightarrow y_i.$$

The objective is to infer the function $f : \mathbb{R}^M \rightarrow \mathbb{R}$ from the observations $\mathbf{y} = (y_1, \ldots, y_N)^{\mathrm{T}}$ at locations given by $\mathbf{x} = (\mathbf{x}_1, \ldots, \mathbf{x}_N)$. The usual NN learning approach is to define a parametric family of functions $\phi_\theta : \mathbb{R}^M \times \theta \rightarrow \mathbb{R}$ that is flexible enough so that $\exists \theta^\star$ such that $\phi(\cdot) \approx \phi_{\theta^\star}(\cdot)$:

$$\{\mathbf{x}_i, y_i\} \rightarrow \boxed{\text{NN Learning: } y_i = \phi(\mathbf{x}_i) \approx \phi_{\theta^\star}(\mathbf{x}_i)} \rightarrow \theta^\star.$$

Deep learning focuses on learning the optimal parameters $\theta^\star$, which can then be used for predicting the output $\hat{y}_j$ for any input $\mathbf{x}_j$:

$$\mathbf{x}_j \rightarrow \boxed{\text{NN: } \phi_{\theta^\star}(\mathbf{x}_j)} \rightarrow \hat{y}_j.$$

In this approach, there is no uncertainty quantification.

The Bayesian deep learning approach can be summarized as follows:

$$\{\mathbf{x}_i, y_i\} \rightarrow \boxed{p(\theta|\mathcal{D}) = \frac{p(\mathcal{D}|\theta)p(\theta)}{p(\mathcal{D})}} \rightarrow p(\theta|\mathcal{D}),$$

$$\mathbf{x}_j \rightarrow \boxed{p(y_j|\mathbf{x}_j, \mathcal{D}) = \int p(y_j|\mathbf{x}_j, \theta, \mathcal{D}) p(\theta|\mathcal{D}) \, \mathrm{d}\theta} \rightarrow \hat{y}_j.$$

As we can see, uncertainties are accounted for in both steps of the parameter estimation and prediction. However, as we will see, the computational costs are important. We need to find solutions to perform fast computation.

3. Bayesian Inference and Approximate Computation

In a general Bayesian framework for NNs and DL, the objective is to infer the parameters θ from the data, $\mathcal{D} : \{\mathbf{x}_i, y_i\}$, using the Bayes rule:

$$p(\theta|\mathcal{D}) = \frac{p(\mathcal{D}|\theta)p(\theta)}{p(\mathcal{D})}, \tag{1}$$

where $p(\theta)$ is the prior, $\ell(\mathcal{D}|\theta) = p(\mathcal{D}|\theta)$ is the likelihood, $p(\theta|\mathcal{D})$ is the posterior, and $p(\mathcal{D}) = \int p(\mathcal{D}|\theta)p(\theta)\,\mathrm{d}\theta$ is called the evidence. We can also write $p(\theta|\mathcal{D}) \propto \ell(\mathcal{D}|\theta)p(\theta)$, where the classical maximum likelihood estimation (MLE) is defined as
$\widehat{\theta} = \arg\max_\theta\{\ell(\mathcal{D}|\theta)\}$.

A particular point of the posterior is of high interest, because we may be interested in maximum a posterior (MAP) estimation: $\widehat{\theta} = \arg\max_\theta\{\ell(\mathcal{D}|\theta)p(\theta)\}$. We may also be interested in mean squared error (MSE) estimation, which is shown that it corresponds to $\widehat{\theta} = \mathrm{E}_{p(\theta|\mathcal{D})}\{\theta\} = \int \theta\, p(\theta|\mathcal{D})\,\mathrm{d}\theta$.

The exact expression of the posterior and the computations of these integrals for great dimensional problems may be very computationally costly. For this reason, we need to perform approximate computation. In the following subsections, we review a few solutions.

3.1. Laplace Approximation

Rewriting the general Bayes rules slightly differently gives us the following:

$$p(\theta|\mathcal{D}) = \frac{1}{p(\mathcal{D})}p(\mathcal{D}|\theta)p(\theta) = \frac{1}{Z}\exp[\mathcal{L}(\theta|\mathcal{D})], \tag{2}$$

the Laplace approximations use a second-order expansion of $\mathcal{L}(\theta|\mathcal{D}) = \ln p(\mathcal{D}|\theta) + \ln p(\theta)$ around $\widehat{\theta}_{MAP}$ to construct a Gaussian approximation of $p(\theta|\mathcal{D})$:

$$\mathcal{L}(\theta) \approx \mathcal{L}(\widehat{\theta}_{MAP}) + \frac{1}{2}(\theta - \widehat{\theta}_{MAP})'\left(\nabla^2_\theta\mathcal{L}(\theta)|_{\widehat{\theta}_{MAP}}\right)(\theta - \widehat{\theta}_{MAP}), \tag{3}$$

where the first-order term vanishes at the $\widehat{\theta}_{MAP}$. This is equivalent to calculating the following approximation:

$$\boxed{\text{Approximate } p(\theta|\mathcal{D}) \text{ by } q(\theta) = \mathcal{N}\left(\theta|\widehat{\theta}_{MAP}, \mathbf{\Sigma} = [\nabla^2_\theta\mathcal{L}(\theta)|_{\widehat{\theta}_{MAP}}]\right).}$$

With this approximation, the evidence $p(\mathcal{D})$ is approximated by the following:

$$Z = p(\mathcal{D}) = (2\pi)^{d/2}|\mathbf{\Sigma}|^{1/2}\exp\left[-\mathcal{L}(\widehat{\theta}_{MAP})\right]. \tag{4}$$

For great dimensional problems such as BDL, the full computation of $\mathbf{\Sigma}$ is very costly. We have still to do more approximations. The following are a few solutions for scalable approximations for BDL:

- Work with the subnetwork or last layer (transfer learning);
- Perform covariance matrix decomposition (low rank, Kronecker-factored approximate curvature (KFAC), Diag);
- Conduct the computation during the hyperparameter tuning using crossvalidation;
- Use approximate predictive computation.

3.2. Approximate Computation: Variational Inference

The main idea then is to perform approximate Bayesian computation (ABC) by approximating the posterior $p(\theta|\mathcal{D})$ using a simpler expression $q(\theta)$. When approximation is performed by minimizing

$$\mathrm{KL}[q(\theta) : p(\theta|\mathcal{D})] = \int q(\theta) \log \frac{q(\theta)}{p(\theta|\mathcal{D})} \,\mathrm{d}\theta, \quad (5)$$

the method is called the variational Bayesian approximation (VBA). When $q(\theta)$ is chosen to be separable in some components of the parameters $q(\theta) = \prod_j q_j(\theta_j)$, the approximation is called *the mean field VBA (MFVBA).*

Let us come back to the general VBA, $\mathrm{KL}[q : p] = \int q \log \frac{q}{p} = \mathrm{E}_q\{\log q\} - \mathrm{E}_q\{\log p\}$, and note using $L = \mathrm{E}_q\{\log p\}$ the expected likelihood and using $S = -\mathrm{E}_q\{\log q\}$ the entropy of q. Then, we have the following:

$$q* = \arg\min_q \{\mathrm{KL}[q : p]\} = \arg\max_q \{E = S + L\}. \quad (6)$$

E is also called the evidence lower bound (ELBO):

$$ELBO = -\mathrm{E}_q\{\log q\} + \mathrm{E}_q\{\log p\}. \quad (7)$$

At this point, it is important to note one main property of the VBA: When $p(\theta|\mathcal{D})$, the posterior probability law p and the approximate probability law q are in the exponential family; then, $\mathrm{E}_p\{\theta\} = \mathrm{E}_q\{\theta\}$.

3.3. VBA and Natural Exponential Family

If q is chosen to be in a *natural exponential family,* $q(\theta|\boldsymbol{\eta}) = \exp[\boldsymbol{\eta}'\theta - A(\boldsymbol{\eta})]$, then it is entirely characterized by its mean $\mathbf{m} = \mathrm{E}_q\{\theta\}$, and if q is conjugate to p, then $q^*(\theta|\boldsymbol{\eta}) = \exp\left[\boldsymbol{\eta}^{*\prime}\theta - A(\boldsymbol{\eta}^*)\right]$, which is entirely characterized by its mean $\mathbf{m}^* = \mathrm{E}_{q^*}\{\theta\}$. We can then define the objective E as a function of $\mathbf{m}$, and the first order condition of the optimality is $\frac{\partial E}{\partial \mathbf{m}}\big|_{\mathbf{m}=\mathbf{m}*} = 0$. From this property, we can obtain a fixed-point algorithm to compute $\mathbf{m}^* = \mathrm{E}_{q^*}\{\theta\}$:

$$\frac{\partial E}{\partial \mathbf{m}}\Big|_{\mathbf{m}=\mathbf{m}*} = 0 \to \frac{\partial E}{\partial \mathbf{m}}\Big|_{\mathbf{m}=\mathbf{m}*} + \mathbf{m} = \mathbf{m} \to \mathbf{M}(\mathbf{m}) = \mathbf{m}.$$

Iterating on this *fixed-point algorithm* gives us the following:

$$\mathbf{M}(\mathbf{m}) = \mathbf{m}^{(k-1)} \quad \text{with} \quad \mathbf{M}(\mathbf{m}) := \frac{\partial E}{\partial \mathbf{m}} + \mathbf{m} \quad (8)$$

which converges to $\mathbf{m}^* = \mathrm{E}_{q^*}\{\theta^*\}$ and is also $\mathbf{m}^* = \mathrm{E}_p\{\theta^*\}$. This algorithm can be summarized as follows:

- Having chosen the prior and likelihood, find the expression of $p(\theta|\mathcal{D}) \propto p(\mathcal{D}|\theta)p(\theta)$;
- Choose a family q and find the expressions of $L = \mathrm{E}_q\{\ln p\}$ and $S = \mathrm{E}_q\{\ln q\}$, which thus yield $E = L + S$ as a function of $\mathbf{m} = \mathrm{E}_q\{\theta\}$;
- Find the expression of the vector operator $\mathbf{M} = \nabla_{\mathbf{m}} E + \mathbf{m}$ and update it $\mathbf{M}(\mathbf{m}) = \mathbf{m}^{(k-1)}$ until convergence, which results in $\mathbf{m}^* = \mathrm{E}_{q^*}\{\theta^*\} = \mathrm{E}_{p^*}\{\theta^*\}$.

At this point, it is important to note that, in this approach, even if the mean is well approximated, the variances or the covariance are underestimated. Some authors who are interested in this approach have proposed solutions to better estimate the covariance. See [14] for one of the solutions.

4. Deep Learning and Bayesian DL

As introduced before, in classical DL, the training and prediction steps can be summarized as follows:

$$\{\mathbf{x}_i, y_i\} \rightarrow \boxed{\text{NN Learning: } \phi_{\theta^*}(\cdot)} \rightarrow \theta^* \qquad \mathbf{x}_j \rightarrow \boxed{\text{NN: } \phi_{\theta^*}(\mathbf{x}_j)} \rightarrow \hat{y}_j$$

In this approach, there is no uncertainty quantification. The Bayesian deep learning can be summarized as follows:

$$\{\mathbf{x}_i, y_i\} \rightarrow \boxed{p(\theta|\mathcal{D}) = \frac{p(\mathcal{D}|\theta)p(\theta)}{p(\mathcal{D})}} \rightarrow p(\theta|\mathcal{D}) \text{ or } q(\theta|\mathcal{D}),$$

$$\mathbf{x}_j \rightarrow \boxed{p(y_j|\mathbf{x}_j, \mathcal{D}) = \int p(y_j|\mathbf{x}_j, \theta, \mathcal{D}) p(\theta|\mathcal{D}) \, \mathrm{d}\theta} \rightarrow p(y_i|\mathbf{x}_j, \mathcal{D}) \text{ or } q(y_i|\mathbf{x}_j, \mathcal{D}).$$

As we can see, uncertainties are accounted for in both steps of the parameter estimation and prediction. However, computational costs are important. We need to find solutions to perform fast computation. As mentioned before, the different possible tools are Laplace and Gaussian approximation, variational inference, and more controlled approximation to design new deep learning algorithms, which can scale up for practical situations.

Exponential Family Approximation

Let us consider the case of general exponential families $q(\theta|\boldsymbol{\lambda}) = \exp[\lambda^T \mathbf{t}(\theta) - F(\lambda)]$, where θ represents the original parameters, λ represents the natural parameters, $\mathbf{t}(\theta)$ represents sufficient statistics, $F(\lambda)$ represents the log partition function, and we define the expectations parameters as $\mu := \mathrm{E}_q\{\mathbf{t}(\theta)\}$. Let us also define the dual function $F^*(\boldsymbol{\eta})$ and the dual parameters $\boldsymbol{\eta}$ via the Legendre transform:

$$G(\boldsymbol{\eta}) = F^*(\boldsymbol{\eta}) = \sup_{\lambda}\{< \boldsymbol{\lambda}, \boldsymbol{\eta} > -F(\boldsymbol{\lambda})\}.$$

Then, we can show the triangular relation between $\theta \in \Theta$, $\lambda \in \Lambda$, and $\mu \in \mathcal{M}$ shown in Figure 1.

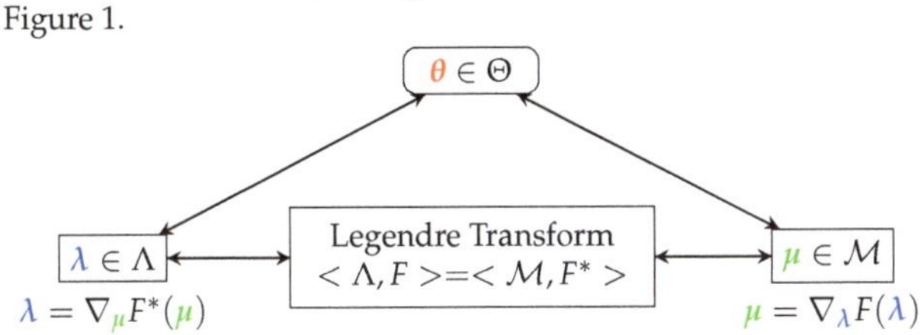

Figure 1. General exponential family, with original parameters θ, natural parameters λ, expectations parameters $\mu := \mathrm{E}_q\{\mathbf{t}(\theta)\}$, and their relations via the Legendre Transform.

With these notations, applying the VBA rule $\min_{q \in \mathcal{Q}} \mathrm{E}_q\{\mathcal{L}(\theta)\} - H(q)$ results in the following updating rule for the natural parameters:

$$\boldsymbol{\lambda} \leftarrow \boldsymbol{\lambda} - \rho \nabla_{\boldsymbol{\mu}}(\mathcal{L}(\theta) - H(q)). \tag{9}$$

For example, by considering the Gaussian case

$$q(\theta) = \mathcal{N}(\theta|\mathbf{m}, \mathbf{S}^{-1}) \propto \exp\left[-\frac{1}{2}(\theta - \mathbf{m})^T \mathbf{S}(\theta - \mathbf{m})\right] \propto \exp\left[(\mathbf{S}\mathbf{m})^T \theta + \mathrm{Tr}(-\frac{\mathbf{S}}{2}\theta\theta^T)\right]. \tag{10}$$

we can identify the natural parameters as $\lambda = [\mathbf{Sm}, -\mathbf{S}/2] \rightarrow \mu := [\mathrm{E}_q\{\theta\}, \mathrm{E}_q\{\theta\theta^T\}]$, and we easily obtain the following algorithm:

$$\begin{cases} \mathbf{Sm} \leftarrow (1-\rho)\mathbf{Sm} - \rho\nabla_{\mathrm{E}_q\{\theta\}}\mathrm{E}_q\{\mathcal{L}(\theta)\}, \\ \mathbf{S} \leftarrow (1-\rho)\mathbf{S} - \frac{\rho}{2}\nabla_{\mathrm{E}_q\{\theta\theta^T\}}\mathrm{E}_q\{\mathcal{L}(\theta)\}. \end{cases} \tag{11}$$

where

$$\begin{cases} \nabla_{\mathrm{E}_q\{\theta\}}\mathrm{E}_q\{\mathcal{L}(\theta)\} = \mathrm{E}_q\{\nabla_\theta\mathcal{L}(\theta)\} - 2\mathrm{E}_q\{\mathcal{L}(\theta)\}, \\ \nabla_{\mathrm{E}_q\{\theta\theta^T\}}\mathrm{E}_q\{\mathcal{L}(\theta)\} = \mathrm{E}_q\{\mathbf{H}(\theta)\}, \end{cases}$$

which results, explicitly, in

$$\begin{cases} \mathbf{m} \leftarrow \mathbf{m} - \rho\mathbf{S}^{-1}\nabla_{\mathbf{m}}\mathcal{L}(\mathbf{m}), \\ \mathbf{S} \leftarrow (1-\rho)\mathbf{S} + \rho\mathbf{H}_{\mathbf{m}}. \end{cases} \tag{12}$$

For a linear model $\mathbf{y} = \mathbf{X}\theta$ and Gaussian priors, we have

$$\mathcal{L}(\theta) = (\mathbf{y} - \mathbf{X}\theta)^T(\mathbf{y} - \mathbf{X}\theta) + \gamma\theta^T\theta = -2\theta^T(\mathbf{X}^T\mathbf{y}) + \mathrm{Tr}\left[\theta\theta^T(\mathbf{X}^T\mathbf{X} + \gamma\mathbf{I})\right], \tag{13}$$

and $\mathrm{E}_q\{\mathcal{L}(\theta)\} = \boldsymbol{\lambda}^T\boldsymbol{\mu}, \nabla_\mu\mathrm{E}_q\{\mathcal{L}(\theta)\} = \boldsymbol{\lambda}$.

It is interesting to note that many classical algorithms for updating the parameters, such as forward–backward, sparse variational inference, and variational message passing, become special cases. A main remark here is that this linear generating function case will link us to the linear inverse problems $\mathbf{g} = \mathbf{H}\mathrm{f} + \epsilon$ if we replace $\mathbf{y}$ with $\mathbf{g}$, $\mathbf{X}$ with $\mathbf{H}$, and θ with f.

5. NN, DL, and Bayesian Inference for Linear Inverse Problems

To show the possibilities of the interaction between inverse problems methods, neural networks and deep learning, the best way is to give a few examples.

5.1. First Example: A Known Linear Forward Model

The first and easiest example is the case of linear inverse problems $\mathrm{g} = \mathbf{H}\mathrm{f} + \epsilon$, where we know the forward model $\mathbf{H}$ and we assume the Gaussian likelihood $p(\mathrm{g}|\mathrm{f}) = \mathcal{N}(\mathrm{g}|\mathbf{H}\mathrm{f}, \sigma_\epsilon^2\mathbf{I})$ and the Gaussian prior $p(\mathrm{f}) = \mathcal{N}(\mathrm{f}|0, \sigma_f^2\mathbf{I})$ are the easiest cases to consider, where we know that the posterior is Gaussian $p(\mathrm{f}|\mathrm{g}) = \mathcal{N}(\mathrm{f}|\widehat{\mathrm{f}}, \widehat{\mathbf{\Sigma}})$ with

$$\widehat{\mathrm{f}} = (\mathbf{H}^t\mathbf{H} + \lambda\mathbf{I})^{-1}\mathbf{H}^t\mathrm{g} = \mathbf{A}\,\mathrm{g} = \mathbf{B}\mathbf{H}^t\,\mathrm{g} \;\text{ or still }\; \widehat{\mathrm{f}} = \mathbf{H}^t(\frac{1}{\lambda}\mathbf{H}\mathbf{H}^t + \mathbf{I})^{-1}\mathrm{g} = \mathbf{H}^t\mathbf{C}\,\mathrm{g},$$

where $\lambda = \sigma_\epsilon^2/\sigma_f^2$, $\mathbf{A} = (\mathbf{H}^t\mathbf{H} + \lambda\mathbf{I})^{-1}\mathbf{H}^t$, $\mathbf{B} = (\mathbf{H}^t\mathbf{H} + \lambda\mathbf{I})^{-1}$, $\mathbf{C} = (\frac{1}{\lambda}\mathbf{H}\mathbf{H}^t + \mathbf{I})^{-1}$, and $\widehat{\mathbf{\Sigma}} = (\mathbf{H}^t\mathbf{H} + \lambda\mathbf{I})^{-1}$.

These relations can be presented schematically as

$$\mathrm{g} \rightarrow \boxed{\quad\mathbf{A}\quad} \rightarrow \widehat{\mathrm{f}}, \qquad \mathrm{g} \rightarrow \boxed{\;\mathbf{H}^t\;} \rightarrow \boxed{\;\mathbf{B}\;} \rightarrow \widehat{\mathrm{f}}, \qquad \mathrm{g} \rightarrow \boxed{\;\mathbf{C}\;} \rightarrow \boxed{\;\mathbf{H}^t\;} \rightarrow \widehat{\mathrm{f}}.$$

We can then consider replacing $\mathbf{A}$, $\mathbf{B}$, and $\mathbf{C}$ with appropriate deep neural networks and apply all the previous BDL methods to them. As we can see, these relations directly induce a linear feed-forward NN structure. In particular, if $\mathbf{H}$ represents a convolution operator, then $\mathbf{H}^t$, $\mathbf{H}^t\mathbf{H}$, and $\mathbf{H}\mathbf{H}^t$ are too, as well as the operators $\mathbf{B}$ and $\mathbf{C}$. Thus, the whole inversion can be modeled using a CNN [15,16].

For the case of computed tomography (CT), the first operation is equivalent to an analytic inversion, the second corresponds to back projection that is first followed by 2D filtering in the image domain, and the third corresponds to the famous filtered back projection (FBP), which is implemented on classical CT scans. These cases are illustrated in Figure 2.

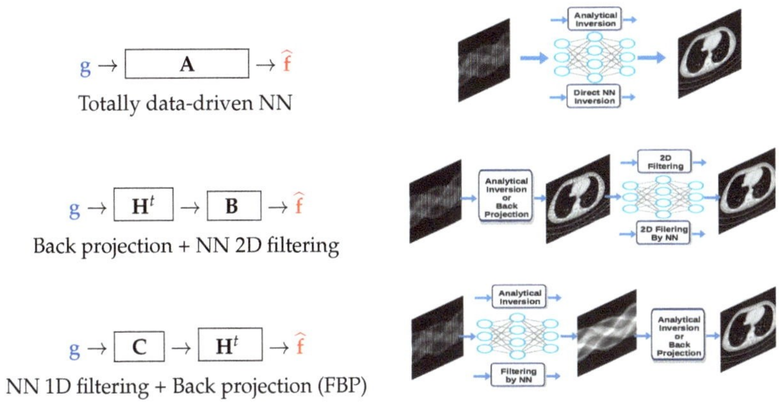

Figure 2. Three linear NN structures, which are derived directly from quadratic regularization inversion method. The right part of this figure is adapted from [16].

5.2. Second Example: A Deep Learning Equivalence of Iterative Gradient-Based Algorithms

One of the classical iterative methods in linear inverse problem algorithms is based on the gradient descent method to optimize $J(\mathbf{f}) = \|\mathbf{g} - \mathbf{H}\mathbf{f}\|^2$:

$$\mathbf{f}^{(k+1)} = \mathbf{f}^{(k)} + \alpha \mathbf{H}^t(\mathbf{g} - \mathbf{H}\mathbf{f}^{(k)}) = \alpha \mathbf{H}^t \mathbf{g} + (\mathbf{I} - \alpha \mathbf{H}^t \mathbf{H})\mathbf{f}^{(k)}, \tag{14}$$

where the solution to the problem is obtained recursively. Everybody knows that when the forward model operator $\mathbf{H}$ is singular or ill-conditioned, this iterative algorithm starts by converging, but it may diverge easily. One of the experimental methods to obtain an acceptable approximate solution is just to stop the iterations after K iterations. This idea can be translated to a deep learning NN by using K layers. Each layer represents one iteration of the algorithm. See Figure 3.

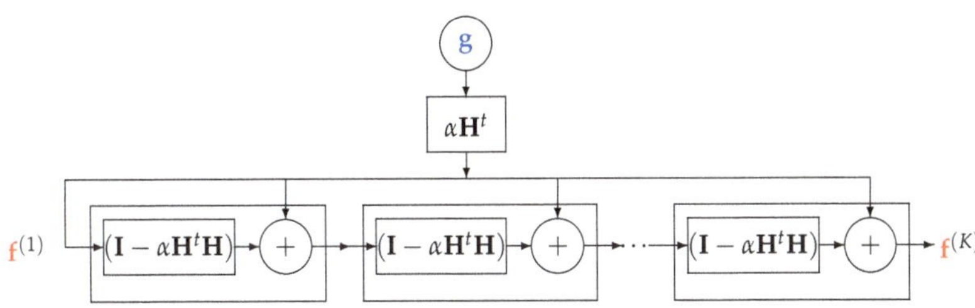

Figure 3. A K layers DL NN equivalent to K iterations of the basic optimization algorithm.

This DL structure can easily be extended to a regularized criterion $J(\mathbf{f}) = \frac{1}{2}\|\mathbf{g} - \mathbf{H}\mathbf{f}\|^2 + \lambda\|\mathbf{D}\mathbf{f}\|^2$, which can also be interpreted as the MAP or posterior mean solution with a Gaussian likelihood and prior. In this case, we have the following:

$$\mathbf{f}^{(k+1)} = \mathbf{f}^{(k)} + \alpha[\mathbf{H}^t(\mathbf{g} - \mathbf{H}\mathbf{f}^{(k)}) - \lambda \mathbf{D}^t \mathbf{D}] = \alpha \mathbf{H}^t \mathbf{g} + (\mathbf{I} - \alpha \mathbf{H}^t \mathbf{H} - \alpha\lambda \mathbf{D}^t \mathbf{D})\mathbf{f}^{(k)}. \tag{15}$$

We just need to replace $(\mathbf{I} - \alpha \mathbf{H}^t \mathbf{H})$ with $(\mathbf{I} - \alpha \mathbf{H}^t \mathbf{H} - \alpha\lambda \mathbf{D}^t \mathbf{D})$.

This structure can also be extended to all the sparsity-enforcing regularization terms such as ℓ_1 and the total variation (TV) using appropriate algorithms such as the ISTA (iterative soft threshold algorithm) or its fast version FISTA by replacing the update expression and by adding an NL operation, much like the ordinary NNs. A simple example is given in the following subsection.

5.3. Third Example: ℓ_1 Regularization and NN

The case of the MAP solution with a Gaussian likelihood and double exponential prior becomes equivalent to the ℓ_1 regularization criterion:

$$J(\mathbf{f}) = \|\mathbf{g} - \mathbf{H}\mathbf{f}\|_2^2 + \lambda\|\mathbf{f}\|_1, \tag{16}$$

where the solution can be obtained with an iterative optimization algorithm, such as ISTA:

$$\mathbf{f}^{(k+1)} = Prox_{\ell_1}\left(\mathbf{f}^{(k)}, \lambda\right) \stackrel{\triangle}{=} \mathcal{S}_{\lambda\alpha}\left(\alpha\mathbf{H}^t\mathbf{g} + (\mathbf{I} - \alpha\mathbf{H}^t\mathbf{H})\mathbf{f}^{(k)}\right), \tag{17}$$

where $\mathcal{S}_\theta$ is a soft threshold operator, and $\alpha \leq |eig(\mathbf{H}^t\mathbf{H})|$ is the Lipschitz constant of the normal operator. When $\mathbf{H}$ is a convolution operator, then

- $(\mathbf{I} - \alpha\mathbf{H}^t\mathbf{H})\mathbf{f}^{(k)}$ can also be approximated by a convolution and thus considered as a filtering operator;
- $\frac{1}{\alpha}\mathbf{H}^t\mathbf{g}$ can be considered as a bias term and is also a convolution operator;
- $\mathcal{S}_{\theta=\lambda\alpha}$ is a nonlinear pointwise operator. In particular, when $\mathbf{f}$ is a positive quantity, this soft threshold operator can be compared to the ReLU activation function of the NN. See Figure 4.

Using the iterative gradient-based algorithm with a fixed number of iterations for computing a GI or a regularized one, as explained in the previous section, can be used to propose a DL structure with K layers, with K being the number of iterations before stopping. Figure 5 shows this structure for a quadratic regularization, which results in a linear NN, and Figure 6 shows the case of ℓ_1 regularization.

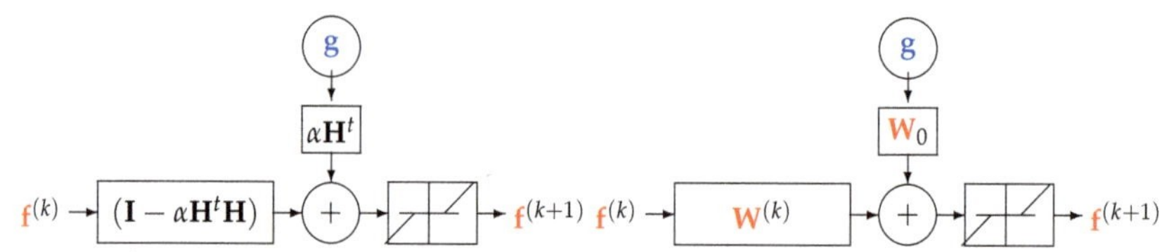

Figure 4. A K layers DL NN equivalent to K iterations of a basic gradient-based optimization algorithm. A quadratic regularization results in a linear NN, while a ℓ_1 regularization results in a classical NN with a nonlinear activation function. Left: supervised case. Right: unsupervised case. In both cases, all the K layers have the same structure.

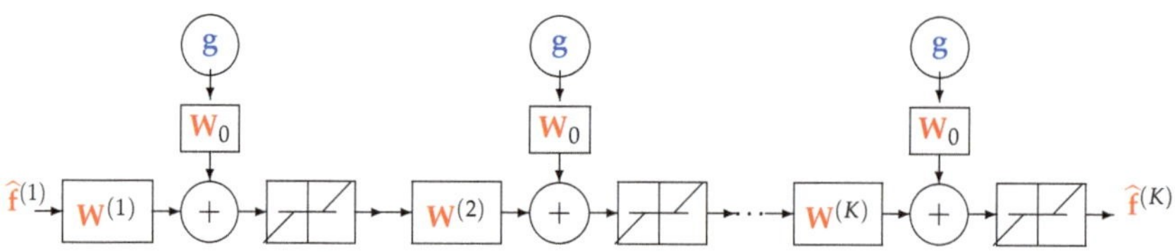

Figure 5. All the K layers of DL NN equivalent to K iterations of an iterative gradient-based optimization algorithm. The simplest solution is to choose $\mathbf{W}_0 = \alpha\mathbf{H}$ and $\mathbf{W}^{(k)} = \mathbf{W} = (\mathbf{I} - \alpha\mathbf{H}^t\mathbf{H})$, $k = 1, \cdots, K$. A more robust, but more costly approach, is to learn all the layers for $\mathbf{W}^{(k)} = (\mathbf{I} - \alpha^{(k)}\mathbf{H}^t\mathbf{H})$, $k = 1, \cdots, K$.

In all these examples, we could directly obtain the structure of the NN from the forward model and known parameters. However, in these approaches, there are some difficulties that consist of the determination of the structure of the NN. For example, in the first example, obtaining the structure of $\mathbf{B}$ depends on the regularization parameter λ. The same difficulty arises in determining the shape and the threshold level of the threshold bloc of the network in the second example. The same need for the regularization parameter, as well as many other hyperparameters, makes it necessary to create the NN structure and weights. In practice, we can decide, for example, on the number and structure of a DL network, but as their corresponding weights depend on many unknown or difficult-to-fix parameters, ML may become of help. In the following, we first consider the training part of a general ML method. Then, we will see how to include the physics-based knowledge of the forward model in the structure of learning [4,10,17,18].

Learning Data $(\mathbf{g}_k, \mathbf{f}_k)_{k=1}^K$ → [Fixed Physics-based $\widetilde{\mathbf{f}}_k = \mathbf{H}^t \mathbf{g}_k$] → $\widetilde{\mathbf{f}}_k$ → [Trainable part $\widehat{\mathbf{B}} = \arg\min_{\mathbf{B}} \left\{ \sum_{k=1}^{K} \|\mathbf{f}_k - \boldsymbol{\phi}(\mathbf{B}\widetilde{\mathbf{f}}_k)\|^2 + \lambda R(\mathbf{B}) \right\}$] → $\widehat{\mathbf{B}}$

New Data $\mathbf{g}_j$ → [Physics-based part $\widetilde{\mathbf{f}}_j = \mathbf{H}^t \mathbf{g}_j$] → $\widetilde{\mathbf{f}}_j$ → [Trained part $\widehat{\mathbf{f}}_j = \boldsymbol{\phi}(\widehat{\mathbf{B}}\widetilde{\mathbf{f}}_j)$] → $\widehat{\mathbf{f}}_j$

Figure 6. Training (**top**) and testing (**bottom**) steps in the first use of physics-based ML approach.

5.4. Decomposition of the NN Structure to Fixed and Trainable Parts

The first easiest and most understandable method consists of decomposing the structure of the network $\mathbf{W}$ in two parts: a fixed part and a learnable part. As the simplest example, we can consider the case of analytical expression of the quadratic regularization $\widehat{\mathbf{f}} = (\mathbf{H}\mathbf{H}^t + \lambda \mathbf{D}\mathbf{D}^t)^{-1}\mathbf{H}^t\mathbf{g} = \mathbf{B}\mathbf{H}^t\mathbf{g}$, which suggests having a two-layer network with a fixed part structure $\mathbf{H}^t$ and a trainable one $\mathbf{B} = (\mathbf{H}\mathbf{H}^t + \lambda\mathbf{D}\mathbf{D}^t)^{-1}$. See Figure 6.

It is interesting to note that in X-ray-computed tomography (CT), the forward operator $\mathbf{H}$ is called *projection*, the adjoint operator $\mathbf{H}^t$ is called *back projection (BP)*, and the $\mathbf{B}$ operator is assimilated to a 2D filtering (convolution).

6. DL Structure and Deterministic or Bayesian Computation

To be able to look at the DNN and analyze it either in a deterministic or Bayesian manner, let us come back to the general notations and consider the following NN with input $\mathbf{x}$, output $\mathbf{y}$, and intermediate hidden variables $\mathbf{z}_1, \mathbf{z}_2, \ldots, \mathbf{z}_L$:

$$\mathbf{x} \rightarrow \boxed{\text{first layer}} \rightarrow \mathbf{z}_1 \rightarrow \boxed{\text{second layer}} \rightarrow \mathbf{z}_2 \cdots \rightarrow \boxed{\text{last layer}} \rightarrow \mathbf{y}.$$

In the deterministic case, each layer is defined by its parameters (W_l, b_l):

$$\mathbf{x} \rightarrow \boxed{f_{W_0,b_0}(\mathbf{x})} \rightarrow \mathbf{z}_1 \rightarrow \boxed{f_{W_1,b_1}(\mathbf{z}_1)} \rightarrow \mathbf{z}_2 \cdots \rightarrow \boxed{f_{W_L,b_L}(\mathbf{z}_{L-1})} \rightarrow \mathbf{y},$$

and we can write:

$$\mathbf{y} = f_{\boldsymbol{\theta}}(\mathbf{x}) = (f_{(W_0,b_0)} \odot f_{(W_1,b_1)} \odot \cdots f_{(W_L,b_L)})(\mathbf{x})$$

with

$$\boldsymbol{\theta} = ((W_0, b_0), (W_1, b_1), \cdots, (W_L, b_L))$$

and

$$\mathbf{y} = f_L(W_L\mathbf{z}_{L-1} + b_L), \quad \mathbf{z}_l = f_l(W_l\mathbf{z}_{l-1} + b_l), \quad l = L, \cdots, 1, \quad \mathbf{z}_0 = \mathbf{x}. \tag{18}$$

6.1. Deterministic DL Computation

In general, during the training steps, the parameters θ are estimated via

$$\theta^* = \arg\min_{\theta} \left\{ \sum_i \ell(y_i, f_\theta(\mathbf{x}_i)) + \sum_k \lambda_k \phi_k(W_k, b_k) \right\}. \quad (19)$$

The main point here is to choose how to choose λ_k and $\phi(.)$, as well as which optimization algorithm to choose for better convergence.

When parameters are obtained (the model is trained), we can use it easily via

$$\mathbf{z}_0 = \mathbf{x}, \quad \mathbf{z}_l = f_l(W_l \mathbf{z}_{l-1} + b_l), \quad \mathbf{y} = f_L(W_L \mathbf{z}_{L-1} + b_L). \quad (20)$$

6.2. Bayesian Deep Neural Network

In Bayesian DL, the question of choosing λ_k and $\phi(.)$ in the previous case becomes the choice of the prior, $p(\theta)$, which can also be assumed to be a priori separable in the components of θ or not. Then, we have to choose the expression of the likelihood (in general Gaussian) and find the expression of the posterior $p(\theta|\mathcal{D})$. As explained extensively before, directly using this posterior is almost impossible. Hopefully, we have a great number of approximate computation methods, such as MCMC sampling, slice sampling, nested sampling, data augmentation, and variational inference, which can still be used in practical situations. However, the training step in the Bayesian DL still stays very costly, particularly if we want to quantify the uncertainties.

Prediction Step

In the prediction step, we again have to consider choosing a probability law $p(\mathbf{x})$ for the class of the possible inputs and for all the outputs $\mathbf{z}_l$ that are conditional to their inputs $\mathbf{z}_{l-1}$. Then, we can consider, for example, the Gibbs sampling scheme. A comparison between deterministic and Bayesian DL is shown here:

$$\begin{array}{lcl} \text{Deterministic:} & & \text{Bayesian:} \\ \left\{ \begin{array}{l} \mathbf{z}_0 = \mathbf{x}, \\ \mathbf{z}_l = f_l(W_l \mathbf{z}_{l-1} + b_l), \quad l = 1, \cdots, L, \\ \mathbf{y} = f_L(W_L \mathbf{z}_{L-1} + b_L). \end{array} \right. & \rightarrow & \left\{ \begin{array}{l} \mathbf{z}_0 \sim p(\mathbf{x}), \\ \mathbf{z}_l \sim p(\mathbf{z}_l|\mathbf{z}_{l-1}), \quad l = 1, \cdots, L, \\ \mathbf{y} \sim p(\mathbf{y}|\mathbf{z}_{L-1}). \end{array} \right. \end{array}$$

If we consider Gaussian laws for the input and all the conditional variables, then we can write the following:

$$\mathbf{z}_0 \sim N(\mathbf{z}_0|\mathbf{x}, \tau_0 \mathbf{I}), \qquad \mathbf{z}_l \sim N(\mathbf{z}_l|f_l(\mathbf{z}_{l-1}), \tau_l \mathbf{I}),\ l = 1, \cdots, L, \qquad \mathbf{y} \sim p(\mathbf{y}|\mathbf{z}_{L-1}). \quad (21)$$

Here too, the main difficulty occurs when there are nonlinear activation functions, particularly in the last layer, where the Gaussian approximation may not be more valid.

7. Application: Infrared Imaging

Infrared (IR) imaging is used to diagnose and to survey the temperature field distribution of sensitive objects in many industrial applications. These images are, in general, low resolution and very noisy. The real values of the temperature also depend on many other parameters, such as emissivity, attenuation, and diffusion, due to the distance of the camera to the sources. To be really useful in practice, we need to reduce the noise, calibrate and increase the resolution, segment for detection of the hot area, and finally survey the temperature values of the different areas during the time to be able to conduct preventive diagnosis and possible maintenance.

Reducing the noise can be accomplished by filtering using the Fourier transform, wavelet transform, or other sparse representations of images. To increase the resolution, we may use deconvolution methods if we can obtain the point spread function (PSF) of the camera, or if not by using blind deconvolution techniques. The segmentation and detection of the hot area and the temperature value estimation at each area are also very important steps in real applications. Any of these steps can be performed separately, but trying to propose global processing using DL or BDL is necessary for real applications. As any of these steps are in fact different inverse problems, and it is difficult to fix the parameters in each step in a robust way, we propose a global process using the BDL. See the global scheme in Figure 7.

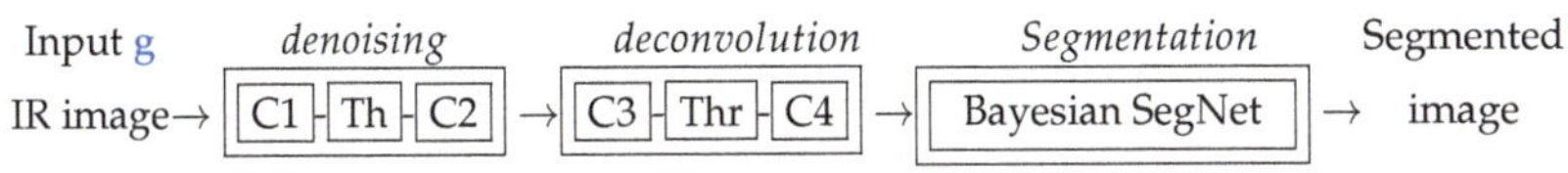

Figure 7. The proposed four groups of layers of NN for denoising, deconvolution, and segmentation of IR images.

In the first step, as the final objective is to segment the image to obtain different levels of temperature (for example, four levels: background, normal, high, and very high), we propose to design an NN that obtains as input a low resolution and noisy image and outputs a segmented image with those four levels and, at the same time, a good estimate of the temperature at each segment. See Figure 8.

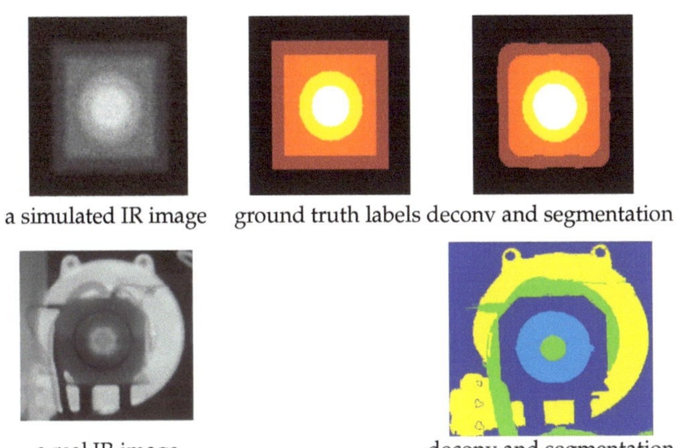

Figure 8. Example of expected results in deterministic methods. First row: a simulated IR image (**left**), its ground truth labels (**middle**), and the result of the deconvolution and segmentation (**right**). Second row: a real IR image (**left**) and the result of its deconvolution and segmentation (**right**).

To train this NN, we can generate different known shaped images to consider as the ground truth and simulate the blurring effects of temperature diffusions via the convolution of different appropriate PSFs. We can also add some noise to generate realistic images. We can also use black body thermal sources and acquire different images at different conditions. All these images can be used for the training of the network. See an example of the obtained result in Figure 9.

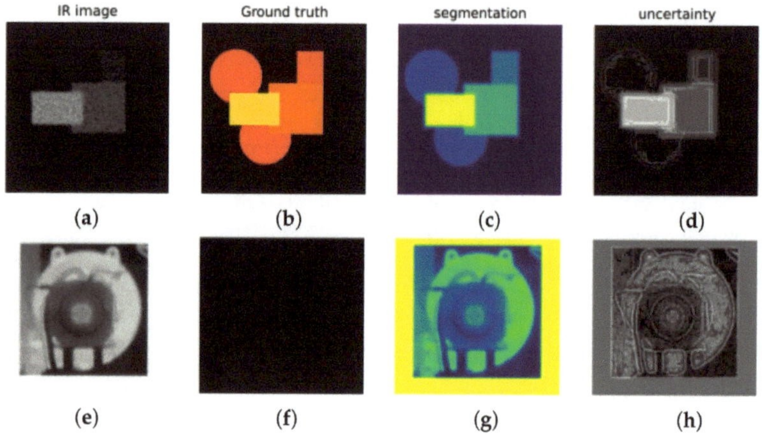

Figure 9. Example of expected results in Bayesian methods. First row from left: (**a**) simulated IR image, (**b**) its ground truth labels, (**c**) the result of the deconvolution and segmentation, and (**d**) uncertainties. Second row: (**e**) a real IR image, (**f**) no ground truth, (**g**) the result of its deconvolution and segmentation, and (**h**) uncertainties.

Author Contributions: Conceptualization and methodology, A.M.-D., N.C., L.W. and L.Y.; software, A.M.-D.; validation, A.M.-D., N.C., L.W. and L.Y.; formal analysis, A.M.-D.; investigation, resources, and data curation, N.C.; writing—original draft preparation, writing—review and editing, visualization and supervision, A.M.-D.; project administration and funding acquisition, N.C. and A.M.-D. All authors have read and agreed to the published version of the manuscript.

Funding: This research received no external funding.

Institutional Review Board Statement: Not applicable.

Informed Consent Statement: Not applicable.

Data Availability Statement: Data sharing is not applicable to this article.

Conflicts of Interest: The authors Ali Mohammad-Djafari and Ning Chu are the scientific researchers of Zhejiang Shangfeng Special Blower Company. The authors declare no conflict of interest.

References

1. Ayasso, H.; Mohammad-Djafari, A. Joint NDT Image Restoration and Segmentation Using Gauss-Markov-Potts Prior Models and Variational Bayesian Computation. *IEEE Trans. Image Process.* **2010**, *19*, 2265–2277. [CrossRef] [PubMed]
2. Tang, J.; Egiazarian, K.; Golbabaee, M.; Davies, M. The Practicality of Stochastic Optimization in Imaging Inverse Problems. *IEEE Trans. Comput. Imaging* **2020**, *6*, 1471–1485. [CrossRef]
3. Zhu, Y.; Zabaras, N. Bayesian deep convolutional encoder–decoder networks for surrogate modeling and uncertainty quantification. *J. Comput. Phys.* **2018**, *366*, 415–447. [CrossRef]
4. McCann, M.T.; Jin, K.H.; Unser, M. A Review of Convolutional Neural Networks for Inverse Problems in Imaging. *Image Video Process.* **2017**, *34*, 85–95. [CrossRef]
5. Fang, Z. A High-Efficient Hybrid Physics-Informed Neural Networks Based on Convolutional Neural Network. *IEEE Trans. Neural Netw. Learn. Syst.* **2021**, *33*, 5514–5526. [CrossRef] [PubMed]
6. Gilton, D.; Ongie, G.; Willett, R. Neumann Networks for Linear Inverse Problems in Imaging. *IEEE Trans. Comput. Imaging* **2020**, *6*, 328–343. [CrossRef]
7. Gong, D.; Zhang, Z.; Shi, Q.; van den Hengel, A.; Shen, C.; Zhang, Y. Learning Deep Gradient Descent Optimization for Image Deconvolution. *IEEE Trans. Neural Netw. Learn. Syst.* **2020**, *31*, 5468–5482. [CrossRef] [PubMed]
8. De Haan, K.; Rivenson, Y.; Wu, Y.; Ozcan, A. Deep-Learning-Based Image Reconstruction and Enhancement in Optical Microscopy. *Proc. IEEE* **2020**, *108*, 30–50. [CrossRef]
9. Aggarwal, H.K.; Mani, M.P.; Jacob, M. MoDL: Model-Based Deep Learning Architecture for Inverse Problems. *IEEE Trans. Med. Imaging* **2019**, *38*, 394–405. [CrossRef] [PubMed]
10. Chen, Y.; Lu, L.; Karniadakis, G.E.; Negro, L.D. Physics-informed neural networks for inverse problems in nano-optics and metamaterials. *Opt. Express* **2020**, *28*, 11618–11633. [CrossRef] [PubMed]

1. Raissi, M.; Perdikaris, P.; Karniadakis, G.E. Physics-informed neural networks: A deep learning framework for solving forward and inverse problems involving nonlinear partial differential equations. *J. Comput. Phys.* **2019**, *378*, 686–707. [CrossRef]
2. Mohammad-Djafari, A. Hierarchical Markov modeling for fusion of X-ray radiographic data and anatomical data in computed tomography. In Proceedings of the IEEE International Symposium on Biomedical Imaging, Washington, DC, USA, 7–10 July 2002; pp. 401–404. [CrossRef]
3. Mohammad-djafari, A. Regularization, Bayesian Inference and Machine Learning methods for Inverse Problems. *Entropy* **2021**, *23*, 1673. [CrossRef] [PubMed]
4. Giordano, R.; Broderick, T.; Jordan, M. Linear response methods for accurate covariance estimates from mean field variational Bayes. In Proceedings of the Advances in Neural Information Processing Systems, Montreal, QC, Canada, 7–12 December 2015.
5. Chun, I.Y.; Huang, Z.; Lim, H.; Fessler, J. Momentum-Net: Fast and convergent iterative neural network for inverse problems. *IEEE Trans. Pattern Anal. Mach. Intell.* **2020**, *45*, 4915–4931. [CrossRef] [PubMed]
6. Lucas, A.; Iliadis, M.; Molina, R.; Katsaggelos, A.K. Using Deep Neural Networks for Inverse Problems in Imaging: Beyond Analytical Methods. *IEEE Signal Process. Mag.* **2018**, *35*, 20–36. [CrossRef]
7. Chang, J.H.R.; Li, C.L.; Poczos, B.; Kumar, B.V.K.V.; Sankaranarayanan, A.C. One Network to Solve Them All—Solving Linear Inverse Problems using Deep Projection Models. In Proceedings of the IEEE International Conference on Computer Vision 2017, Venice, Italy, 22–29 October 2017.
8. Liang, D.; Cheng, J.; Ke, Z.; Ying, L. Deep Magnetic Resonance Image Reconstruction: Inverse Problems Meet Neural Networks. *IEEE Signal Process. Mag.* **2020**, *37*, 141–151. [CrossRef] [PubMed]

physical sciences forum

MDPI

Proceeding Paper

Physics-Consistency Condition for Infinite Neural Networks and Experimental Characterization †

Sascha Ranftl * and Shaoheng Guan

Institute of Theoretical Physics-Computational Physics, Graz University of Technology, Petersgasse 16, 8010 Graz, Austria

* Correspondence: ranftl@tugraz.at

† Presented at the 42nd International Workshop on Bayesian Inference and Maximum Entropy Methods in Science and Engineering, Garching, Germany, 3–7 July 2023.

Abstract: It has previously been shown that prior physics knowledge can be incorporated into the structure of an artificial neural network via neural activation functions based on (i) the correspondence under the infinite-width limit between neural networks and Gaussian processes if the central limit theorem holds and (ii) the construction of physics-consistent Gaussian process kernels , i.e., specialized covariance functions that ensure that the Gaussian process fulfills a priori some linear (differential) equation. Such regression models can be useful in many-query problems, e.g., inverse problems, uncertainty quantification or optimization, when a single forward solution or likelihood evaluation is costly. Based on a small set of training data, the learned model or "surrogate" can then be used as a fast approximator. The bottleneck is then for the surrogate to also learn efficiently and effectively from small data sets while at the same time ensuring physically consistent predictions. Based on this, we will further explore the properties of so-constructed neural networks. In particular, we will characterize (i) generalization behavior and (ii) the approximation quality or Gaussianity as a function of network width and discuss (iii) extensions from shallow to deep NNs.

Keywords: neural networks; Gaussian process; physics-consistent machine learning; infinite-width limit; surrogate modeling

1. Introduction

The impressive achievements of neural networks (NNs) in computer science have sparked a growing interest also in physics and scientific computing. Here, we explore their utilization as surrogate models in computationally intensive tasks like simulation-based optimization or uncertainty quantification [1,2]. For this, the surrogates' predictions should be consistent with physics, which typically cannot be achieved with data alone as such models often generalize poorly, i.e., they learn the physics implicitly and well only in regimes where (enough) data are available.

Here, we pursue a previously proposed approach [3] as follows: (i) linear differential equations can be inherently integrated into a Gaussian process (GP) model by carefully selecting the covariance function or kernel [4–8]. The GP defined by the so-constructed kernel may be referred to as a "physics-consistent GP" [6]. (ii) Neal [9] established that Bayesian NNs converge toward a GP in an infinite-width limit using the central limit theorem. By combining these two insights, we infer that at least some laws of physics can also be a priori embedded into a suitable infinite NN model by choosing neural architecture and activation functions in alignment with the corresponding GP's kernel, i.e., the corresponding infinite NN has the same physics-consistency property as the physics-consistent GP. Note that this approach fundamentally differs from "physics-informed" learning machines [10], where the residual of the equation (of the physical law) is included into the loss function independent of NN architecture. Departing from the foundations introduced in [3], we here consider generalization outside of the training data regime, and Gaussianity, i.e., convergence of the infinite-width limit approximation.

2. Background

This section provides a concise overview of the relevant literature on Gaussian processes (GPs), neural networks (NNs), and the infinite-width correspondence between the two. Additionally, related works on physically inspired GPs and NNs are discussed.

2.1. Physics-Consistent Gaussian Processes

Gaussian process regression, also known as "Kriging", has been extensively studied for several decades [11]. GPs have become one of the most popular classes of machine learning [12] and Bayesian models [13], and found wide application in the field of "Uncertainty Quantification" [14,15]. The defining element of a Gaussian process is its covariance function, also referred to as the kernel. Consequently, significant efforts have been dedicated to the design and selection of appropriate kernel functions [16].

As a matter of fact, a linear (differential) operator applied to a GP results in another GP with a modified kernel [5–7,17]. This allows one to derive a "physics-consistency condition" for the kernel, allowing the construction of kernels to define a GP that satisfies physical laws expressed in terms of said differential equations a priori, i.e., before training. Examples include divergence-free fields [18] and the Helmholtz equation [6]. More general linear constraints and boundary conditions can be considered in an analogous manner [17,19–21]. Note that it is also possible to construct physics-based or "weakly" physics-informed GPs through training [22] given some (non necessarily physical) base kernel, or through Green's functions [23,24].

While this approach was generally believed to be limited to linear equations, it has recently been extended even toward non-linear partial differential equations (PDEs) [25].

2.2. Infinite Neural Networks

Neural networks (NNs) are arguably one of the most popular classes of machine learning models. Similar to GPs and kernels, the design and selection of activation functions play a crucial role in NNs. Numerous activation functions have been proposed [26,27], but few principled approaches are known on how to choose an appropriate neural activation.

The infinite-width correspondence relates to the fact that Bayesian NNs converge toward a GP under the infinite-width limit [9]. This infinite-width correspondence has been used to compute kernels for GPs corresponding to specific neural activations and architectures [28–32], which may be regarded as a form of the "kernel trick". However, the inverse approach, i.e., computing activations from kernels, has received little attention. Here, we aim to take this route due to the beneficial circumstance that we can make GPs physics-consistent. Note that the term "physics-informed neural network" relates to training based on a physics-loss term composed of the differential equation residual, and was popularized by Raissi et al. [33], although the idea was around earlier [34,35].

Quite importantly, it has recently been shown that a very large class of NN architectures converge toward a GP [36], including deep NNs [29], convolutional NNs [37], recurrent NNs [38] and transformers [39]. Furthermore, the infinite-width limit provides valuable insight into the training dynamics of NNs as well [40]. Note that random Fourier features [41] also share the underlying mechanism of the central limit theorem. As for interactions arising from the breaking of CLT through either a finite number of neurons or non-independence of neurons, i.e., the emergence of a non-Gaussian field, see [42].

3. Infinite Neural Networks Converging to Physics-Consistent Gaussian Processes

In this section, we give a brief review of the notions introduced previously in [3]. We introduce two *sets* of functions (function spaces) $V, W : X \to Y \subseteq \mathbb{R}^Q$. From the first *set* V, we denote two particular elements (i.e., functions from this space) as $g(x) \in V,\ x \mapsto g(x)$ and $f(x) \in V,\ x \mapsto f(x)$, $x \in X \subseteq \mathbb{R}^D$, where D, Q denote the input/output dimension.

We will later see that GPs are *distributions* for functions; hence, we need sets of functions accordingly. We aim to learn these functions from data related to some physical law expressed in terms of a linear operator $\hat{\mathcal{O}} : V \to W$, $\hat{\mathcal{O}} \circ f(x) \mapsto \tilde{f}(x)$. In words, an operator maps a function from the set (function space) V to another function from the set (function space) W. Linearity requires $\hat{\mathcal{O}} \circ (\alpha \cdot f + \beta \cdot g) = \alpha \cdot \hat{\mathcal{O}} \circ f + \beta \cdot \hat{\mathcal{O}} \circ g$ for some constants $\alpha, \beta \in \mathbb{R}$. Let the physical law be expressed in the form $\hat{\mathcal{O}} f(x) = 0$, where, from now on, we omit the $\circ$-sign for simpler notation, e.g., if $\hat{\mathcal{O}}$ were the D-dimensional Laplace-operator $\hat{\mathcal{O}} \triangleq \nabla^2 := \sum_{k=1}^{D} \frac{\partial^2}{\partial x_k^2}$, x_k denoting the $k-th$ element of D-dimensional vector $x = (x_1, \ldots, x_D)^T$, then V would be at least the set of (continuous) functions twice-differentiable and $\hat{\mathcal{O}} : C^2(X) \to C^0(X) \supset C^2(X)$, i.e., $W \subset V$.

In the following, g will denote a GP and f will denote an NN.

3.1. Physics-Consistent Gaussian Processes

We will introduce GPs along the lines of [7]. A GP is a distribution on the set of functions $\mathbb{R}^D \to \mathbb{R}^Q$ such that the function values $g(x^{(1)}), \cdots, g(x^{(n)})$ at $x^{(i)} \in \mathbb{R}^D$ have a joint Gaussian distribution. Note the consistent marginalization property that follows from this definition. The distribution is specified by a mean function $\mu : \mathbb{R}^D \to \mathbb{R}^Q : x \mapsto \langle g(x) \rangle$ and a positive semi-definite covariance function $k : \mathbb{R}^D \oplus \mathbb{R}^D \mapsto \mathbb{R}^{Q \times Q}_{\succeq 0} : (x, x') \mapsto \langle (g(x) - \mu(x)) (g(x') - \mu(x'))^T \rangle$, where $\langle \cdot \rangle$ denotes expectations. k is also known as *kernel*, and the common notation is $g(x) \sim \mathcal{GP}(\mu(x), k(x, x'))$. For simplicity, we assume g to be a zero-mean GP,

$$g(x) \sim \mathcal{GP}(0, k(x, x')) \tag{1}$$

We further require that g be consistent with a physical law expressed in terms of a linear differential operator $\hat{\mathcal{O}}_x$, i.e., differential with regard to x, i.e., that it admits the following equation:

$$\hat{\mathcal{O}}_x g(x) = 0\,, \tag{2}$$

In other words, we use a GP as an ansatz for solving Equation (2). As mentioned before, this is a differential operator defining a differential equation. Then, due to the linearity of $\hat{\mathcal{O}}$,

$$\hat{\mathcal{O}}_x g(x) \sim \mathcal{GP}(0, \hat{\mathcal{O}}_x \hat{\mathcal{O}}^{\dagger}_{x'} k(x, x')) \tag{3}$$

where $\dagger$ denotes the adjoint, acting on the *second* argument of the kernel function x'. We have introduced the subscript x to clarify which argument of the kernel the differential operator is acting on. At this point, we restrict ourselves to linear operators, but recent extensions to non-linear differential PDEs have been fairly successfully [25]. The above relationship can be used to formulate a number of equivalent physics-consistency conditions for the GP (or its kernel, respectively), e.g., from [5]:

$$\hat{\mathcal{O}}_x \hat{\mathcal{O}}_{x'}\, k(x, x')\Big|_{x=x'} \overset{!}{=} 0\,. \tag{4}$$

The solution for k in Equation (4) then defines a GP that satisfies Equation (3) a priori, i.e., *before* training. In other words, these physics-consistency conditions can be used to find and design physics-consistent kernels. For this, Mercer's theorem can be especially useful, in that it states that a kernel function (i.e., an element from a reproducing kernel Hilbert space) can always be expressed in terms of a set of basis functions $\{\phi_i\}$ and vice versa [43]. Formally, this means that

$$k(x, x') = \lim_{P \to \infty} \sum_{i,j}^{P} \phi_i(x) M_{ij} \phi_j(x') \tag{5}$$

with a suitable prior covariance matrix M for the function weights. Those basis functions could, e.g., be constructed from Green's functions as proposed by [6]. Exploiting linearity and requiring uniform convergence, the substitution of Equation (5) into Equation (4) would lead the operators to act on the basis functions in the sum directly. Note that [7] introduced a symbolic, algebraic algorithm program for constructing such kernels.

3.2. *Infinite Neural Networks*

Let us now consider an NN f with a *single* hidden layer with N neurons with non-linear but bounded activation functions h,

$$f(x) = \sum_{k=1}^{N} \mathrm{v}_k h_k \Big(\mathrm{w}_k \cdot x + b_k \Big) \,, \tag{6}$$

where the index $k = 1, \ldots, N$ sums over the N neurons in the single hidden layer. w_k are the input-to-latent weight matrices, b are bias terms and v_k are the latent-to-output weights. Under the infinite-width limit, $N \to \infty$, Equation (6) converges to a Gaussian process, i.e., a Gaussian distribution $\forall x$, using the central limit theorem if the activation functions h_k are bounded and weights and biases are i.i.d. [9]. Figure 1 shows an illustration of the correspondence between GPs and infinite NNs.

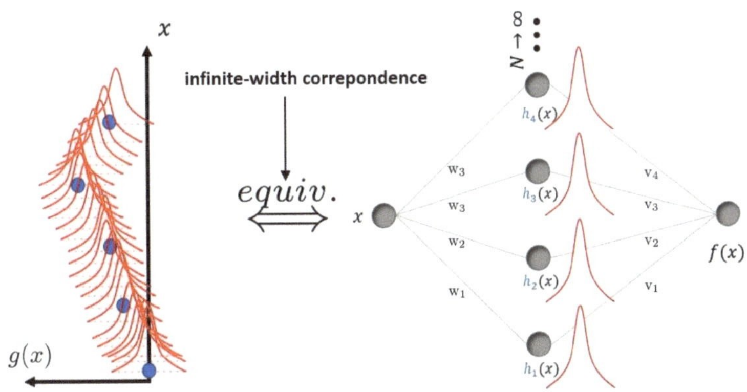

Figure 1. Schematic illustration of convergence of infinite NNs toward GPs for building intuition. The right shows a simple single-layer NN with grey disks being the neurons, where we keep adding neurons in the hidden (latent) layer in the middle ad infinitum. The distribution curves next to the neurons should emphasize that each of the outputs of the neurons (i.e., the resulting values of non-linear activation after linear transform) are random variables with some probability distribution. The left side shows a GP, i.e., a stochastic process with a Gaussian distribution on *every* point on the real line of x (this is essentially the definition of a GP). The correlation between all those Gaussians is defined by the covariance function $k(x, x')$. The blue disks denote data points to which the GP has been adjusted (i.e., we are looking at a posterior GP trained on data, but we may as well illustrate a prior GP centered, e.g., around $g = 0$). The x-axis of the GP has been rotated and mirrored for visualization purposes such that the orientation of the bell curves on the GP x-axis align with the distributions over the neurons r.h.s.

The same finding holds for a considerably larger class of models [36]. Then, the NN converges to a GP with kernel $k(x, x') = \langle f(x) f(x') \rangle$, i.e., the covariance of NN predictions must satisfy the covariance of the GP, which is a sensible requirement. This gives rise to the following condition for equivalence between a certain GP and a certain NN:

$$
\begin{aligned}
k(x,x') &\stackrel{!}{=} \langle f(x)f(x')\rangle \\
&\propto \lim_{N\to\infty}\int \sum_{k=1}^{N}\sum_{k'=1}^{N} \mathrm{v}_k\mathrm{v}_{k'}h_k(\mathrm{w}_k x+b_k)h_{k'}(\mathrm{w}_{k'}x'+b_{k'})p(\theta)dV_\theta
\end{aligned} \tag{7}
$$

where we used the linearity of the expectation and $p(\theta)$ is the probability density function of the weights and biases $\theta = \{\mathrm{v},\mathrm{w},b\}$. Note that we exchanged limit and integration, which is (typically) permissible if convergence is uniform and the limit function is integrable [44]. This integral can be solved numerically and, in some cases, analytically [28]. Note that Equation (7) is essentially determined by the covariance of the activations.

3.3. Physics-Consistency Condition for Neural Networks

We could now take the route of finding a physics-consistent kernel k on the left-hand side of Equation (7) and then try to find "physics-consistent" activations, i.e., activations such that the infinite NN becomes a physics-consistent GP. This approach may be regarded as an "inverse kernel trick", especially if the Mercer representation is used.

An alternative route would be to find physics-consistent activations directly. Substituting the NN-GP consistency condition Equation (7) into the physics-consistency condition for the GP Equation (4) (or variants thereof) allows us to formulate a physics-consistency condition for the NN directly. Reciting uniform convergence and integrable limits as above used for Equation (7) [44], we switch the order of limit, expectation and differentiation:

$$
\left(\lim_{N\to\infty}\int p(\theta)dV_\theta \sum_{k,k'=1}^{N} \mathrm{v}_k\mathrm{v}_{k'}\left[\hat{\mathcal{O}}h_k\right](\mathrm{w}_k x+b_k)\cdot\left[\hat{\mathcal{O}}h_{k'}\right](\mathrm{w}_{k'}x'+b_{k'})\right)\Bigg|_{x=x'} \stackrel{!}{=} 0\,. \tag{8}
$$

where the chain rule has to be applied, giving rise to differentiated activations $x \mapsto \left[\hat{\mathcal{O}}h_k\right](x)$ applied to their original argument $(\mathrm{w}_k x + b_k)$, where, if $h_k(x) \in V$, then $\left[\hat{\mathcal{O}}h_k\right] \in W$. This gives rise to the problem of finding pairs of probability density functions $p(\theta)$ and activations h such that the physics-consistency condition for the NN Equation (8) is satisfied. This implies that the choice of activation and prior cannot be simply treated separately. The choice of prior $p(\theta)$ and activation h is then reminiscent of the choice of the prior covariance matrix M and basis functions ϕ in the Mercer representation of the GP kernel in Equation (5).

Through equating Equation (8) l.h.s. with the kernel's Mercer representation, Equation (5), and recalling that Green's functions can admit kernels (i.e., applying the kernel trick to Green's functions), it is easy to see through the choices $M_{ij} = \sigma_{\mathrm{v}_i}\delta_{ij}$, Gaussian priors and the limit $p(\mathrm{w}_k) = \lim_{\sigma_{\mathrm{w}_k}\to 0}\mathcal{N}(0,\sigma^2_{\mathrm{w}_k}\Sigma)$ (where Σ is diagonal) that $h \to \phi$. This means that the activations become the Green's functions [3].

3.4. Training and Regularization with Physics-Information

Given pairs of input–output data $D = \{x^{(d)}, y^{(d)}\}_{d=1}^{N_d}$, where $y^{(d)} = f(x^{(d)}) + \varepsilon$ are noisy observations with noise ε, we ought to choose an optimization criterion or loss function for the learning, e.g., here,

$$
\theta^* = \arg\min_\theta \left[\sum_{d=1}^{N_d}\left(y^{(d)} - f(x^{(d)})\right)^2 + \lambda\sum_{p=1}^{N_p}\left\|\left(\hat{\mathcal{O}}_x f(x)\right)\Big|_{x=x^{(p)}}\right\|\right], \tag{9}
$$

which essentially consists of the likelihood and a prior defined via the residual of Equation (3). The last term denotes the differential to be applied to f and then evaluated at pivot points $x = x^{(p)}$ that may be distinct from the data points $x^{(d)}$. In more detail, for the second term (the "prior"), a network f is substituted as an ansatz function into Equation (3) and evaluated at pivot points (or sometimes called collocation points) $x^{(p)}$. This evaluation generally yields non-zero values for a generic NN (as opposed to the zero right-hand side in Equation (3), i.e., the equation at this point is not actually satisfied). The sum of these (absolute) values may be called the residual. In the physics-informed training approach, this residual is to be minimized by adjusting the function parameters (in an NN, the weights and biases). This minimization can be performed conjointly with minimizing the data misfit (i.e., the first term in Equation (9)), where the trade-off between the physics-residual and the data misfit term is controlled by a tunable regularization parameter λ. The choice of λ and the set $\{x^{(p)}\}_{p=1}^{N_p}$ is generally non-trivial [45].

4. Experiments

4.1. Generalization

We consider a Helmholtz equation

$$(\nabla^2 - c^2)f(x) = 0 \tag{10}$$

where $c = 6.0$ is a constant parameter related to the wavenumber. In the one-dimensional case, the solution to this equation is a sinusoidal function. The physics-consistent GP kernel for this physical law in 1D has the form $k(x, x') = \cos\left(\alpha(x - x')\right)$ [46]. As derived in [3], one particular activation that defines an infinite NN with the same physics-consistent covariance is a sinusoidal activation function, $h(\cdot) := \sin(\cdot)$.

We used the function $y = A\sin(\omega x + \phi)$ to generate the training data. A noise of normal distribution $\mathcal{N}(0, 0.04^2)$ was added to the training sets. Next, we tried to learn an NN function $f(x)$ from this data, and compared the performance of three networks: all three networks are single-layer with the structure introduced before with $N = 100$ neurons in the hidden layer. We denote as "vanilla" a simple NN with an ReLU$(\cdot)$ activation, and training with $L2$-loss. We denote as "physics-informed" the same network where the training is additionally regularized through a physics-loss term as in Equation (9). For the physics-informed neural network, the loss function specifically is $\mathcal{L} = \mathcal{L}_d + \mathcal{L}_p = |y - \hat{y}|^2 + \lambda|\nabla^2 y + c^2 y|$. Note that in the physics-consistent NN, there are no training difficulties due to higher-order derivatives and the training is as straightforward and fast as with the vanilla NN.

Note the distinction between training error, validation error and testing error. The training error is the prediction error on the training sets. Additional validation sets are used for early stopping to avoid overfitting, where training is stopped if there is no improvement in validation errors for a number of epochs. The validation sets are distributed within the same range/domain as the training sets, and test sets are distributed in a different range. The test sets are typically used to investigate the generalization behavior of an NN.

One reason for the wide-spread popularity of physics-informed NNs (PINNs) is that they typically generalize better than NNs trained on the data alone. PINNs achieve this by introducing a weak constraint that penalizes un-physical behavior. In our physics-consistent approach, in contrast, we enforce a hard constraint instead. In particular, this avoids the inconsistency of the multiple training objectives or the Pareto front as discussed in [47].

We compare the generalization behavior of the three NNs in Figure 2.

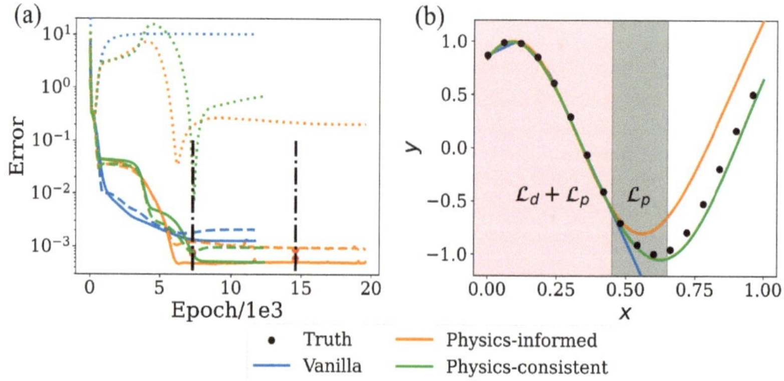

Figure 2. Generalization of neural network models. (**a**) Evolution of loss during training. (**b**) Predictions of different networks. Solid line: training error; dashed line: validation error; dotted line: test error. Generalization is typically measured in terms of the gap between training and test error. Black vertical lines indicate optimal model according to validation set.

4.2. Convergence to Gaussian as a Function of Network Width

The Gaussianity of an NN is only exact under the infinite-width limit, whereas, on a computer, NNs are always finite. This begs the question of how good the approximation is as a function of NN width, or how fast the NN converges to a GP. Here, we use a single-hidden-layer neural network with prior distributions all i.i.d. normal with zero-mean. The variances are $\sigma_w = 5$, $\sigma_b = 5$, $\sigma_v = 1$, $\sigma_a = 0.1$.

For pedagogical illustration, we show scatter plots in Figure 3 of predictions collected from simple Monte Carlo samples from the prior on the weights and biases, $p(\theta)$ similar to [9] with a single-layer NN and tanh$(\cdot)$-activations. $NN(-0.2)$ and $NN(0.4)$ denote the outputs of the network corresponding to the inputs $x = -0.2$ and $x = 0.4$, respectively.

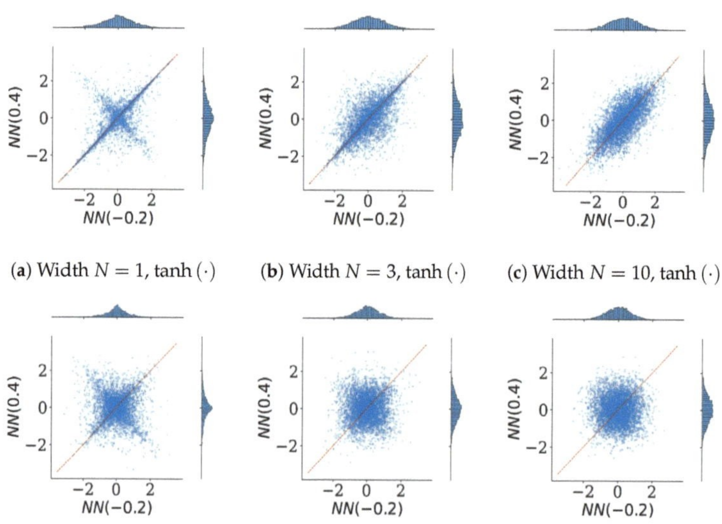

(**a**) Width $N = 1$, tanh $(\cdot)$ (**b**) Width $N = 3$, tanh $(\cdot)$ (**c**) Width $N = 10$, tanh $(\cdot)$

(**d**) Width $N = 1$, sin $(\cdot)$ (**e**) Width $N = 3$, sin $(\cdot)$ (**f**) Width $N = 10$, sin $(\cdot)$

Figure 3. Scatter plots of MC samples with different single-layer network widths (number of neurons, N) evaluated at fixed NN inputs x. Ordinate and abscissa values and then mean values of NN prediction for two (distinct) given inputs. Upper row: tanh-activation. Lower row: sin-activation.

The NN's limit kernel is defined by Equation (7). For some NNs, this kernel can be computed analytically, while more often it has to be computed numerically, e.g., through Monte Carlo integration. For a single-layer NN with ReLU-activation, we compare the analytical kernel to the estimated kernel in Figure 4, which shows that such estimates can converge quickly and easily with only few neurons and samples.

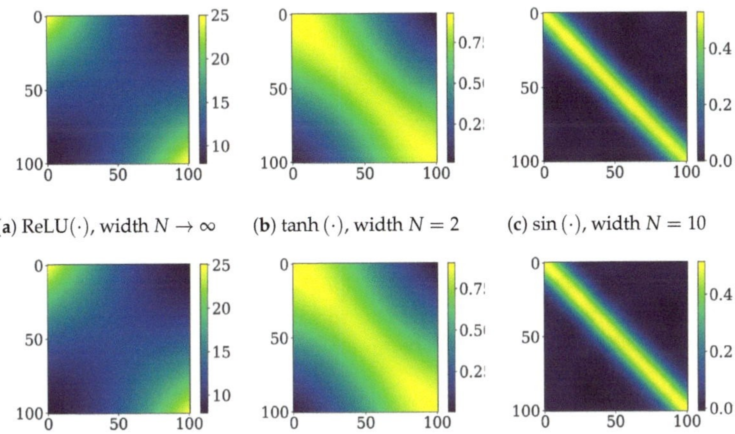

(**a**) ReLU$(\cdot)$, width $N \to \infty$ (**b**) tanh $(\cdot)$, width $N = 2$ (**c**) sin $(\cdot)$, width $N = 10$

(**d**) ReLU$(\cdot)$, width $N = 5000$ (**e**) tanh $(\cdot)$, width $N = 5000$ (**f**) sin $(\cdot)$, width $N = 5000$

Figure 4. Comparing kernels of NNs with ReLU (left), tanh (middle column) and physics-consistent sinus (right) activation for varying widths (rows). (**a**) Infinite NN limit kernel as computed by [28]. (**b–f**) Monte Carlo estimates with 5000 samples. N denotes number of neurons. Ordinate and abscissa mean (pairs of) input indices x, x'.

This, however, is only an estimate of the second moments, and does not tell "how Gaussian" the distribution itself is. One suitable measure for this would be the Kullback–Leibler divergence, measuring the (dis-)similarity between two probability distributions. By defining a Gaussian distribution $Q = \mathcal{N}(0, K_Q)$ defined by the computed or estimated covariance matrix K, e.g., from Figure 4, the Kullback–Leibler divergence allows us to measure the dissimilarity between the distribution of NN predictions $P\big((f(x)\big)$, i.e., the histograms of NN realizations collected from prior samples fed through the NN as in the scatter plots in Figure 3, and the (Gaussian) limit distribution Q. Note that alternative measures would be possible, such as the Wasserstein distance.

The Kullback–Leibler divergence is defined as:

$$\mathcal{D}_{KL}(P \| Q) = \sum_{x \in \mathcal{X}} P(f) \log\left(\frac{P(f)}{Q(f)}\right) \tag{11}$$

which can be estimated, e.g., through Monte Carlo. We plot the KL-divergence between the finite NN and its infinite-limit Gaussian as a function of the finite network's width in Figure 5. The result confirms the intuition from Figure 3, and the finite NN behaves similar to a Gaussian already with just 10^2 neurons.

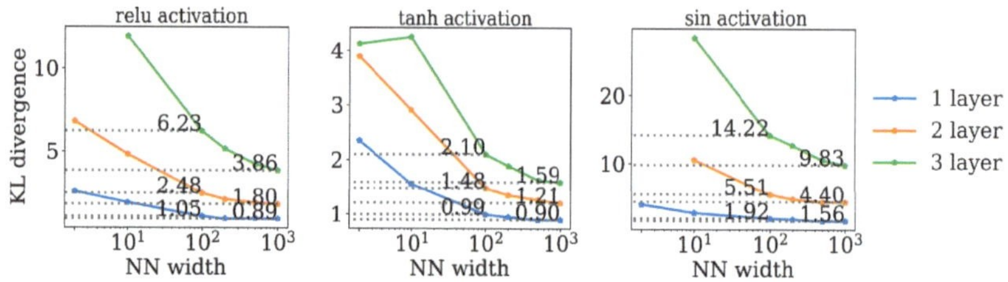

Figure 5. Kullback–Leibler divergence between finite NN and GP (i.e., infinite NN) for three layer depths, two generic activations and the physics-consistent activation (*sin*) from before.

5. Conclusions

We have derived a physics-consistency condition for NNs by exploiting the physics-consistency condition for GPs along with infinite NNs converging to GPs using the central limit theorem. We quantify experimentally that finite NNs with only order $\sim 10^2$ neurons already behave like GPs, i.e., that the approximation is sensible. We further show that engineering NNs to be physics-consistent can have advantages over both purely data-driven approaches and physics-based approaches with a similar intent, e.g., PINNs, in terms of convergence rate and generalization.

Author Contributions: Conceptualization, S.R.; methodology, S.R.; software, S.G.; validation, S.R. and S.G.; formal analysis, S.R.; investigation, S.R. and S.G.; resources, S.R.; data curation, N/A; writing—original draft preparation, S.R.; writing—review and editing, S.R. and S.G.; visualization, S.G. and S.R.; supervision, S.R.; project administration, S.R.; funding acquisition, S.R. All authors have read and agreed to the published version of the manuscript.

Funding: The authors gratefully acknowledge funding by UFO—Unkonventionelle Forschung of Das Land Steiermark (Styrian Government), Abt. 12 Referat für Wissenschaft & Forschung, PN32.

Institutional Review Board Statement: Not applicable.

Informed Consent Statement: Not applicable.

Data Availability Statement: All information is contained in the manuscript. Codes are available at https://github.com/guanshaoheng/gaussian_nn_tugraz (accessed on 1 July 2023).

Acknowledgments: S.R. carried out parts of the research during a visit at the Munich Institute for Astro-, Particle and BioPhysics (MIAPbP), funded by the Deutsche Forschungsgemeinschaft (DFG, German Research Foundation), Germany's Excellence Strategy—EXC-2094—390783311.

Conflicts of Interest: The authors declare no conflicts of interest.

References

1. Ranftl, S.; von der Linden, W. Bayesian Surrogate Analysis and Uncertainty Propagation. *Phys. Sci. Forum* **2021**, *3*, 6. [CrossRef]
2. Ranftl, S.; Rolf-Pissarczyk, M.; Wolkerstorfer, G.; Pepe, A.; Egger, J.; von der Linden, W.; Holzapfel, G.A. Stochastic modeling of inhomogeneities in the aortic wall and uncertainty quantification using a Bayesian encoder–decoder surrogate. *Comput. Methods Appl. Mech. Eng.* **2022**, *401*, 115594. [CrossRef]
3. Ranftl, S. A Connection between Probability, Physics and Neural Networks. *Phys. Sci. Forum* **2022**, *5*, 11. [CrossRef]
4. Dong, A. Kriging Variables that Satisfy the Partial Differential Equation ΔZ = Y. In *Geostatistics: Proceedings of the Third International Geostatistics Congress September 5–9, 1988*; Armstrong, M., Ed.; Springer: Dordrecht, The Netherlands, 1989; Volume 4, pp. 237–248. [CrossRef]
5. van den Boogaart, K.G. Kriging for processes solving partial differential equations. In Proceedings of the Conference of the International Association for Mathematical Geology (IAMG), Cancun, Mexico, 6–12 September 2001.
6. Albert, C.G. Gaussian processes for data fulfilling linear differential equations. *Proceedings* **2019**, *33*, 5. [CrossRef]
7. Lange-Hegermann, M. Algorithmic linearly constrained Gaussian processes. In Proceedings of the Advances in Neural Information Processing Systems 31 (NeurIPS 2018), Montreal, QC, Canada, 3–8 December 2018.
8. Härkönen, M.; Lange-Hegermann, M.; Raiţă, B. Gaussian Process Priors for Systems of Linear Partial Differential Equations with Constant Coefficients. *arXiv* **2022**, arXiv:2212.14319.
9. Neal, R.M. Bayesian Learning for Neural Networks. Chapter 2: Priors on Infinite Networks. Ph.D. Thesis, University of Toronto, Toronto, ON, Canada, 1996. [CrossRef]

10. Karniadakis, G.E.; Kevrekidis, I.G.; Lu, L.; Perdikaris, P.; Wang, S.; Yang, L. Physics-informed machine learning. *Nat. Rev. Phys.* **2021**, *3*, 422–440. [CrossRef]
11. O'Hagan, A. Curve Fitting and Optimal Design for Prediction. *J. R. Stat. Soc. Ser. B (Methodol.)* **1978**, *40*, 1–24. [CrossRef]
12. Rasmussen, C.E.; Williams, C.K. *Gaussian Processes for Machine Learning*; The MIT Press: Cambridge, MA, USA, 2006.
13. Eberle, V.; Frank, P.; Stadler, J.; Streit, S.; Enßlin, T. Efficient Representations of Spatially Variant Point Spread Functions with Butterfly Transforms in Bayesian Imaging Algorithms. *Phys. Sci. Forum* **2022**, *5*, 33. [CrossRef]
14. Bilionis, I.; Zabaras, N.; Konomi, B.A.; Lin, G. Multi-output separable Gaussian process: Towards an efficient, fully Bayesian paradigm for uncertainty quantification. *J. Comput. Phys.* **2013**, *241*, 212–239. [CrossRef]
15. Schöbi, R.; Sudret, B.; Wiart, J. Polynomial-chaos-based Kriging. *Int. J. Uncertain. Quantif.* **2015**, *5*, 171–193. [CrossRef]
16. Duvenaud, D.K. Automatic Model Construction with Gaussian Processes. Ph.D. Thesis, University of Cambridge, Cambridge, UK, 2014. [CrossRef]
17. Swiler, L.P.; Gulian, M.; Frankel, A.L.; Safta, C.; Jakeman, J.D. A Survey of Constrained Gaussian Process Regression: Approaches and Implementation Challenges. *J. Mach. Learn. Model. Comput.* **2020**, *1*, 119–156. [CrossRef]
18. Jidling, C.; Wahlstrom, N.; Wills, A.; Schön, T.B. Linearly Constrained Gaussian Processes. In Proceedings of the Advances in Neural Information Processing Systems 30 (NIPS 2017), Long Beach, CA, USA, 4–9 December 2017. [CrossRef]
19. Graepel, T. Solving Noisy Linear Operator Equations by Gaussian Processes: Application to Ordinary and Partial Differential Equations. In Proceedings of the ICML, Washington, DC, USA, 21–24 August 2003.
20. Gulian, M.; Frankel, A.L.; Swiler, L.P. Gaussian process regression constrained by boundary value problems. *Comput. Methods Appl. Mech. Eng.* **2022**, *388*, 114117. [CrossRef]
21. Särkkä, S. Linear Operators and Stochastic Partial Differential Equations in Gaussian Process Regression. In Proceedings of the Artificial Neural Networks and Machine Learning—ICANN, Espoo, Finland, 14–17 June 2011; Springer: Berlin, Germany, 2011; pp. 151–158. [CrossRef]
22. Raissi, M.; Perdikaris, P.; Karniadakis, G.E. Machine learning of linear differential equations using Gaussian processes. *J. Comput. Phys.* **2017**, *348*, 683–693. [CrossRef]
23. Álvarez, M.A.; Luengo, D.; Lawrence, N.D. Linear latent force models using gaussian processes. *IEEE Trans. Pattern Anal. Mach. Intell.* **2013**, *35*, 2693–2705. [CrossRef] [PubMed]
24. López-Lopera, A.F.; Durrande, N.; Álvarez, M.A. Physically-inspired Gaussian process models for post-transcriptional regulation in Drosophila. *IEEE/ACM Trans. Comput. Biol. Bioinform.* **2021**, *18*, 656–666. [CrossRef]
25. Chen, Y.; Hosseini, B.; Owhadi, H.; Stuart, A.M. Solving and learning nonlinear PDEs with Gaussian processes. *J. Comput. Phys.* **2021**, *447*, 110668. [CrossRef]
26. Apicella, A.; Donnarumma, F.; Isgrò, F.; Prevete, R. A survey on modern trainable activation functions. *Neural Netw.* **2021**, *138*, 14–32. [CrossRef] [PubMed]
27. Jagtap, A.D.; Karniadakis, G.E. How important are activation functions in regression and classification? A survey, performance comparison, and future directions. *J. Mach. Learn. Model. Comput.* **2022**, *4*, 21–75. [CrossRef]
28. Williams, C.K. Computing with infinite networks. In Proceedings of the Advances in Neural Information Processing Systems 9 (NIPS 1996), Denver, CO, USA, 2–5 December 1996.
29. Tsuchida, R.; Roosta-Khorasani, F.; Gallagher, M. Invariance of weight distributions in rectified MLPs. In Proceedings of the ICML, Stockholm, Sweden, 10–15 July 2018.
30. Cho, Y.; Saul, L.K. Kernel methods for deep learning. In Proceedings of the Advances in Neural Information Processing Systems 22 (NIPS 2009), Vancouver, BC, Canada, 7–10 December 2009.
31. Pearce, T.; Tsuchida, R.; Zaki, M.; Brintrup, A.; Neely, A. Expressive priors in Bayesian neural networks: Kernel combinations and periodic functions. In Proceedings of the UAI, Tel Aviv, Israel, 22–25 July 2019; PMLR: New York, NY, USA, 2020.
32. Hazan, T.; Jaakkola, T. Steps Toward Deep Kernel Methods from Infinite Neural Networks. *arXiv* **2015**, arXiv:1508.05133. [CrossRef]
33. Raissi, M.; Perdikaris, P.; Karniadakis, G.E. Physics-informed neural networks: A deep learning framework for solving forward and inverse problems involving nonlinear partial differential equations. *J. Comput. Phys.* **2019**, *378*, 686–707. [CrossRef]
34. Dissanayake, M.W.M.G.; Phan-Thien, N. Neural-network-based approximations for solving partial differential equations. *Commun. Numer. Methods Eng.* **1994**, *10*, 195–201. [CrossRef]
35. Lagaris, I.E.; Likas, A.; Fotiadis, D.I. Artificial neural networks for solving ordinary and partial differential equations. *IEEE Trans. Neural Netw.* **1998**, *9*, 987–1000. [CrossRef] [PubMed]
36. Yang, G. Tensor Programs I: Wide feedforward or recurrent neural networks of any architecture are gaussian processes. In Proceedings of the Advances in Neural Information Processing Systems 32 (NeurIPS 2019), Vancouver, BC, Canada, 8–14 December 2019.
37. Novak, R.; Xiao, L.; Lee, J.; Bahri, Y.; Yang, G.; Hron, J.; Abolafia, D.A.; Pennington, J.; Sohl-Dickstein, J. Bayesian deep convolutional networks with many channels are Gaussian processes. In Proceedings of the ICLR, New Orleans, LA, USA, 6–9 May 2019.
38. Sun, X.; Kim, S.; Choi, J.i. Recurrent neural network-induced Gaussian process. *Neurocomputing* **2022**, *509*, 75–84. [CrossRef]
39. Hron, J.; Bahri, Y.; Sohl-Dickstein, J.; Novak, R. Infinite attention: NNGP and NTK for deep attention networks. In Proceedings of the ICML, Online, 12–18 July 2020.

40. Jacot, A.; Gabriel, F.; Hongler, C. Neural Tangent Kernel: Convergence and Generalization in Neural Networks. In Proceeding of the Advances in Neural Information Processing Systems 31 (NeurIPS 2018), Montreal, QC, Canada, 3–8 December 201 [CrossRef]
41. Rahimi, A.; Recht, B. Random Features for Large-Scale Kernel Machines. In Proceedings of the Advances in Neural Informatio Processing Systems 20 (NIPS 2007), Vancouver, BC, Canada, 3–6 December 2007.
42. Demirtas, M.; Halverson, J.; Maiti, A.; Schwartz, M.D.; Stoner, K. Neural Network Field Theories: Non-Gaussianity, Actions, an Locality. *arXiv* **2023**, arXiv:2307.03223.
43. Schaback, R.; Wendland, H. Kernel techniques: From machine learning to meshless methods. *Acta Numer.* **2006**, *15*, 543–63 [CrossRef]
44. Kamihigashi, T. Interchanging a limit and an integral: Necessary and sufficient conditions. *J. Inequalities Appl.* **2020**, *2020*, 24 [CrossRef]
45. Wang, S.; Yu, X.; Perdikaris, P. When and why PINNs fail to train: A neural tangent kernel perspective. *J. Comput. Phys.* **202** *449*, 110768. [CrossRef]
46. Albert, C.G. Physics-informed transfer path analysis with parameter estimation using Gaussian Processes. *Int. Congr. Acous* **2019**, *23*, 459–466. [CrossRef]
47. Rohrhofer, F.M.; Posch, S.; Geiger, B.C. On the Pareto Front of Physics-Informed Neural Networks. *arXiv* **2021**, arXiv:2105.0086 [CrossRef]

physical sciences forum

MDPI

Proceeding Paper

Quantum Measurement and Objective Classical Reality †

Vishal Johnson [1,2,*], Philipp Frank [1] and Torsten Enßlin [1,2]

1 Max Planck Institute for Astrophysics, 85748 Garching, Germany; philipp@mpa-garching.mpg.de (P.F.); ensslin@mpa-garching.mpg.de (T.E.)
2 Faculty of Physics, Ludwig-Maximilians-Universität, 80539 Munich, Germany
* Correspondence: vishal@mpa-garching.mpg.de
† Presented at the 42nd International Workshop on Bayesian Inference and Maximum Entropy Methods in Science and Engineering, Garching, Germany, 3–7 July 2023.

Abstract: We explore quantum measurement in the context of Everettian unitary quantum mechanics and construct an explicit unitary measurement procedure. We propose the existence of prior correlated states that enable this procedure to work and therefore argue that correlation is a resource that is consumed when measurements take place. It is also argued that a network of such measurements establishes a stable objective classical reality.

Keywords: quantum measurement; quantum entanglement; classical reality; measurement problem; unitary quantum mechanics

1. Introduction

The usual treatment of quantum mechanics is to consider two distinct kinds of dynamics for the quantum wavefunction: reversible unitary dynamics of the Schrödinger equation and irreversible wavefunction collapse. Similar to the approach of Hugh Everett [1], we assume a completely unitary dynamics for quantum mechanics and explore its consequences. The aim is to explore the extent to which the results of quantum experiments, and thereafter our own subjective experience of our world, can be explained starting from a hypothesis of unitarity (or equivalently, reversible dynamics).

In a related publication [2], we explain the details of how to define a measurement procedure within the context of unitary quantum mechanics. The current article explores a particular instructive example in detail and sets it in the context of explaining the objective classical reality we are familiar with.

A consequence of the assumption of unitarity is that redundant information about the measurement is required in order to correct for the inevitable environmental influence. As a result, we propose the existence of a "correlated environment" that provides this information. This is related to "spectrum broadcast structures" described in [3]. We thereafter explain the emergence of an objective classical reality following the arguments in [4].

The article is structured as follows. The limits that unitarity of quantum mechanics places on the measurement procedure and its consequences are explored in Section 2. The motivation of a "correlated environment" and explanation of how it is a conserved resource follow in Section 3. We then explain how these considerations help explain the familiar objective classical reality in Section 4. A discussion follows in Section 5.

2. Limits of Unitary Quantum Mechanics

In this section, we discuss how the assumption of unitary quantum mechanics limits the dynamics of quantum measurement. The arguments presented here are quite similar to those discussed in [1].

Consider a quantum system being measured,

$$|\psi\rangle_s = \psi_\uparrow|\uparrow\rangle_s + \psi_\downarrow|\downarrow\rangle_s, \quad (1)$$

where we use the mental picture of an electron spin being measured, which is labelled s for "signal". The electron spin could be either in the $|\uparrow\rangle_s$ or $|\downarrow\rangle_s$ state and generically it is in a superposition, such as in Equation (1).

The quantum system is measured by an observer. As we assume that quantum mechanics is a complete description of our universe, the observer is also described using a quantum state. Let us assume that the observer is in the arbitrary state

$$|\phi\rangle_o = \phi_\uparrow|\uparrow\rangle_o + \phi_\downarrow|\downarrow\rangle_o. \quad (2)$$

The states $|\uparrow,\downarrow\rangle_o$ for the observer indicate its physical configurations that correspond to recording a measurement of $|\uparrow,\downarrow\rangle_s$, respectively, for the signal. Note that, due to time reversibility, it is not required to label the states with a time label; the state of the observer when the signal is measured corresponds to a well-defined state when the measurement procedure is begun.

A specific initial state for the observer is assumed in usual approaches to Everettian quantum mechanics and this leads, implicitly or explicitly, to the issue of preferred bases [5]. In our approach, an arbitrary state for the observer is assumed (Equation (2)) so that the observer need not be in any predetermined initial state to allow the measurement procedure to take place. Nevertheless, our approach does not completely resolve the issue, as the basis dependence comes in implicitly at the level of the Hamiltonian (or unitary) enabling the measurement procedure; this is discussed in Section 5.

In order for the measurement procedure to be unitary, it is required that the information in the observer (Equation (2)) be transferred to another system so that the no-deletion theorem [6] is not violated. It is therefore required to consider a (large enough) third system into which the observer's prior state can be dumped. We call this the environment and consider for its state

$$|\chi\rangle_e = |\uparrow\rangle_e. \quad (3)$$

In this case, the state is not in a superposition because we assume an environment state that is well-known. The extension to an arbitrary environment state is considered later in Section 2.2. Figure 1 summarizes the setup we consider here.

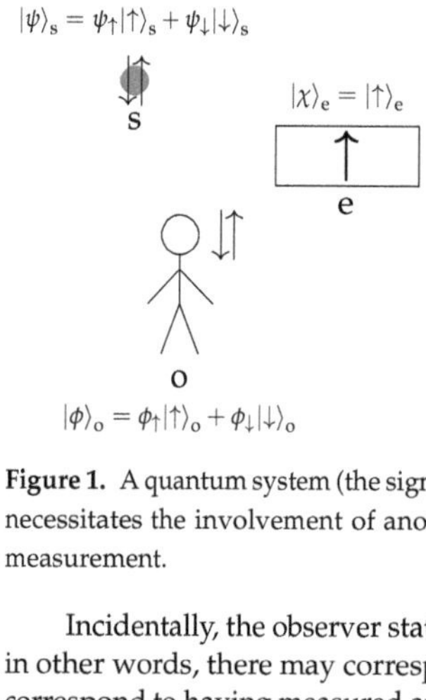

Figure 1. A quantum system (the signal) is measured by an observer. Unitarity of quantum mechanics necessitates the involvement of another system (the environment) in order to facilitate quantum measurement.

Incidentally, the observer state itself might live in a larger Hilbert space than the signal; in other words, there may correspond physical configurations of the observer that do not correspond to having measured any state of the signal. This is not an issue, as any arbitrary initial observer state can be dumped into an environment which is large enough.

2.1. Quantum Measurement

We define quantum measurement as a process that establishes perfect correlation between the signal and the observer. We restrict ourselves to the simplified and idealized case where the outcomes of the quantum measurement are orthogonal states for both the system and observer; the case of generalized measurements is considered in [2]. In this case, perfect correlation means that the different states of the observer correspond bijectively to states of the signal; cases where the "basis" for the observer differs slightly from that of the signal are considered in [2]. Therefore, starting with the three disconnected subsystems of Figure 1, the goal is to reach a state where the signal and observer completely agree with one another.

$$\begin{aligned} |\psi\rangle_s|\phi\rangle_o|\chi\rangle_e &= (\psi_\uparrow|\uparrow\rangle+\psi_\downarrow|\downarrow\rangle)_s(\phi_\uparrow|\uparrow\rangle+\phi_\downarrow|\downarrow\rangle)_o|\uparrow\rangle_e \\ &\to (\psi_\uparrow|\uparrow\uparrow\rangle+\psi_\downarrow|\downarrow\downarrow\rangle)_{so}(\phi_\uparrow|\uparrow\rangle+\phi_\downarrow|\downarrow\rangle)_e \\ &=: |\Psi\rangle_{so}|\phi\rangle_e; \end{aligned} \tag{4}$$

the perfectly correlated signal–observer state is denoted as $|\Psi\rangle_{so} = \psi_\uparrow|\uparrow\uparrow\rangle_{so} + \psi_\downarrow|\downarrow\downarrow\rangle_{so}$.

At the level of the individual basis vectors, the action is

$$\begin{aligned} |\uparrow\uparrow\uparrow\rangle_{soe} &\to |\uparrow\uparrow\uparrow\rangle_{soe} \\ |\uparrow\downarrow\uparrow\rangle_{soe} &\to |\uparrow\uparrow\downarrow\rangle_{soe} \\ |\downarrow\uparrow\uparrow\rangle_{soe} &\to |\downarrow\downarrow\uparrow\rangle_{soe} \\ |\downarrow\downarrow\uparrow\rangle_{soe} &\to |\downarrow\downarrow\downarrow\rangle_{soe}. \end{aligned} \tag{5}$$

2.2. Generic Environment

In case the environment is not in the state $|\uparrow\rangle_e$ but $|\downarrow\rangle_e$, by the requirement of unitarity, the subspace $|XY\downarrow\rangle_{soe}$ can only be mapped to $\mathrm{span}\{|\uparrow\downarrow Z\rangle_{soe}, |\downarrow\uparrow Z\rangle_{soe}\}$ so that the measurement procedure necessarily goes wrong; that is, the signal and observer are no longer correlated but anticorrelated.

In terms of the basis vectors, a natural extension of the action of Equation (5) is

$$\begin{aligned} |\uparrow\uparrow\downarrow\rangle_{soe} &\to |\uparrow\downarrow\uparrow\rangle_{soe} \\ |\uparrow\downarrow\downarrow\rangle_{soe} &\to |\uparrow\downarrow\downarrow\rangle_{soe} \\ |\downarrow\uparrow\downarrow\rangle_{soe} &\to |\downarrow\uparrow\uparrow\rangle_{soe} \\ |\downarrow\downarrow\downarrow\rangle_{soe} &\to |\downarrow\uparrow\downarrow\rangle_{soe} \end{aligned} \tag{6}$$

where the environment absorbs the state the observer initially is in. Therefore, the measurement procedure proceeds as

$$|\psi\rangle_s|\phi\rangle_o|\downarrow\rangle_e \to (\psi_\uparrow|\uparrow\downarrow\rangle+\psi_\downarrow|\downarrow\uparrow\rangle)_{so}|\phi\rangle_e =: |\bar{\Psi}\rangle_{so}|\phi\rangle_e \tag{7}$$

where $|\bar{\Psi}\rangle_{so} = (\psi_\uparrow|\uparrow\downarrow\rangle+\psi_\downarrow|\downarrow\uparrow\rangle)_{so}$ is the state of the signal and observer where they are anticorrelated.

In case the environment state is in a generic superposition, such as $|\chi\rangle_e = \chi_\uparrow|\uparrow\rangle + \chi_\downarrow|\downarrow\rangle$, the measurement proceeds as

$$|\psi\rangle_s|\phi\rangle_o(\chi_\uparrow|\uparrow\rangle+\chi_\downarrow|\downarrow\rangle)_e \to (\chi_\uparrow|\Psi\rangle_{so}+\chi_\downarrow|\bar{\Psi}\rangle_{so})|\phi\rangle_e, \tag{8}$$

where the signal–observer is correlated in the branch beginning with $|\uparrow\rangle_e$ and anticorrelated in the other branch, $|\downarrow\rangle_e$.

2.3. Local Operations

The results of Section 2.2 can be obtained using only local operators. The local operators codify interactions that take place between systems which are physically close to each other. Using local operators between observer–environment and signal–environment, the measurement procedure of Equation (8) can be carried out; there is no requirement for the signal and observer to directly interact.

An imprint operation is defined as

$$\begin{aligned} \mathcal{I}_{\mathrm{a}\to\mathrm{b}}|\uparrow\uparrow\rangle_{\mathrm{ab}} &= |\uparrow\uparrow\rangle_{\mathrm{ab}} \\ \mathcal{I}_{\mathrm{a}\to\mathrm{b}}|\uparrow\downarrow\rangle_{\mathrm{ab}} &= |\uparrow\downarrow\rangle_{\mathrm{ab}} \\ \mathcal{I}_{\mathrm{a}\to\mathrm{b}}|\downarrow\uparrow\rangle_{\mathrm{ab}} &= |\downarrow\downarrow\rangle_{\mathrm{ab}} \\ \mathcal{I}_{\mathrm{a}\to\mathrm{b}}|\downarrow\downarrow\rangle_{\mathrm{ab}} &= |\downarrow\uparrow\rangle_{\mathrm{ab}} \end{aligned} \tag{9}$$

and is similar to the CNOT gate used in quantum computation. A swap operation is defined as

$$\begin{aligned} \mathcal{S}_{\mathrm{a}\leftrightarrow\mathrm{b}}|\uparrow\uparrow\rangle_{\mathrm{ab}} &= |\uparrow\uparrow\rangle_{\mathrm{ab}} \\ \mathcal{S}_{\mathrm{a}\leftrightarrow\mathrm{b}}|\uparrow\downarrow\rangle_{\mathrm{ab}} &= |\downarrow\uparrow\rangle_{\mathrm{ab}} \\ \mathcal{S}_{\mathrm{a}\leftrightarrow\mathrm{b}}|\downarrow\uparrow\rangle_{\mathrm{ab}} &= |\uparrow\downarrow\rangle_{\mathrm{ab}} \\ \mathcal{S}_{\mathrm{a}\leftrightarrow\mathrm{b}}|\downarrow\downarrow\rangle_{\mathrm{ab}} &= |\downarrow\downarrow\rangle_{\mathrm{ab}} \end{aligned} \tag{10}$$

which is similar to the swap gate in quantum computing. It can now readily be seen that $\mathcal{S}_{o\leftrightarrow e} \circ \mathcal{I}_{s\to e}$ achieves the required outcome of Equation (8),

$$\mathcal{S}_{o\leftrightarrow e} \circ \mathcal{I}_{s\to e} \quad \left[|\psi\rangle_{\mathrm{s}}|\phi\rangle_{\mathrm{o}}(\chi_\uparrow|\uparrow\rangle + \chi_\downarrow|\downarrow\rangle)_{\mathrm{e}}\right] \to (\chi_\uparrow|\Psi\rangle_{\mathrm{so}} + \chi_\downarrow|\bar{\Psi}\rangle_{\mathrm{so}})|\phi\rangle_{\mathrm{e}}. \tag{11}$$

The environmental influence on the measurement procedure is evidenced by the fact that there are multiple branches of the wavefunction (Equation (8)) depending on which branch of the environment the overall system began in. The signal and observer are perfectly correlated in one of the branches and anticorrelated in the other. In the following, Section 3, we explore how this environmental influence can be corrected for. The measurement procedure so far is summarized in Figure 2.

Figure 2. Measurement procedure without environmental correction. The environment influences the measurement procedure as indicated by the term $k \circ i$. The clouds indicate systems which are correlated with each other. Two clouds touching indicates that the quantum systems they are part of are correlated (and thereby entangled). The clouds are colored to distinguish between the different quantum systems participating in the measurement procedure.

3. Correlation as a Resource

In this section, we discuss how the influence of the environment on the measurement procedure can be corrected for. There must be a redundant record of the environment in order for this correction to work. We argue that a correlated environment of the form

$$|\chi\rangle_{\mathrm{e}} = \sum_k \chi_k |k\rangle_{\mathrm{e}_1}|k\rangle_{\mathrm{e}_2}|k\rangle_{\mathrm{e}_3}\cdots|k\rangle_{\mathrm{e}_N} \tag{12}$$

be used to correct for the environmental influence.

3.1. Measurement Procedure

The measurement procedure proceeds as follows. The observer first swaps its state with the environment using a swap operation. This makes the observer now part of the correlated environment. Therefore, the environmental state can be corrected to obtain a "0 state" $|\uparrow\rangle_{\mathrm{o}}$, thereby allowing the usual imprint operation by the signal onto the observer to result in a perfectly correlated state. Note that this decoherence of the observer into the network of the environment could take place long before the signal is measured. It enables the observer to measure the signal and also sets the basis of measurement. The resulting state is as in Figure 3.

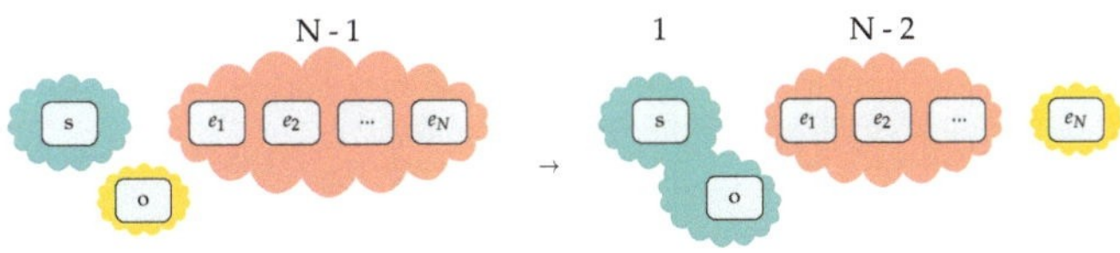

Figure 3. Measurement procedure with environmental correction. The redundant information provided by the correlated environment is used for this correction. Also indicated are the measures of correlation in the involved subsystems.

The mathematics works as follows. The swap operation enables the observer to become a part of the correlated environment:

$$\mathcal{S}_{\mathrm{o}\to\mathrm{e_N}}\quad \begin{array}{l} \mathrm{s}: |i\rangle_{\mathrm{s}} \\ \mathrm{o}: |\phi\rangle_{\mathrm{o}} \\ \mathrm{e}: |k\rangle_{\mathrm{e_1}}|k\rangle_{\mathrm{e_2}}\cdots|k\rangle_{\mathrm{e_N}} \end{array} \quad\to\quad \begin{array}{l} \mathrm{s}: |i\rangle_{\mathrm{s}} \\ \mathrm{o}: |k\rangle_{\mathrm{o}} \\ \mathrm{e}: |k\rangle_{\mathrm{e_1}}|k\rangle_{\mathrm{e_2}}\cdots|\phi\rangle_{\mathrm{e_N}}. \end{array} \tag{13}$$

The environmental influence can now be corrected for:

$$\begin{array}{ccc} \mathcal{I}^{-1}_{\mathrm{e_2}\to\mathrm{e_1}}\ \begin{array}{l} \mathrm{s}: |i\rangle_{\mathrm{s}} \\ \mathrm{o}: |k\rangle_{\mathrm{o}} \\ \mathrm{e}: |k\rangle_{\mathrm{e_1}}|k\rangle_{\mathrm{e_2}}\cdots|\phi\rangle_{\mathrm{e_N}} \end{array} & \to & \mathcal{I}_{\mathrm{s}\to\mathrm{e_1}}\ \begin{array}{l} \mathrm{s}: |i\rangle_{\mathrm{s}} \\ \mathrm{o}: |k\rangle_{\mathrm{o}} \\ \mathrm{e}: |\uparrow\rangle_{\mathrm{e_1}}|k\rangle_{\mathrm{e_2}}\cdots|\phi\rangle_{\mathrm{e_N}} \end{array} \\ & & \downarrow \\ \begin{array}{l} \mathrm{s}: |i\rangle_{\mathrm{s}} \\ \mathrm{o}: |i\rangle_{\mathrm{o}} \\ \mathrm{e}: |k\rangle_{\mathrm{e_1}}|k\rangle_{\mathrm{e_2}}\cdots|\phi\rangle_{\mathrm{e_N}} \end{array} & \leftarrow & \mathcal{S}_{\mathrm{s}\leftrightarrow\mathrm{e_1}}\ \begin{array}{l} \mathrm{s}: |i\rangle_{\mathrm{s}} \\ \mathrm{o}: |k\rangle_{\mathrm{o}} \\ \mathrm{e}: |i\rangle_{\mathrm{e_1}}|k\rangle_{\mathrm{e_2}}\cdots|\phi\rangle_{\mathrm{e_N}} \end{array} \end{array} \tag{14}$$

and perfect correlation between signal and observer results. $\mathcal{I}^{-1}_{\mathrm{a}\to\mathrm{b}}$ is the inverse of the imprint operation; this action undoes an imprint, $\mathcal{I}_{\mathrm{a}\to\mathrm{b}}$, that system a might have made on system b. In case of the CNOT gate of Equation (9), the inverse imprint is the same as the imprint.

3.2. Correlation as a Resource

We propose that there is a cost associated with correlating a signal–observer pair as the environment loses some of its own correlation; it is a resource used up in the measurement procedure. By associating a well-motivated measure, it can be seen that the resource of correlation is conserved over the measurement procedure, as described in Figure 3.

For every signal–observer qubit pair that a correlated environment in turn correlates, it must lose at least one qubit of correlation. This is because of the basic requirement of the observer having to dump its arbitrary initial state somewhere due to constraints of the no-deletion theorem [6]. Therefore, assuming a completely efficient procedure, the cost of correlating one signal–observer pair is one qubit of correlation.

Proceeding this way, it can be seen that a correlated environment, as in Equation (12), can correlate a further N-1 signal–observer pairs before it runs out of all its ability to correlate. Therefore, for a perfectly efficient procedure, the measure of correlation of a correlated environment state, as in Equation (12), is N-1.

A newly correlated signal–observer pair can itself sacrifice its correlation to correlate a different signal–observer pair. Therefore, the resource of correlation does not vanish, but is conserved and transferred from the environment to the newly correlated signal–observer pair.

4. Objective Classical Reality

In this section, we explore how the objective classical reality we are all familiar with emerges from the considerations in the previous Sections 2 and 3.

4.1. Multiple Observers

If multiple observers perform the above measurement procedure as in Figure 3, the result is a network of observers:

$$\begin{aligned} & \left(\sum_i \psi_i |i\rangle_s\right) |\phi_1\rangle_{o_1} |\phi_2\rangle_{o_2} |\phi_3\rangle_{o_3} \cdots \\ \overset{o_1 \text{ measures}}{\rightarrow} & \left(\sum_i \psi_i |i\rangle_s |i\rangle_{o_1}\right) |\phi_2\rangle_{o_2} |\phi_3\rangle_{o_3} \cdots (\text{environment suppressed}) \\ \overset{o_2 \text{ measures}}{\rightarrow} & \left(\sum_i \psi_i |i\rangle_s |i\rangle_{o_1} |i\rangle_{o_2}\right) |\phi_3\rangle_{o_3} \cdots \\ \overset{\cdots}{\rightarrow} & \left(\sum_i \psi_i |i\rangle_s |i\rangle_{o_1} |i\rangle_{o_2} |i\rangle_{o_3} \cdots\right). \end{aligned} \tag{15}$$

This enables the repeatability of experiment and, therefore, for multiple observers to agree with one another on what the state of reality of a signal is. This is the definition of objectivity used in the current article, multiple entities all in agreement with one another about the state of a signal (in agreement with [3]).

Furthermore, if each of these observers o_i are themselves classical objects that decohere, what obtains is a highly branched network of states that all agree with one another (Figure 4). This lends stability to the decohered network of states, as in order to delete the information about the state of the signal, all the systems it has interacted with have to come together to conspire to undo this correlation. In the following Section 4.2, we consider what happens when measurements are performed in different bases. It turns out that an objective classical reality does not then emerge, as the different observers do not agree with one another on what the state of a signal is. The network of redundant information in Figure 4 is exactly what protects against such a loss of objectivity.

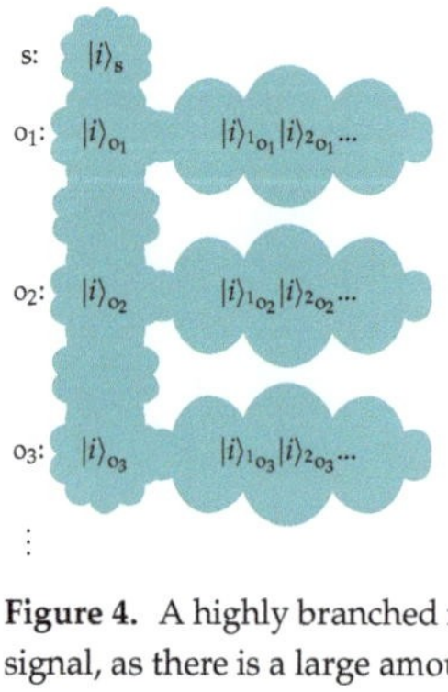

Figure 4. A highly branched network of decohered states lends stability to the measurement of the signal, as there is a large amount of redundancy in the information. Also, for this information to be deleted, all the involved systems must conspire to come together and undo this correlation.

4.2. Different Basis

In the discussion of the measurement procedure in Sections 2 and 3, most of the analysis was (made) basis independent. None among the signal, observer, and environment need be in any particular state. The swap operation of Equation (10) can also be shown to be basis-independent. Explicit dependence on a particular basis only makes its way in Equation (9) through the imprint operation. In this subsection, we explore what happens in case the measurement is performed in a different basis. It shall be seen that the measurement procedure need no longer give rise to an objective classical reality; different observers may disagree with one another on the state of a signal.

Consider a signal s that is first measured by an observer o_1 in a particular basis $\{|\uparrow\rangle, |\downarrow\rangle\}$. Let it further be observed by a different observer o_2 in a different basis

$$|\rightarrow\rangle = \frac{1}{\sqrt{2}}(|\uparrow\rangle + |\downarrow\rangle) \qquad |\leftarrow\rangle = \frac{1}{\sqrt{2}}(|\uparrow\rangle - |\downarrow\rangle). \tag{16}$$

Let observer o_1 further be measured by a third observer o_3' in the new basis. The resulting expressions are (ignoring normalization of the quantum state)

$$\begin{aligned} & \psi_\uparrow|\uparrow\rangle_s + \psi_\downarrow|\downarrow\rangle_s \\ \rightarrow\; & \psi_\uparrow|\uparrow\rangle_s|\uparrow\rangle_{o_1} + \psi_\downarrow|\downarrow\rangle_s|\downarrow\rangle_{o_1} \\ \rightarrow\; & |\rightarrow\rightarrow\rangle_{so_2}\left(\psi_\uparrow|\uparrow\rangle_{o_1} + \psi_\downarrow|\downarrow\rangle_{o_1}\right) + |\leftarrow\leftarrow\rangle_{so_2}\left(\psi_\uparrow|\uparrow\rangle_{o_1} - \psi_\downarrow|\downarrow\rangle_{o_1}\right) \\ \rightarrow\; & (\psi_\uparrow + \psi_\downarrow)|\rightarrow\rightarrow\rangle_{so_2}|\rightarrow\rightarrow\rangle_{o_1o_3'} + (\psi_\uparrow - \psi_\downarrow)|\leftarrow\leftarrow\rangle_{so_2}|\rightarrow\rightarrow\rangle_{o_1o_3'} \\ & +(\psi_\uparrow - \psi_\downarrow)|\rightarrow\rightarrow\rangle_{so_2}|\leftarrow\leftarrow\rangle_{o_1o_3'} + (\psi_\uparrow + \psi_\downarrow)|\leftarrow\leftarrow\rangle_{so_2}|\leftarrow\leftarrow\rangle_{o_1o_3'} \end{aligned} \tag{17}$$

and it is seen that there are instances (those highlighted in red in Equation (17)) where the different observers o_2 and o_3' do not agree on the state of the signal. Objectivity is lost! Figure 5 summarizes the argument.

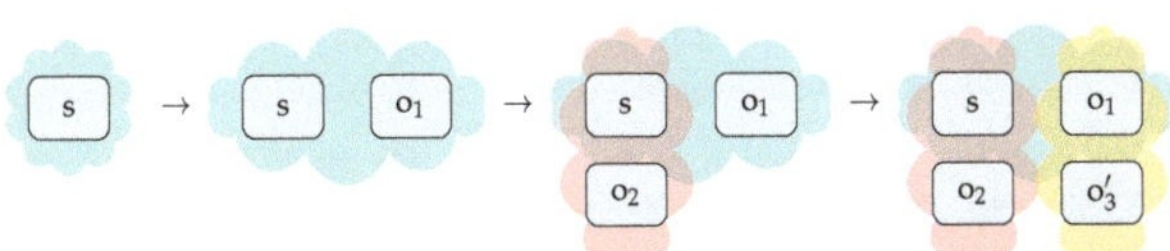

Figure 5. Measurement in a different basis is no longer objective. Different observers may disagree on what consititutes reality. Transparent clouds over the same quantum system indicate its being observed by observers in different bases.

However, in case observer o_1 is classical, it is further observed by other observers, and this creates a network of observers, which allows recovery of the information regarding measurement of the signal (Figure 6). The mathematics is worked out in Appendix A.

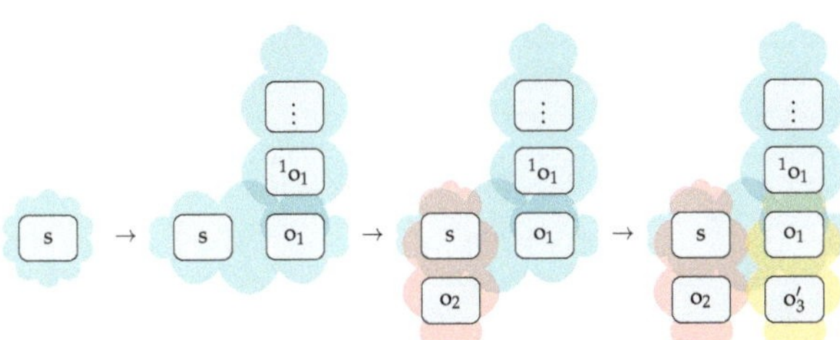

Figure 6. Due to the redundant information stored in the environment, it is possible to recover information about the environment and thereby allow for the signal to once again be determined objectively.

It could be argued that what is made objective is not the state of the signal but rather the measurement of the signal by the observer. In this sense, it is the whole network of observer states that is assigned the adjective "objective". This, of course, implies that the state of the signal is also objective in the sense that there is now a redundant record of the measurement, as in Figure 4, such that all observers within that network agree with one another on its state. Moreover, this information is safeguarded against measurements of the signal in different bases due to the redundancy of information stored in the network of observers. In this sense, the network of Figure 5 is unstable, whereas that of Figure 6 is stable.

5. Discussion

We begin with an assumption of purely unitary quantum dynamics in Section 2 and explain the limits it imposes on the quantum measurement procedure. The assumption of unitarity along with that of an arbitrary initial state for signal and observer necessitates the use of an environment into which the observer can dispose its initial state. Extending the arbitrariness of the initial state to the environment (and along with the usual assumption of unitarity), it is seen that the environment can influence the measurement procedure and leave a trace therein (Equation (8)).

This implies the necessity of redundant information about the environment so that the influence of the environment can be undone. We therefore describe a correlated environment (Equation (12)) which is redundantly encoded and which can help correct for the environmental influence. With this addition, we describe a corrected measurement procedure (Equations (13) and (14)), which results in perfect correlation between signal and observer. It is also seen that one can come up with a measure of this correlation and that it is a conserved resource used up in the quantum measurement procedure. These considerations make up Section 3.

Finally, in Section 4, we describe how, using the considerations of the previous sections, the emergence of an objective classical reality can be argued for. It is seen that redundancy of information about the measurement procedure is key, as it is the agreement of different observers regarding the state of affairs of the world that is described as objective classical reality [2–4]. Stability of this classical reality is argued to be due to the formation of highly branched structures, such as in Figure 4, which makes it difficult to undo this correlation.

The current article discusses a particular example in detail in order to highlight the main ideas of our approach. A related article [2] is a more abstract analysis and addresses other issues in more detail. One of the issues we do not address in either article is how a correlated environment such as in Equation (12) emerges in the first place. While one might obtain a hint of an answer in Section 4, detailed analysis is required in order to properly address this issue. The issue of basis independence is also something we gloss over. While the state of signal, observer, and environment are all basis-independent and so is the swap operation (Equation (10)), the imprint operation (Equation (9)) is basis-dependent. Therefore, the issue of preferred basis brought up in Section 2 is not completely addressed. How/whether the emergence of basis dependence in the imprint operation relates to the emergence of a correlated environment is another interesting question for future research.

Author Contributions: Conceptualization, V.J., P.F. and T.E.; validation, V.J., P.F. and T.E.; formal analysis, V.J.; investigation, V.J.; writing—original draft preparation, V.J.; writing—review and editing, V.J., P.F. and T.E; visualization, V.J. and T.E.; supervision, T.E. All authors have read and agreed to the published version of the manuscript.

Funding: Philipp Frank acknowledges funding through the German Federal Ministry of Education and Research for the project ErUM-IFT: Informationsfeldtheorie für Experimente an Großforschungsanlagen (Förderkennzeichen: 05D23EO1).

Institutional Review Board Statement: Not applicable.

Informed Consent Statement: Not applicable.

Data Availability Statement: No new data was generated.

Acknowledgments: We thank our co-author in [2], Reimar Leike, for his discussion and advice on the background work leading up to this article.

Conflicts of Interest: The authors declare no conflict of interest.

Appendix A

The mathematics following Figure 6 proceeds as follows:

$$\psi_\uparrow|\uparrow\rangle_s+\psi_\downarrow|\downarrow\rangle_s \to \psi_\uparrow|\uparrow\rangle_s \overset{\overset{\vdots}{|\uparrow\rangle_{1_{o_1}}}}{|\uparrow\rangle_{o_1}} + \psi_\downarrow|\downarrow\rangle_s \overset{\overset{\vdots}{|\downarrow\rangle_{1_{o_1}}}}{|\downarrow\rangle_{o_1}} \tag{A1}$$

where state $\overset{\overset{\vdots}{|i\rangle_{1_{o_1}}}}{|i\rangle_{o_1}}$ indicates that observer o_1 is a highly branched state which is observed by several consequent observers. The observation in different bases then proceeds as follows:

$$\to |\rightarrow\rightarrow\rangle_{so_2}\Big(\psi_\uparrow \overset{\overset{\vdots}{|\uparrow\rangle_{1_{o_1}}}}{|\uparrow\rangle_{o_1}} + \psi_\downarrow \overset{\overset{\vdots}{|\downarrow\rangle_{1_{o_1}}}}{|\downarrow\rangle_{o_1}}\Big) + |\leftarrow\leftarrow\rangle_{so_2}\Big(\psi_\uparrow \overset{\overset{\vdots}{|\uparrow\rangle_{1_{o_1}}}}{|\uparrow\rangle_{o_1}} - \psi_\downarrow \overset{\overset{\vdots}{|\downarrow\rangle_{1_{o_1}}}}{|\downarrow\rangle_{o_1}}\Big) \tag{A2}$$

$$\begin{aligned}\to\ & \Big(\overset{\overset{\vdots}{|\uparrow\rangle_{1_{o_1}}}}{\psi_\uparrow} + \overset{\overset{\vdots}{|\downarrow\rangle_{1_{o_1}}}}{\psi_\downarrow}\Big)|\rightarrow\rightarrow\rangle_{so_2}|\rightarrow\rightarrow\rangle_{o_1o_3'} + \Big(\overset{\overset{\vdots}{|\uparrow\rangle_{1_{o_1}}}}{\psi_\uparrow} - \overset{\overset{\vdots}{|\downarrow\rangle_{1_{o_1}}}}{\psi_\downarrow}\Big)|\leftarrow\leftarrow\rangle_{so_2}|\rightarrow\rightarrow\rangle_{o_1o_3'} \\ & +\Big(\overset{\overset{\vdots}{|\uparrow\rangle_{1_{o_1}}}}{\psi_\uparrow} - \overset{\overset{\vdots}{|\downarrow\rangle_{1_{o_1}}}}{\psi_\downarrow}\Big)|\rightarrow\rightarrow\rangle_{so_2}|\leftarrow\leftarrow\rangle_{o_1o_3'} + \Big(\overset{\overset{\vdots}{|\uparrow\rangle_{1_{o_1}}}}{\psi_\uparrow} + \overset{\overset{\vdots}{|\downarrow\rangle_{1_{o_1}}}}{\psi_\downarrow}\Big)|\leftarrow\leftarrow\rangle_{so_2}|\leftarrow\leftarrow\rangle_{o_1o_3'}.\end{aligned} \tag{A3}$$

In this case, observers o_2 and o_3' continue to disagree about the measured state of the signal. However, the redundant record maintained in states $|i\rangle_{1_{o_1}}\cdots$ allows other independent observers to query and infer the previously measured state of the signal. Therefore, the redundancy in recording the measurement outcomes for observer o_1 allows the formation of a stable objective classical reality for the measurement of signal s despite the disagreement between observers o_2 and o_3'. The network of states $|i\rangle_{1_{o_1}}\cdots$ contains redundant information about the measurement of signal s by observer o_1. Redundancy is key!

References

1. Everett, H. "Relative State" Formulation of Quantum Mechanics. *Rev. Mod. Phys.* **1957**, *29*, 454–462. [CrossRef]
2. Johnson, V.; Leike, R.; Frank, P.; Enßlin, T. Measurement in a Unitary World. *arXiv* **2023**, arXiv:2212.03829.
3. Horodecki, R.; Korbicz, J.K.; Horodecki, P. Quantum origins of objectivity. *Phys. Rev. A* **2015**, *91*, 032122. [CrossRef]
4. Zurek, W.H. Quantum Darwinism, classical reality, and the randomness of quantum jumps. *Phys. Today* **2014**, *67*, 44–50. [CrossRef]
5. Hemmo, M.; Shenker, O. The Preferred Basis Problem in the Many-Worlds Interpretation of Quantum Mechanics: Why Decoherence Does Not Solve It. *Synthese* **2022**, *200*, 261. [CrossRef]
6. Pati, A.K.; Braunstein, S.L. Impossibility of deleting an unknown quantum state. *Nature* **2000**, *404*, 164–165. [CrossRef]

physical sciences forum

MDPI

Proceeding Paper

Snowballing Nested Sampling †

Johannes Buchner

Max Planck Institute for Extraterrestrial Physics, Giessenbachstrasse, 85748 Garching, Germany; jbuchner@mpe.mpg.de

† Presented at the 42nd International Workshop on Bayesian Inference and Maximum Entropy Methods in Science and Engineering, Garching, Germany, 3–7 July 2023.

Abstract: A new way to run nested sampling, combined with realistic MCMC proposals to generate new live points, is presented. Nested sampling is run with a fixed number of MCMC steps. Subsequently, snowballing nested sampling extends the run to more and more live points. This stabilizes the MCMC proposal of later MCMC proposals, and leads to pleasant properties, including that the number of live points and number of MCMC steps do not have to be calibrated, that the evidence and posterior approximation improve as more compute is added and can be diagnosed with convergence diagnostics from the MCMC community. Snowballing nested sampling converges to a "perfect" nested sampling run with an infinite number of MCMC steps.

Keywords: nested sampling; Markov Chain Monte Carlo

1. Introduction

Nested sampling [1] is a popular [2] Monte Carlo algorithm for Bayesian inference, estimating the posterior distribution and the Bayesian evidence. A set of K live points are initially drawn from the prior, and their likelihoods are evaluated. In each iteration of the algorithm, the live point with the lowest likelihood is discarded and becomes a dead point. It is replaced with a new point sampled from the prior, under the constraint that the likelihood of the discarded point is exceeded. This likelihood-restricted prior sampling (LRPS) can be achieved in various ways (see Buchner [3] for an extensive review).

Skilling [1] originally proposed Markov Chain Monte Carlo (MCMC) random walk algorithms for the LRPS. This is still the most popular choice for high-dimensional problems, including slice sampling variants (e.g., [4]), recently compared for efficiency and reliability in [5]. Such algorithms start from a randomly chosen live point and make M transitions to yield the new live point. To avoid a collapse of the live point set, and sufficiently explore the likelihood-restricted prior, M has to be chosen large enough. However, as Salomone et al. [6] discusses, mild correlations between live points are acceptable. Thus, a suitably efficient LRPS algorithm has to be chosen by the user, as well as the size of the live points, K, which regulates the resolution of the sampling and the uncertainty on the outputs. Since M is not known beforehand, to obtain reliable outputs [7] recommended to repeatedly run nested sampling with ever-doubling M until the Bayesian evidence stabilizes. However, this may discard information built with substantial computational resources.

This work highlights that there is an interplay between M and K: If the number of live points is large, nested sampling discards only a small fraction of the prior in each iteration ($1/K$ of its probability mass), and a large set of live points is representing the likelihood distribution of the restricted prior. The higher K, and the slower nested sampling traverses the likelihood, the more acceptable correlations between points should be.

At termination, nested sampling provides posterior samples from the dead points. Each sampled point is weighted by the likelihood times the prior volume discarded at that iteration. Because nested sampling iterates with a fixed speed through the prior, the number of samples with non-negligible weight may be low. Two solutions have been proposed for this: (1) to keep intermediate "phantom" points [4] from the MCMC chains to represent the dead point, instead of a single estimate. A drawback here is memory storage. (2) to dynamically widen K in a selected likelihood interval (dynamic nested sampling, [8]). The latter can rerun likelihood intervals where most of the posterior weight is, bulking the posterior samples. The uncertainty on the evidence is dominated by the stochastic sampling of a limited number of K live points up to this point. Therefore, for enhanced precision on the evidence, all iterations have to be extended to more live points.

In this work, we present a relatively derivative form of nested sampling which (1) avoids choosing M and K carefully beforehand, (2) does not discard previous computations, (3) achieves an arbitrarily large effective number of posterior samples as compute increases, and (4) can be diagnosed for convergence despite an imperfect LRPS.

2. Method

Instead of improving nested sampling convergence with a limited Markov Chain by increasing the number of steps M, we increase the number of live points K.

2.1. Basic Algorithm

The basic algorithm proposed here, Snowballing Nested Sampling, is listed below (Algorithm 1).

Algorithm 1: Basic Snowballing Nested Sampling

```
K ← 1
Loop:
    run Nested Sampling with K live points
    K ← K + 1
```

At each iteration, this algorithm produces nested sampling results (evidence, posterior samples). "Running nested sampling" means running with a MCMC LRPS with a fixed number of steps, M, until termination. For example, one can adopt a nested sampling termination criterion that makes the integral contribution remaining in the live points negligible (e.g., $Z_{\text{live}}/Z_i < 10^{-6}$).

2.2. Convergence Analysis

The rate of compression in (vanilla) nested sampling is $\frac{K-1}{K}$ [9]. Thus, after t iterations, the likelihood-restricted prior has volume $\left(\frac{K-1}{K}\right)^t$. The same restricted prior volume can be reached after $t' = 2 \times t$ iterations with $K' = 1 + \sqrt{1 - 1/K} \approx 2$ times the number of live points. With twice the number of iterations, there are twice as many dead points. Loosely speaking, each dead point from a run with K is split into two dead points from a run with K' live points. The noise in the nested sampling estimate due to more dead points is also reduced, because the variance is inversely proportional to K [10,11]. However, here, we focus on analyzing the convergence requirements of the LRPS.

We assume an MCMC random walk with a fixed number of steps M is adopted as an LRPS method. With the restricted prior compressing twice as slowly, and each dead point being split into two dead points, naturally, we find the following equivalence: A nested sampling run with M steps and K live points is comparable to a run with $K \times 2$ live points but $M/2$ steps, in terms of correlations between live points that could be problematic for nested sampling. Based on this, $K \to \infty$ with fixed M achieves the properties of $M \to \infty$ with fixed K, i.e., a perfect LRPS algorithm for vanilla nested sampling. For the latter case, Ref. [9] proved unbiasedness and convergence of Z and the posterior distribution. A perfect LRPS algorithm is reached with a proposal kernel that is unbounded (or has the support of the prior), as $M \to \infty$. Therefore, the results of [9] also hold for snowballing NS, as $K \to \infty$.

2.3. Improved Algorithm

For practical purposes, the algorithm above can be improved slightly. Some LRPS methods need at least a few times the dimensionality of the parameter to build effective transition kernels. Therefore, the iterations with the lowest K can be skipped, and we start at K_0. Since there may be some computation overhead beyond model evaluations in nested sampling, K can be incremented in steps of K_{inc}.

It is also not necessary to discard runs from previous iterations and start from scratch. Firstly, a subsequent nested sampling run can be initialized with the same initial prior samples as in the previous iterations, and K_{inc} additional ones. Subsequently in each nested sampling iteration, a live point is discarded, and replaced with a new point. In continuous likelihood functions without plateaus, the call to the LRPS at each dead point, creating a new prior sample based on a threshold L_{min}, has a uniquely identifiable input, L_{min}. Therefore, the LRPS call can be memoized. That is, the result of the function is stored with the argument, L_{min}, as its identifier, and when later the function evaluation with the same argument is needed, the result is returned based on a look-up. Only if the identifier is not stored yet does the LRPS call at that L_{min} needs to be made. See [5,7,12,13] for previous work on resuming or combining nested sampling runs.

Put together, the improved algorithm is listed below as Algorithm 2:

Algorithm 2: Basic Snowballing Nested Sampling

$K \leftarrow K_0$
run Nested Sampling with K0 live points
Loop:
  $K \leftarrow K + K_{\mathrm{inc}}$
  resume Nested Sampling with K live points

3. Results

We test the algorithm on the Rosenbrock function over a flat prior from -10 to +10 in $d = 20$ dimensions. The likelihood is:

$$\ln \mathcal{L}(\theta) = -2 \times \sum_i \left(100 \times \left(\theta_{i+1} - \theta_i^2\right)^2 + (1 - \theta_i)^2\right)$$

With $K_0 = 20$, and $K_{\mathrm{inc}} = 20$, and $M = 20$, we observe that $\ln(Z)$ evolves as shown in Figure 1. From an initial high value, the estimate declines approximately proportional to $1/(M \times K)$ towards the (presumably true) value if nested sampling was run with $K = \infty$.

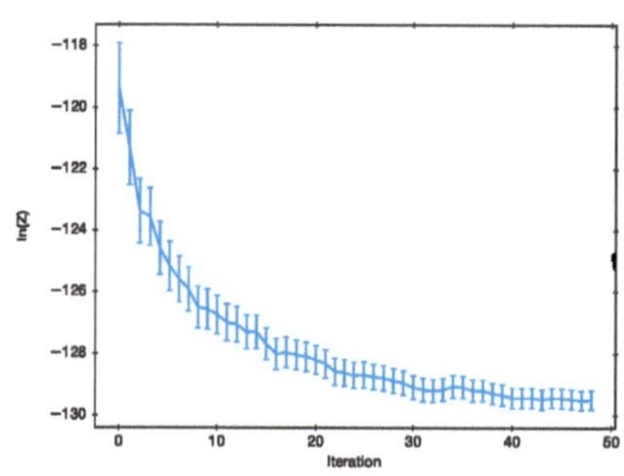

Figure 1. Estimate of $\ln(Z)$ for each algorithm iteration.

4. Discussion

We have presented an iterative form of applying nested sampling. It can be understood as a kind of dynamic nested sampling. The proposed algorithm has several amicable properties:

1. the number of live points and the number of steps do not have to be chosen carefully,
2. an initial quick look result is available,
3. the evidence estimate improves as more compute is added,
4. in multi-modal settings, additional peaks are discovered as more compute is added,
5. the convergence can be observed and tested with conventional methods based on MCMC,
6. the posterior sample size increases proportionally to the compute,
7. it is easy to implement.

The convergence as more compute is added, the revisiting and further exploration of missed peaks, and the possibility to use conventional MCMC convergence diagnostics, are properties shared with diffusive nested sampling [14]. That algorithm is based on likelihood levels. Instead of monotonically switching to higher likelihoods, the sampler can randomly go up or down in levels, and add more samples (live points) there. Unlike diffusive nested sampling, snowballing nested sampling provides an uncertainty estimate on the evidence Z.

Finally, we make some remarks on adaptive MCMC algorithms. Salomone et al. [6] pointed out that on-the-fly adaptation of MCMC algorithms is problematic, and a warm-up nested sampling run should be performed, the adapted proposal kernel (parameters) at each iteration stored, and applied without adaptation in a second, fresh nested sampling run. If the MCMC transition kernel is created based on estimating a sample covariance matrix of the live point population, this estimate stabilizes as K increases. Newly added K_{inc} live points only mildly alter the estimate as $K \to \infty$. Another popular adaptation is to alter the proposal scale in Metropolis random walks to target a certain acceptance rate. If the proposal scale is chosen based on the acceptance rate of the previous K points and their proposal scale, through a weighted mean targeting an acceptance rate of 23.4 percent, then we have a situation of a vanishing adaptation see [15] with convergence to the true posterior. Furthermore, as the sampling is more faithful, the live points are also more faithfully distributed, which in turn improves the sampling, leading to a positive feedback effect. This accelerates the convergence further.

Funding: This research received no external funding.

Institutional Review Board Statement: Not applicable.

Informed Consent Statement: Not applicable.

Data Availability Statement: Data and code are available on request.

Acknowledgments: I thank Robert Salomone and Nicolas Chopin for feedback on an early draft, and the two anonymous referees for helpful comments.

Conflicts of Interest: The author declares no conflict of interest.

References

1. Skilling, J. Nested sampling. *AIP Conf. Proc.* **2004**, *735*, 395. [CrossRef]
2. Ashton, G.; Bernstein, N.; Buchner, J.; Chen, X.; Csányi, G.; Fowlie, A.; Feroz, F.; Griffiths, M.; Handley, W.; Habeck, M.; et al. Nested sampling for physical scientists. *arXiv* **2022**, arXiv:2205.15570.
3. Buchner, J. Nested Sampling Methods. *arXiv* **2021**, arXiv:2101.09675.
4. Handley, W.J.; Hobson, M.P.; Lasenby, A.N. POLYCHORD: Nested sampling for cosmology. *MNRAS* **2015**, *450*, L61–L65. [CrossRef]
5. Buchner, J. Comparison of Step Samplers for Nested Sampling. *Phys. Sci. Forum* **2022**, *5*, 46. [CrossRef]
6. Salomone, R.; South, L.F.; Drovandi, C.C.; Kroese, D.P. Unbiased and Consistent Nested Sampling via Sequential Monte Carlo. *arXiv* **2018**, arXiv:1805.03924.
7. Higson, E.; Handley, W.; Hobson, M.; Lasenby, A. NESTCHECK: Diagnostic tests for nested sampling calculations. *MNRAS* **2019**, *483*, 2044–2056. [CrossRef]
8. Higson, E.; Handley, W.; Hobson, M.; Lasenby, A. Dynamic nested sampling: An improved algorithm for parameter estimation and evidence calculation. *arXiv* **2017**, arXiv:1704.03459.
9. Chopin, N.; Robert, C.P. Properties of nested sampling. *Biometrika* **2010**, *97*, 741–755. [CrossRef]

10. Chopin, N.; Robert, C. *Contemplating Evidence: Properties, Extensions of, and Alternatives to Nested Sampling*; Technical repor Technical Report 2007-46; CEREMADE, Université Paris Dauphine: Paris, France, 2007.
11. Skilling, J. Nested sampling's convergence. *AIP Conf. Proc.* **2009**, *1193*, 277–291. [CrossRef]
12. Skilling, J. Nested sampling for general Bayesian computation. *Bayesian Anal.* **2006**, *1*, 833–859. [CrossRef]
13. Speagle, J.S. DYNESTY: A dynamic nested sampling package for estimating Bayesian posteriors and evidences. *MNRAS* **202** *493*, 3132–3158. [CrossRef]
14. Brewer, B.J.; Pártay, L.B.; Csányi, G. Diffusive nested sampling. *Stat. Comput.* **2011**, *21*, 649–656. [CrossRef]
15. Andrieu, C.; Thoms, J. A tutorial on adaptive MCMC. *Stat. Comput.* **2008**, *18*, 343–373. [CrossRef]

physical sciences forum

MDPI

Proceeding Paper

A BRAIN Study to Tackle Image Analysis with Artificial Intelligence in the ALMA 2030 Era †

Fabrizia Guglielmetti [1,*], Michele Delli Veneri [2], Ivano Baronchelli [3,4], Carmen Blanco [5], Andrea Dosi [6], Torsten Enßlin [5], Vishal Johnson [5], Giuseppe Longo [6], Jakob Roth [5,7], Felix Stoehr [1], Łukasz Tychoniec [1] and Eric Villard [1]

1 European Southern Observatory, Karl-Schwarzschild-Straße 2, 85748 Garching, Germany; Lukasz.Tychoniec@eso.org (Ł.T.); evillard@eso.org (E.V.)
2 National Institute of Nuclear Physics, Section of Naples (INFN), Via Cinthia, 80126 Naples, Italy; delliven@na.infn.it
3 INAF—Institute of Radio Astronomy, Via Gobetti 101, 40129 Bologna, Italy; ivano.baronchelli@inaf.it
4 Italian ALMA Regional Centre, Via Gobetti 101, 40129 Bologna, Italy
5 Max Planck Institute for Astrophysics, Karl-Schwarzschild-Straße 1, 85748 Garching, Germany; carmenbp26@gmail.com (C.B.); ensslin@mpa-garching.mpg.de (T.E.); vishal@mpa-garching.mpg.de (V.J.); roth@mpa-garching.mpg.de (J.R.)
6 Department of Physics Ettore Pancini, Federico II University, Via Cinthia 26, 80126 Naples, Italy; andrea.dosi@unina.it (A.D.); giulongo@unina.it (G.L.)
7 Department for Computer Science, Technical University of Munich, Boltzmannstr. 3, 85748 Garching, Germany
* Correspondence: fgugliel@eso.org
† Presented at the 42nd International Workshop on Bayesian Inference and Maximum Entropy Methods in Science and Engineering, Garching, Germany, 3–7 July 2023.

Abstract: An ESO internal ALMA development study, BRAIN, is addressing the ill-posed inverse problem of synthesis image analysis, employing astrostatistics and astroinformatics. These emerging fields of research offer interdisciplinary approaches at the intersection of observational astronomy, statistics, algorithm development, and data science. In this study, we provide evidence of the benefits of employing these approaches to ALMA imaging for operational and scientific purposes. We show the potential of two techniques, RESOLVE and DeepFocus, applied to ALMA-calibrated science data. Significant advantages are provided with the prospect to improve the quality and completeness of the data products stored in the science archive and the overall processing time for operations. Both approaches evidence the logical pathway to address the incoming revolution in data rates dictated by the planned electronic upgrades. Moreover, we bring to the community additional products through a new package, ALMASim, to promote advancements in these fields, providing a refined ALMA simulator usable by a large community for training and testing new algorithms.

Keywords: inverse problems; inference methods; Bayesian probability theory; machine learning; image analysis; radio astronomy

1. Introduction

Since its inception in 2011, the Atacama Large Millimeter/submillimeter Array (ALMA) (Figure 1) has been at the forefront of astronomical research (see e.g., [1]), consistently delivering groundbreaking scientific discoveries with its versatile observatory capabilities. In 2018, the ALMA Development Roadmap provided a forward-thinking perspective to increase the observatory's capabilities out to 2030 (ALMA2030) [2,3]. Scientifically motivated by fostering the understanding of the origin of galaxies, the origin of chemical complexity, and the origins of planets, ALMA is going to be enhanced to address (1) larger bandwidths and better receiver sensitivity, (2) improvements to the ALMA Science Archive, and (3) longer baselines. In addition, the observatory is currently expanding the frequency coverage to 35–950 GHz with two new receivers (band 1 and band 2) to capture a broader range of celestial objects and phenomena.

Figure 1. ALMA antennas on the Chajnantor plateau. Credit: ESO.

The ALMA2030 Roadmap [2] encourages innovative technical concepts rooted in astronomy, even if they are not explicitly listed within the document. In this sense, the ESO internal ALMA development study BRAIN fits into the original strategic vision of ALMA2030. In this study, synthesis imaging algorithms capable of learning from ALMA data are developed and tested. The powerful attribute of modern artificial intelligence systems enables algorithms to make data-driven decisions, discover insights, and provide predictions. We include the exploration of algorithms that can leverage graphics processing units (GPUs) to maximize computational power, achieving high performance. GPUs allow for more efficient and cost-effective computing solutions and open up new possibilities for solving complex problems with large data volumes. Moreover, we provide to the community a database and software to create ALMA-simulated data to promote the development of new synthesis imaging algorithms.

Data Rate and Data Volume Incrementing with ALMA2030

The currently top priority of the ALMA2030 vision is to broaden the receiver's intermediate frequency (IF) bandwidth by a factor of 2–4 (currently $\sim$8 GHz) while upgrading the associated electronics and the correlator. A wider bandwidth will allow the observatory to capture more data in a shorter amount of time, accelerating data collection and analysis. Upgrades to the receivers and electronics improve the sensitivity, with the aim to detect fainter signals and gather more precise data. Efforts are made to reduce system noise, interference, and instrumental effects in order to refine the signal-to-noise ratio. These enhancements have been officially designated as the ALMA2030 Wideband Sensitivity Upgrade (WSU) [4].

The development of the ALMA second-generation correlator, the improved receivers, the wider bandwidth, and the transmission system will enable a substantial increase in data volume during a single observation and up to 10 times more observing speed efficiency. When ALMA collected 1.7 PB of visibility data and products until Cycle 9, it is expected that ALMA2030 might augment the visibility data by up to a factor of 100. Secondly, the ALMA Science Archive requires improvements in order to handle the data from the increased bandwidth in addition to enhancement of the user-friendliness of the archive and data products. Thirdly, the maximum baselines may be increased by a factor of 2–3, and the number of antennas will grow up to 16%. Consequently, the ALMA resolution and UV coverage will be enriched from today's several possible solutions because of the longest baselines and more efficient observational strategies.

The expanded bandwidth directly leads to enhancements in both the continuum sensitivity and spectral coverage. While the exact capabilities that can be offered are not yet known, an increase in the data rate of between 10 and 100 times is envisaged. Due to the increased sensitivity and data volume, continuum identification can become an awkward procedure, especially for the potential incrementing of spectral line detection. During the ramp-up years of the WSU, observations will occur with a lower number of antennas than employed today. In this time frame, sparse sampling may affect the data. To handle the increased volume and complexity of the data, data processing and analysis tools have to be strategically developed.

2. Methods and Results: Artificial Intelligence for Synthesis Imaging with BRAIN

The ESO internal ALMA development study named BRAIN was proposed in late 2019 and will conclude by the end of 2023. BRAIN stands for Bayesian Reconstruction through Adaptive Image Notion. Alternatives to commonly used applications in image processing have been sought and tested [5,6]. This study explores the astrostatistics and astroinformatics approaches suited to the ALMA2030 era. Astrostatistics and astroinformatics are essential for advancing our understanding of the universe and for making sense of the enormous volumes of data collected by astronomical instruments and observatories [7].

Astrostatistics involves the application of statistical analysis, data modeling, and probability theory to solve problems, perform predictions, and provide robust uncertainty quantification. By extracting meaningful insights from astronomical data, astrostatistics enhances the quality and reliability of astronomical research by providing quantitative and data-driven approaches.

Astroinformatics focuses on the management, analysis, and interpretation of large and complex astronomical datasets, combining elements of astronomy, computer science, and information technology. This field plays a crucial role in executing operations of large observatories in real time. It can support operations for sanity checks on the health of the observatory, as well as real-time image processing.

In this study, the information field theory [8] approach RESOLVE [9–11] and the astroinformatics technique DeepFocus [12] are applied to ALMA-calibrated data. A companion package of DeepFocus, named ALMASim, was developed with the additional scope to provide simulated data as well as open software for the scientific community. ALMASim enables training and testing supervised machine learning (ML) algorithms as well as reliability and quality assessments and comparisons of several techniques.

2.1. Resolve

The algorithm works on ALMA-calibrated measurement sets in the UV space. The input data d are modeled as a combination of log-celestial signals s corrupted by a dirty beam R (i.e., the spatial response pattern of the telescope array during an observation) and the noise n (systematic and random errors), where $d = Re^s + n$. The process of synthesizing an image involves estimating the posterior probability density function (pdf) of potential true sky signal configurations using variational inference, where $P(s|d) = \frac{e^{-H(s,d)}}{Z(d)}$. For more details on the RESOLVE algorithm, see [9–11] and the references therein. The RESOLVE algorithm outputs posterior samples of sky images and power spectra. From these posterior samples, summary statistics such as a mean sky map or an uncertainty map can be computed. The first application of RESOLVE to ALMA data is shown in [10]. In Figure 2, an application of RESOLVE on a protoplanetary disk sample, Elias 27, from the ALMA Disk Substructures at High Angular Resolution Project (DSHARP) [13,14] is shown. In Figure 2, image (A) is the fiducial image of the continuum detection of Elias 27 from the DSHARP data release [14,15], with a self-calibrated image employing CASA [16]. Images (B) and (C) are the outcomes of RESOLVE's application to Elias 27 continuum ALMA data, showing the mean sky map and the uncertainty map, respectively. The DSHARP project was aimed at detecting a large number of protoplanetary disks with a high resolution (35 mas) to reveal the amplitudes of small-scale substructures in the distributions of the disk material and understand their relation to the planet-forming process.

RESOLVE is currently applied to other science cases such as as self-calibration, full polarization, single-dish imaging, as well as VLBI-based filming of evolving black hole environments [17,18]. Those applications should be extended to ALMA data.

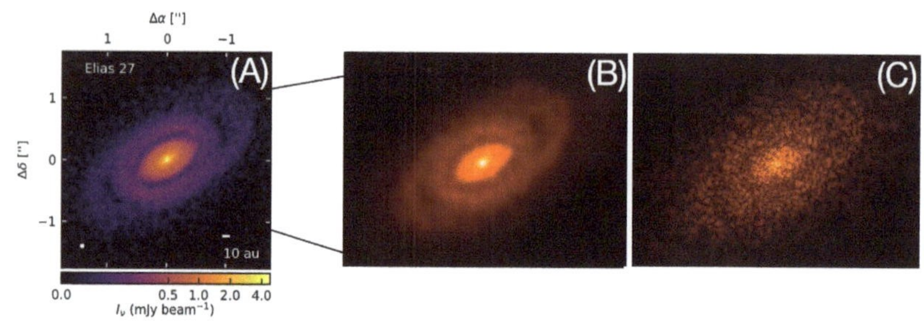

Figure 2. Application of RESOLVE to Elias 27 from the DSHARP ALMA project at 240 GHz (1.25 mm) continuum. (**A**) The fiducial image as given by the DSHARP team [14]. (**B**) RESOLVE mean sky map of Elias 27. (**C**) RESOLVE uncertainty map representation.

2.2. DeepFocus

The deep learning pipeline DeepFocus was described in [6,12]. It performs deconvolution, source detection, and characterization. From the original development targeted to detect faint compact objects, the deconvolution algorithm has been improved to detect extended emissions with a new ML model, the meta-learner, which is capable of exploring different architectures (e.g., CAE-VAE, U-Net, and Res-Net) and assisting in model selection to predict the best-performing architecture given a task and a set of interferometric data. A Bayesian optimization algorithm is used for model selection. The procedure is supported by taxonomy. This framework allows one to include several model architectures, hyperparameters, and evaluation metrics that are specific to the problem, data, and desired performance criteria. Multiple parameter realizations are tested in parallel, and a subsample of the original problem is used to measure the performance. Once the optimal architecture is chosen according to the image data, the deep learning pipeline performs deconvolution and denoising, focusing, and classification tasks [12].

The Bayesian optimization algorithm is supported by a parameter search employing surrogate models. The best set of parameters for a given objective is efficiently found while reducing the number of expensive model evaluations. The expensive objective function (deep learning model) is approximated by its surrogate model, as described by Gaussian processes (GPs), where $y = GP(\mu, \Sigma)$ with $\mu(x)$ and $\Sigma(x, x')$ being mean and covariance functions, respectively. The Matérn kernel is chosen as a covariance function to capture smoothness and correlations in the data, where $\Sigma = K(x, x') = \sigma^2(1 + \frac{(x-x')^2}{2\alpha l})^{-\alpha}$ with α and l representing the smoothness and length scale parameters, respectively. The surrogate model is trained on a limited number of initial evaluations of the objective function. In order to target promising regions in the parameter space, an acquisition function is used to decide which set of parameters is evaluated next. In this development, the expected improvement (EI) is used as an acquisition function, where $EI(x) = E[\max(f(x) - f(x'), 0)]$ with $f(x)$ and $f(x')$ indicating the predicted mean of the objective function at point x, based on the GP, and the best value observed thus far in the optimization process, respectively. The EI quantifies the potential gain in performance. It prefers models for which $\mu > f(x')$ (exploitation) or where the standard deviation $\sigma(x)$ is high (exploration). As already shown in [19], the key advantage of using surrogate models in Bayesian parameter searching is that it reduces the number of costly evaluations of the objective function.

Benchmark on Archived ALMA Data Cubes

The European ALMA Regional Centre (EU ARC) cluster is frequently utilized for processing the standard pipeline and generating official products delivered to principal investigators. Using the EU ARC cluster, Deep Focus was executed on 29×10^3 archived ALMA cubes taken during the three most recent cycles (7–9; see Figure 3). We observed an average processing time of 1.13 min per cube and an average compute throughput of 140 MB/s. The same data were cleaned by the tCLEAN algorithm [16,20] (see also [5,10]).

The tCLEAN algorithm was executed with the parameter niter = 1000, corresponding to the number of cleaning iterations, employing parallel computing. The average computing throughput with tCLEAN was 0.56 MB/s, showing a rate of performance 250 times lower for calculations with respect to the ML algorithm DeepFocus. The image processing speed improvement achieved by DeepFocus varied significantly depending on the image size, ranging from 280 fold to a remarkable 5500 fold increase in speed. When employing algorithms such as DeepFocus, the image deconvolution process may take only few minutes on large cubes. GPUs, which are commonly used by ML algorithms, provide a large benefit for synthesis image analysis. DeepFocus demonstrated high image fidelity and high-performance computing for image reconstruction on the ALMA data cubes.

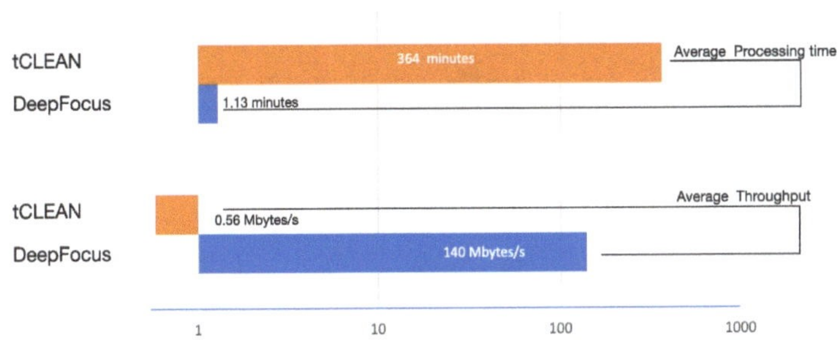

Figure 3. Comparison of processing time and computing throughput with tCLEAN and DeepFocus on 29×10^3 archived cube data from cycles 7, 8, and 9. This represents a rough estimate because at this stage of development, it is challenging to make a robust comparison between the two techniques.

2.3. *A Refined ALMA Simulator: ALMASim*

In [6,12], it was shown that DeepFocus is applied to ALMA dirty cubes, learning from the input data of the celestial sources, the noise, and the instrumental point spread function (beam). The algorithm allows for extreme data compression by leveraging both spatial and frequency information. With DeepFocus being a supervised ML algorithm, the process to build and evaluate the models is essential. To support the needed workflow, the ALMASim package was developed.

ALMASim is an extension of the CASA Simulator package [16]. We increased the capabilities of this simulator to be tailored toward the development of ML imaging algorithms and for quality and reliability assessments. The novel package builds upon the CASA PiP Wheels, the MARTINI Package [21], and the Illustris Python Package [22] to facilitate the creation of observations involving high-redshift point-like sources as well as nearby extended sources across the entire spectrum of ALMA configurations.

Following the CASA Simulator, the observational input parameters (precipitable water vapor, band, antenna configuration, bandwidth, integration time, and scan time) were provided with the source properties (signal-to-noise ratio and peak brightness). ALMASim introduces several optional models for the science target brightness distribution: diffuse, point-like, Gaussian, and extended (Figure 4). For the diffuse emission model, the Numerical Information Field Theory package (NIFTy) [23] was used. The simulated science target distributions entered into the MARTINI package [21] to transform the input data into radio observations. Mock observations of simulated celestial sources (such as high-redshift point sources, low-redshift galaxies, extended structures, and both continuum and emission lines in ALMA cubes) were produced efficiently. In the output, the novel package generated sky model cubes and dirty ALMA cubes, with both in FITS format. The sky models represent the simulated real sky without any noise or instrumental artifacts added to the image. The dirty ALMA cubes correspond to the Fourier inversion of the observed visibilities corrupted by noise (e.g., instrumental or thermal). ALMASim can optionally provide outputs such as the ALMA dirty beam, the measurement sets files, and plots of the 2D integrated cubes and 1D spectra for all simulated data.

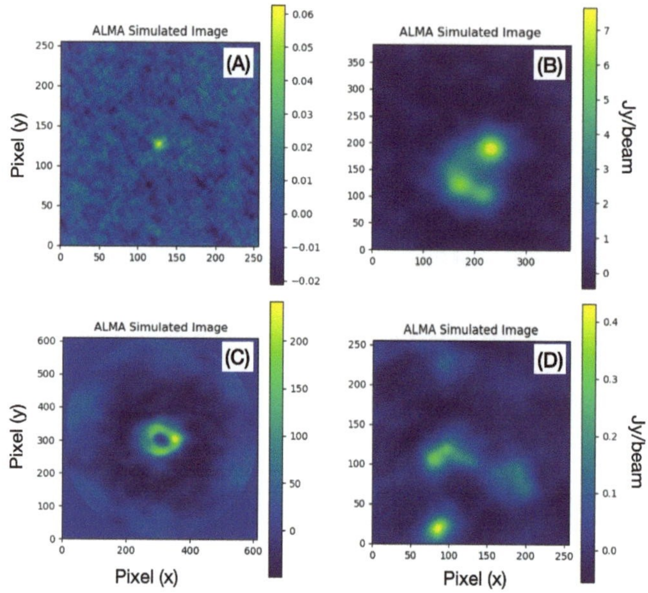

Figure 4. Example of ALMA-simulated sources (dirty images) created with the ALMASim package: (**A**) point-like, (**B**) Gaussian shape, (**C**) extended, and (**D**) diffuse emissions.

We aim at having this service available to the scientific community as an open-source tool. A number of preconfigured parameter sets will be available to generate identical datasets on which to test algorithms. ALMASim empowers users in generating synthetic datasets designed for the development of deconvolution and source detection models. Each user will be able to generate a dataset on its local machine. The package is engineered to harness the usage of MPI parallel computing, particularly on contemporary HPC clusters, to efficiently produce thousands of ALMA data cubes and their visibilities. Users have the choice to define the input parameters or to opt for a completely randomized configuration and sources.

2.3.1. Sharpening Mock Data with Realistic Noise Characteristics

We want to ensure that the simulated data closely resemble actual ALMA observations. The generation of synthetic noise has to reproduce the noise in real ALMA observations and introduce those characteristics in the artificial images. Two approaches are used: (1) adding noise to the generated ALMA measurement sets and (2) empirical noise modeling. Realistic noise characteristics in artificial ALMA images are commonly added to the generated ALMA measurement sets, which contain the visibility data and calibration information from the simulated observations. Synthetic noise can be generated from the CASA Simulator that is capable of corrupting the visibility data with thermal noise and atmospheric attenuation, even with leakages (cross-polarization). Corruption with atmospheric turbulence or adding gain fluctuations or drift are also possible within the CASA Simulator. However, correlated noise (e.g., due to antenna placement) is not accounted for in a straightforward manner, but this is possible while corrupting the measurement sets. Once the several components are introduced, the final noisy image is verified such that the noise levels in the synthetic image match the noise characteristics of the desired simulated observations. Alternatively, the empirical approach to noise modeling is not bounded to any theoretical model, allowing the approach to be applicable at other observatories and wavelengths.

Empirical Noise Modeling

Generation of synthetic noise is limited in reproducing all complexities encountered in real image data, though we wanted the simulated ALMA images to closely resemble actual ALMA observations. Commonly, quality assurance for a principal investigator's parameter of interest (such as sensitivity or resolution) occurs in the image space. We investigated the possibility to learn the noise patterns from real ALMA data with the goal of understanding the noise properties, such as the spatial and spectral distribution of noise, and possibly considering correlated noise. Therefore, rather than relying solely on artificial noise added to a simulated real sky, an empirical approach to noise modeling was used and applied to ALMASim to encompass a broader range of noise components. Noise components were characterized at larger scales and at scales similar to and smaller than the ALMA beam size from the observed data. By extracting high spatial frequency noise patterns present in the image, we tried to capture the complexity of the noise environment. When replicated to create the new simulated image, we took care to preserve the statistical properties of the flux distribution on individual pixels, matching those observed in the original image (except in regions originally occupied by the astronomical sources of interest). In Figure 5, a simulated ALMA image characterized by noise is shown and extracted from a real ALMA image. The real ALMA image displays a quasi-stellar object with a jet-like structure. The empirical approach assists in reinforcing the creation of realistic ALMA simulations. Extracting noise from ALMA observations and integrating it into simulated sky data improves the fidelity of noise modeling, leading to a more precise representation of the intricate noise characteristics often encountered in ALMA observations.

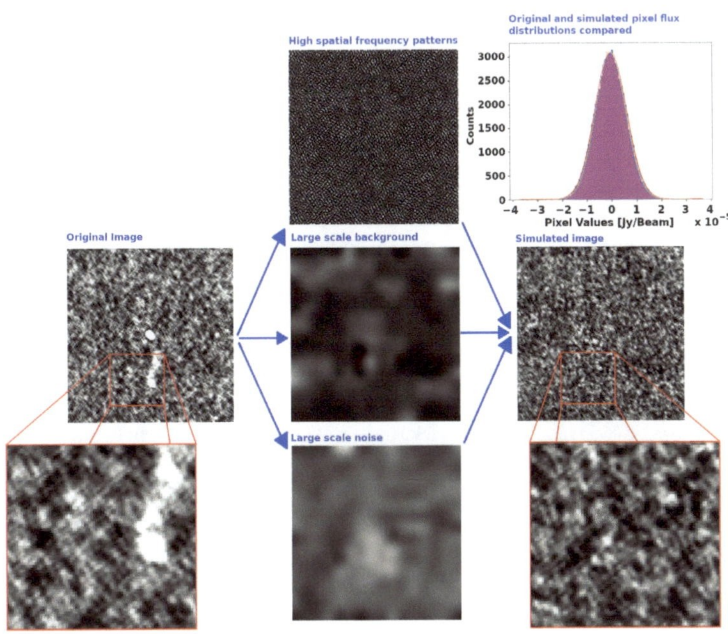

Figure 5. Simplified visual explanation of the empirical approach to noise modeling. Different background and noise components measured at scales larger and shorter than the typical beam scale (central panels) are isolated from a real ALMA image (e.g., an ALMA calibrator (left panel)) and then added to a simulated image (right panel). In this example, we considered local fluctuations (center, bottom panel), a large-scale background (central panel), and high spatial frequency patterns (center, top panel). Instead of using a theoretical model to simulate noise and instrumental response, the same effects were directly measured from real observations obtained in comparable situations (telescope configuration and atmospheric conditions).

3. Outlook and Conclusions

The ALMA2030 roadmap is bringing about a significant transformation due to the incrementing of data rates and volume. The ESO internal ALMA development study BRAIN provided an exploration of concepts for advanced imaging techniques. RESOLVE and DeepFocus are applicable to large data volumes while requiring the least amount of human intervention.

While RESOLVE stands out for its exceptional capacity to detect extended emissions and distinguish them from point sources, it will naturally outclass other options at combining data from various measurement sets for group-level imaging. Group-level imaging allows us to combine data from multiple arrays (12 m, 7 m, and total power) to capture both high-resolution and low-resolution features in a single observation. RESOLVE appears to be emerging as the preferred tool in the scientific community. Currently, the RESOLVE algorithm is in development to enhance its efficiency in processing ALMA cube images. On the other hand, DeepFocus is well-equipped with essential features to be used in operations, such as real-time processing and cube imaging of all spectral windows for large datasets. DeepFocus is actively advancing in its capability to detect diverse source morphologies, and it is planned to learn from freshly archived data. Its companion package, ALMASim, opens doors to the creation of tailored imaging algorithms specifically designed to address unique scientific cases. Although we still do not have a clear design for ALMA2030, it is foreseeable for DeepFocus to perform image processing soon after the calibrated data are reduced in real-time processing. If freshly observed data are calibrated and ingested in the archive, then the DeepFocus algorithm can efficiently perform the imaging of the calibrated data within few minutes. A real-time imaging service from a science platform of an ALMA archive is realistic. During the ALMA imaging process, if structures are present in the images caused by, for example, a glitch, then an alert system can be sent to the system astronomers to perform the needed investigations and take actions for the following ALMA observations, or an alert system can be sent to the next-generation CASA team [24]. If the archive is the central system for acquiring the ingested calibrated data, then the imaging algorithms of choice can be employed to perform the imaging of the measurement set even at the group level when available for each project.

Enhancing the efficiency of the observatory and establishing a comprehensive archive in terms of products will greatly enrich data mining opportunities for the scientific community.

Author Contributions: Conceptualization, F.G.; methodology, F.G., E.V., T.E., G.L., M.D.V., I.B., Ł.T., F.S. and J.R.; software, M.D.V., V.J., C.B., A.D. and J.R.; validation, F.S. and Ł.T.; formal analysis, M.D.V. and V.J.; resources, F.G., T.E., G.L. and M.D.V.; writing—original draft preparation, F.G.; writing—review and editing, T.E., F.S., E.V. and M.D.V. All authors have read and agreed to the published version of the manuscript.

Funding: This research was supported by an ESO internal ALMA development study investigating interferometric image reconstruction methods. The authors would like to acknowledge the Max Planck Computing and Data Facility (MPCDF) for providing computational resources and support for this research. J.R. acknowledges financial support from the German Federal Ministry of Education and Research (BMBF) under grant 05A20W01 (Verbundprojekt D-MeerKAT).

Institutional Review Board Statement: Not applicable.

Informed Consent Statement: Not applicable.

Data Availability Statement: Publicly available datasets were analyzed in this study. The data can be found by querying the ALMA Archive: https://almascience.eso.org/aq/.

Conflicts of Interest: The authors declare no conflict of interest.

Abbreviations

The following abbreviations are used in this manuscript:

ALMA	Atacama Large Millimeter/submillimeter Array
ALMA2030	ALMA Development Roadmap
BRAIN	Bayesian Reconstruction through Adaptive Image Notion
DSHARP	Disk Substructures at High Angular Resolution Project
EI	Expected improvement
ESO	European Southern Observatory
EU ARC	European ALMA Regional Centre
IF	Intermediate frequency bandwidth
FITS	Flexible Image Transport System
GP	Gaussian process
GPU	Graphics processing unit
HPC	High-performance computing
ML	Machine learning
MPI	Message Passing Interface
NIFTy	Numerical Information Field Theory
pdf	Probability density function
WSU	Wideband Sensitivity Upgrade

References

1. Wootten, A.; Thompson, A.R. The Atacama Large Millimeter/Submillimeter Array. *Proc. IEEE* **2009**, *97*, 1463–1471. [CrossRef]
2. Carpenter, J.; Iono, D.; Testi, L.; Whyborn, N.; Wootten, A.; Evans, N. The ALMA Development Roadmap. 2018. Available online: https://www.eso.org/sci/facilities/alma/developmentstudies/ALMA_Development_Roadmap_public.pdf (accessed on 23 November 2023).
3. Carpenter, J.; Iono, D.; Kemper, F.; Wootten, A. The ALMA Development Program: Roadmap to 2030. *arXiv* **2020**, arXiv:2001.11076.
4. Carpenter, J.; Brogan, C.; Iono, D.; Mroczkowski, T. ALMA Memo 621: The ALMA2030 Wideband Sensitivity Upgrade. *arXiv* **2022**, arXiv:2211.00195.
5. Guglielmetti, F.; Villard, E.; Fomalont, E. Bayesian Reconstruction through Adaptive Image Notion. *Proceedings* **2019**, *33*, 21. [CrossRef]
6. Guglielmetti, F.; Arras, P.; Delli Veneri, M.; Enßlin, T.; Longo, G.; Tychoniec, L.; Villard, E. Bayesian and Machine Learning Methods in the Big Data Era for Astronomical Imaging. *Phys. Sci. Forum* **2022**, *5*, 50.
7. Siemiginowska, A.; Eadie, G.; Czekala, I.; Feigelson, E.; Ford, E.B.; Kashyap, V.; Kuhn, M.; Loredo, T.; Ntampaka, M.; Stevens, A.; et al. Astro2020 Science White Paper: The Next Decade of Astroinformatics and Astrostatistics. *arXiv* **2019**, arXiv:1903.06796.
8. Enßlin, T.A.; Frommert, M.; Kitaura, F.S. Information field theory for cosmological perturbation reconstruction and nonlinear signal analysis. *Phys. Rev. D* **2009**, *80*, 105005. [CrossRef]
9. Junklewitz, H.; Bell, M.R.; Selig, M.; Enßlin, T.A. RESOLVE: A new algorithm for aperture synthesis imaging of extended emission in radio astronomy. *Astron. Astrophys.* **2016**, *586*, A76. [CrossRef]
10. Tychoniec, Ł.; Guglielmetti, F.; Arras, P.; Enßlin, T.; Villard, E. Bayesian Statistics Approach to Imaging of Aperture Synthesis Data: RESOLVE Meets ALMA. *Phys. Sci. Forum* **2022**, *5*, 52.
11. Roth, J.; Arras, P.; Reinecke, M.; Perley, R.A.; Westermann, R.; Enßlin, T.A. Bayesian Radio Interferometric imaging with direction-dependent calibration. *Astron. Astrophys.* **2023**, *678*, A177. [CrossRef]
12. Delli Veneri, M.; Tychoniec, Ł.; Guglielmetti, F.; Longo, G.; Villard, E. 3D detection and characterization of ALMA sources through deep learning. *Mon. Not. R. Astron. Soc.* **2023**, *518*, 3. [CrossRef]
13. Andrews, S.M.; Huang, J.; Pérez, L.M.; Isella, A.; Dullemond, C.P.; Kurtovic, N.T.; Guzmán, V.V.; Carpenter, J.M.; Wilner, D.J.; Zhang, S.; et al. The Disk Substructures at High Angular Resolution Project (DSHARP). I. Motivation, Sample, Calibration, and Overview. *Astrophys. J. Lett.* **2018**, *869*, L41. [CrossRef]
14. Huang, J.; Andrews, S.M.; Pérez, L.M.; Zhu, Z.; Dullemond, C.P.; Isella, A.; Benisty, M.; Bai, X.; Birnstiel, T.; Carpenter, J.M.; et al. The Disk Substructures at High Angular Resolution Project (DSHARP). III. Spiral Structures in the Millimeter Continuum of the Elias 27, IM Lup, and WaOph 6 Disks. *Astrophys. J. Lett.* **2018**, *869*, L43. [CrossRef]
15. Disk Substructures at High Angular Resolution Project (DSHARP). Available online: https://almascience.eso.org/almadata/lp/DSHARP/ (accessed on 23 November 2023).
16. The CASA Team. CASA, the Common Astronomy Software Applications for Radio Astronomy. *Publ. Astron. Soc. Pac.* **2022**, *134*, 1041. [CrossRef]
17. Arras, P.; Frank, P.; Haim, P.; Knollmüller, J.; Leike, R.; Reinecke, M.; Enßlin, T. Variable structures in M87* from space, time and frequency resolved interferometry. *Nat. Astron.* **2022**, *6*, 259–269. [CrossRef]
18. Knollmüller, J.; Arras, P.; Enßlin, T. Resolving Horizon-Scale Dynamics of Sagittarius A*. *arXiv* **2023**, arXiv:2310.16889.

19. Preuss, R.; von Toussaint, U. Global Optimization Employing Gaussian Process-Based Bayesian Surrogates. *Entropy* **2018**, *20*, 2([CrossRef] [PubMed]
20. Högbom, J.A. Aperture synthesis with a non-regular distribution of interferometer baselines. *Astron. Astrophys. Suppl.* **197** *15*, 417.
21. Oman, K.A. *MARTINI: Mock Spatially Resolved Spectral Line Observations of Simulated Galaxies*; Bibcode: 2019ascl.soft11005 Astrophysics Source Code Library: College Park, MD, USA, 2019.
22. The Illustris Collaboration. 2018. Available online: https://github.com/illustristng/illustris_python (accessed on 23 Nover ber 2023).
23. The Nifty Team. 2021. Available online: https://gitlab.mpcdf.mpg.de/ift/nifty (accessed on 23 November 2023).
24. CNGI. Available online: https://cngi-prototype.readthedocs.io/en/latest/_api/autoapi/cngi/index.html (accessed on 2 November 2023).

physical sciences forum

MDPI

Proceeding Paper

Inferring Evidence from Nested Sampling Data via Information Field Theory †

Margret Westerkamp [1,2,*], Jakob Roth [1,2,3], Philipp Frank [1], Will Handley [4,5] and Torsten Enßlin [1,2]

1 Max Planck Institute for Astrophysics, 85748 Garching, Germany
2 Faculty for Physics, Ludwig-Maximilians-Universität, 80539 Munich, Germany
3 Department for Computer Science, Technical University Munich, 85748 Munich, Germany
4 Astrophysics Group, Cavendish Laboratory, J. J. Thomson Avenue, Cambridge CB3 0HE, UK
5 Kavli Institute for Cosmology, Madingley Road, Cambridge CB3 0EZ, UK
* Correspondence: margret@mpa-garching.mpg.de
† Presented at the 42nd International Workshop on Bayesian Inference and Maximum Entropy Methods in Science and Engineering, Garching, Germany, 3–7 July 2023.

Abstract: Nested sampling provides an estimate of the evidence of a Bayesian inference problem via probing the likelihood as a function of the enclosed prior volume. However, the lack of precise values of the enclosed prior mass of the samples introduces probing noise, which can hamper high-accuracy determinations of the evidence values as estimated from the likelihood-prior-volume function. We introduce an approach based on information field theory, a framework for non-parametric function reconstruction from data, that infers the likelihood-prior-volume function by exploiting its smoothness and thereby aims to improve the evidence calculation. Our method provides posterior samples of the likelihood-prior-volume function that translate into a quantification of the remaining sampling noise for the evidence estimate, or for any other quantity derived from the likelihood-prior-volume function.

Keywords: nested sampling; Bayesian evidence; information field theory

1. Introduction

Nested sampling is a computational technique for Bayesian inference developed by [1]. Whereas previous statistical sampling algorithms were primarily designed to sample the posterior, the nested sampling algorithm focuses on computing the evidence by estimating how the likelihood function relates to the prior. As discussed in [2], Bayesian inference consists of parameter estimation and model comparison. In Bayesian parameter estimation, the model parameters $\theta_{\mathcal{M}}$ for a given model $\mathcal{M}$ and data d are inferred via Bayes' theorem,

$$\mathcal{P}(\theta_{\mathcal{M}}|d,\mathcal{M}) = \frac{\mathcal{P}(d|\theta_{\mathcal{M}},\mathcal{M})\mathcal{P}(\theta_{\mathcal{M}}|\mathcal{M})}{\mathcal{P}(d|\mathcal{M})}. \quad (1)$$

Here, $\mathcal{P}(\theta_{\mathcal{M}}|d,\mathcal{M})$ is the posterior probability for the model parameters $\theta_{\mathcal{M}}$ given the data d. The likelihood $\mathcal{P}(d|\theta_{\mathcal{M}},\mathcal{M})$ describes the measurement process, which generated the data d, and the prior $\mathcal{P}(\theta_{\mathcal{M}}|\mathcal{M})$ encodes our prior knowledge of the parameters within the given model. The normalization of the posterior,

$$Z = \mathcal{P}(d|\mathcal{M}) = \int d\theta_{\mathcal{M}}\mathcal{P}(\theta_{\mathcal{M}},d|\mathcal{M}), \quad (2)$$

is called the evidence and is the focus of this study. In Bayesian parameter estimation, it is common to work with not normalized posteriors. Thus, in this scenario, the computation of the evidence is less critical. In contrast, when comparing different Bayesian models, estimating the evidence for different models is very important. In this case, the aim is to find the most probable model $\mathcal{M}_i$ given the data,

$$\mathcal{P}(\mathcal{M}_i|d) = \frac{\mathcal{P}(d|\mathcal{M}_i)\mathcal{P}(\mathcal{M}_i)}{\mathcal{P}(d)}. \tag{3}$$

Assuming a uniform prior for all arbitrary models $\mathcal{P}(\mathcal{M}_i) = \text{const}$, this turns out to be equivalent to choosing the model with the highest evidence.

In nested sampling, the possibly multidimensional integral of the posterior in Equation (2) is transformed into a one-dimensional integral by directly using the prior mass X. In particular, by transforming the problem into a series of nested spaces, nested sampling provides an elegant way to compute the evidence. The algorithm starts by drawing N samples from the prior, called the live points. For each of these points, the likelihood is calculated and the live point with the lowest likelihood is removed from the set of live points and added to another set, called the dead points. A new live point is then sampled that has a higher likelihood value than the last added dead point. This type of sampling is commonly referred to as likelihood-restricted sampling. However, the specific methods associated with likelihood-restricted sampling are not discussed further in this paper. As a consequence of the procedure, the prior volume shrinks from one to zero, contracting around the peak of the posterior. The prior mass X contained in the parameter space volume with likelihood values larger than L can be computed by

$$X(L) = \int_{\mathcal{P}(d|\theta_{\mathcal{M}},\mathcal{M})>L} \mathcal{P}(\theta_{\mathcal{M}}|\mathcal{M})\, d\theta_{\mathcal{M}}. \tag{4}$$

Thus, Equation (2) simplifies to a one-dimensional integral,

$$Z = \int_0^1 L(X)dX, \tag{5}$$

where $L(X)$ is the inverse of Equation (4). Accordingly, this integral can be approximated by the weighted sum over all m dead points

$$Z \approx \sum_{i=1}^{m} \omega_i L_i. \tag{6}$$

As proposed in [1], we calculate the weights via $\omega_i = \frac{1}{2}(X_{i-1} - X_{i+1})$ assuming $X_0 = 1$ and $X_{m+1} = 0$. Adding dead points to their set and adjusting the evidence accordingly continues until the remaining live points occupy a tiny prior volume that would contribute little to the weighted sum in Equation (6).

For the calculation in Equation (6) not only the known live and dead contours of the likelihood are needed but also the corresponding prior volumes encoded in ω_i, which are not precisely known. According to [1] there are two different approaches to approximate the prior volumes X_i, a stochastic scheme and a deterministic scheme. In the stochastic scheme the change of volume due to each removed shell i is a stochastic process characterised by a Beta distributed random variable t_i,

$$X_0 = 1, \quad X_i = t_i X_{i-1}, \quad \mathcal{P}(t_i) = \text{Beta}(t_i|1, N) = N t_i^{N-1}, \tag{7}$$

where we assume a constant number of live points N. Approaches with a varying number of live points were i.a. introduced in dynamic nested sampling by [3,4] and extend beyond the boundaries of this research until the present moment. This probabilistic description of the prior volume evolution allows to draw several samples of prior volumes X, according to the likelihood values L, and to thereby get uncertainty estimates on the evidence calculation (Equation (6)). In the deterministic scheme the logarithmic prior volume is estimated via,

$$\ln(X_i) = -\frac{i}{N}, \tag{8}$$

at the ith iteration. This estimate is derived from the fact that the expectation value of the logarithmic volume changes is $\langle \ln t_i \rangle = -1/N$. However, this estimate does not take the uncertainties in the evidence calculation [5] into account and differs from unbiased approaches introduced and analysed in [6–8]. In any case, the imprecise knowledge of the prior volume introduces probing noise that can potentially hinder the accurate calculation of the evidence. In order to improve the accuracy of the posterior integration, we aim to reconstruct the likelihood-prior-volume function given certain a priori assumptions on the function itself using Bayesian inference. Here, we introduce a prior and likelihood model for the reconstruction of the likelihood-prior-volume function, which we will call the reconstruction prior and the reconstruction likelihood to avoid confusion with likelihood contour and prior volume information obtained from nested sampling.

The left side of Figure 1 illustrates the nested sampling likelihood dead contours generated by the software package anesthetic [9] for the simple Gaussian example discussed in Section 2 and two live points (N = 2) as a function of prior volume. In the following, we call the likelihood values of the dead points the likelihood data $\vec{d}_L$ and the prior volume, approximated by Equation (8), the prior volume data $\vec{d}_X$. Additionally, the analytical solution of the likelihood-prior-volume function, which we call the ground truth, is plotted.

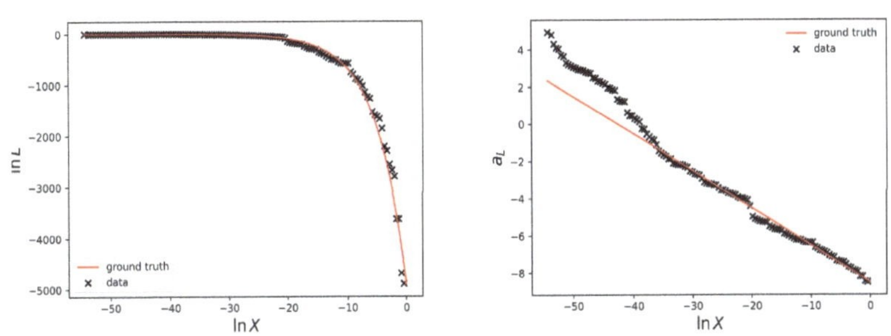

Figure 1. (**Left**): Visualisation of the nested sampling dead point logarithmic likelihoods, $\vec{d}_L$, as a function of logarithmic prior mass data, $\vec{d}_X$, for the normalized simple Gaussian in Equation (14) ($\sigma_X = 0.01,\ D = 10$). The corresponding data was generated by the software package anesthetic [9]. (**Right**): Visualisation of the reparametrised nested sampling dead point logarithmic likelihoods according to Equation (13) as a function of logarithmic prior mass for the same case as shown left.

In accordance with the here considered example, we assume the likelihood-prior-volume function to be smooth for most real-world applications of nested sampling. In this study, we propose an approach that incorporates this assumption of a-priori-smoothness and enforces monotonicity. In particular, we use Information Field Theory (IFT) [10] as a versatile mathematical tool to reconstruct a continuous likelihood-prior-volume function from a discrete dataset of likelihood contours and to impose the prior knowledge on the function.

As noted in [11], the time complexity of the nested sampling algorithm depends on several factors. First, the time complexity depends on the information gain of the posterior over the prior, which is equal to the shrinkage of the prior required to reach the bulk of the posterior. This is described mathematically by the Kullback-Leibler divergence (KL) [12],

$$\mathcal{D}_{\mathrm{KL}} = \int d\theta_{\mathcal{M}} \mathcal{P}(\theta_{\mathcal{M}}|d,\mathcal{M}) \ln \frac{\mathcal{P}(\theta_{\mathcal{M}}|d,\mathcal{M})}{\mathcal{P}(\theta_{\mathcal{M}}|\mathcal{M})}. \tag{9}$$

Second, the time complexity increases with the number of live points N, which defines the shrinkage per iteration. Furthermore, the time for evaluating the likelihood $L(\theta)$, $\mathcal{T}_L$, and the time for sampling a new live point in the likelihood restricted volume, $\mathcal{T}_{\text{samp}}$, contribute to the time complexity. Accordingly, in [13] the time complexity of the nested sampling algorithm T and the error σ_Z have been characterised via,

$$T \propto N \times \langle \mathcal{T}_L \rangle \times \langle \mathcal{T}_{\text{samp}} \rangle \times \mathcal{D}_{\text{KL}} \tag{10}$$

$$\sigma_Z \propto \sqrt{\mathcal{D}_{\text{KL}}/N}. \tag{11}$$

Upon examining the error, σ_Z, it becomes evident that reducing the error by increasing the number of live points leads to significantly longer execution times. Accordingly, by inferring the likelihood-prior-volume function, we aim to reduce the error in the log-evidence for a given $\mathcal{D}_{\text{KL}}$ and a fixed number of live points, N, avoiding a significant increase in time complexity.

The rest of the paper is structured as follows. In Section 2, the description of the reconstruction prior of the likelihood-prior-volume curve is discussed. The model for the reconstruction likelihood and the inference of the likelihood-prior-volume function and the prior volumes using IFT is described in Section 3. The corresponding results for a Gaussian example and the impact on the evidence calculation are shown in Section 4. And eventually, the conclusion and outlook for future work are given in Section 5.

2. The Reconstruction Prior Model for the Likelihood-Prior-Volume Function

A priori we assume that the likelihood-prior-volume function is smooth and monotonically decreasing. This is achieved by representing the negative rate of change of the logarithmic prior volume, $\ln X$, with a monotonic function of the likelihood a_L as a log-normal process,

$$-\frac{d \ln X}{d a_L} = e^{\tau(\ln X)}. \tag{12}$$

In the words of IFT, we model the one-dimensional field τ, which assigns to each logarithmic prior volume a value, as a Gaussian process with $\mathcal{P}(\tau) = \mathcal{G}(\tau, T)$. Thereby, we do not assume a fixed power spectrum for the Gaussian process, but reconstruct it simultaneously with τ itself. An overview of this Gaussian process model is given in Appendix A. The details can be found in [14].

In the most relevant volume for the evidence, the peak region of the posterior is expected to be similar to a Gaussian in a first order approximation. Therefore, the function a_L is chosen such that τ is a constant for the Gaussian case. Deviations from the Gaussian are reflected in deviations of τ from the constant. Accordingly, we define,

$$a_L = -\ln\left(-\ln\left(\frac{L}{L_{\text{max}}}\right)\right), \tag{13}$$

with L_{max} being the maximal likelihood. We consider the simple Gaussian example proposed by [1],

$$L_{\text{gauss}}(X) = \frac{1}{C(\sigma_X)} \exp\left(-\frac{X^{2/D}}{2\sigma_X^2}\right), \tag{14}$$

where D is the dimension and σ_X is the standard deviation. We find that the function $a_L(\ln X)$, defined in Equation (13), becomes linear in this case,

$$a_{\text{gauss}}(\ln X) = \ln(2\sigma_X^2) - \frac{2}{D}\ln X. \tag{15}$$

Figure 1 illustrates the data and the ground truth on log-log-scale on the left and the linear relation $a_L(\ln X)$ on the right. According to the log-normal process defined in

Equation (12), we define the function $a_L(\ln X)$, for arbitrary likelihoods, which is able to account for deviations from the Gaussian case,

$$a_L(\ln X) = a_0 - \int_0^{\ln X} e^{-\tau(z)} dz. \tag{16}$$

By inverting Equation (13) we then get the desired likelihood-prior-volume function. The logarithmic prior volume values given the likelihood contours are obtained by inversion of Equation (16). In Figure 2 several prior samples given the model for the reconstruction prior according to Equation (16) are shown.

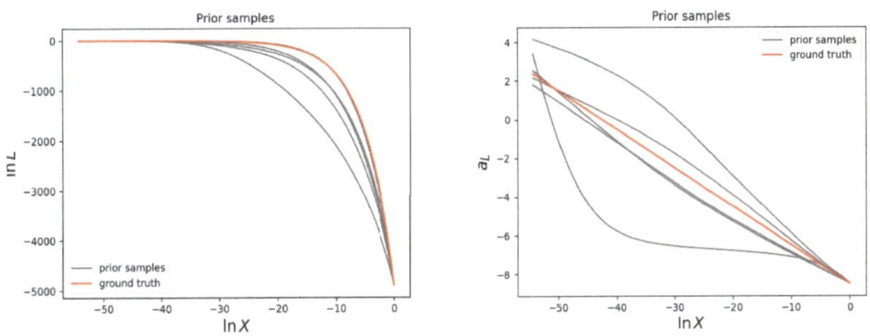

Figure 2. Reconstruction prior samples of the likelihood-prior-volume function plotted together with the ground truth. (**Left**): Log-log-scale. (**Right**): Parametrisation according to Equation (13).

However, often the maximum log-likelihood, $\ln L_{\max}$, is not known. In [15], the calculation of the maximum Shannon entropy $\mathcal{I} = \ln\left(\frac{\mathcal{P}(\theta|d)}{\mathcal{P}(\theta)}\right)$ is given. Using this approach, we can calculate the logarithmic maximum likelihood $\ln L_{\max}$ and thus calculate a_L for unknown likelihoods.

$$\mathcal{I}_{\max} = \mathcal{D}_{\mathrm{KL}} + \frac{D}{2} \tag{17}$$

$$\rightarrow \ln L_{\max} = \langle \ln L \rangle_{\mathcal{P}(\theta_{\mathcal{M}}|d,\mathcal{M})} + \frac{D}{2} \tag{18}$$

Hence, based on the likelihood contours obtained from the nested sampling run, we calculate the data based evidence, Z_d, using the approximated prior volumes according to Equation (8). This allows us to obtain an estimate of the maximum log-likelihood, $\ln L_{\max}$, of the model for reparametrisation,

$$Z_d \approx \sum_{i=1}^{m} d_{L_i} \times (d_{X_{i-1}} - d_{X_i}) \tag{19}$$

$$\rightarrow \ln L_{\max} \approx \sum_{i=1}^{m} \frac{d_{L_i}}{Z_d}(d_{X_{i-1}} - d_{X_i}) \ln d_{L_i} + \frac{D}{2}. \tag{20}$$

3. The Reconstruction Likelihood Model and Joint Inference

In this section we will derive a model for the reconstruction likelihood for the joint inference of the likelihood-prior-volume function and the changes of logarithmic prior volume according to the likelihood data $\vec{d}_L$. Here, IFT and the software package NIFTy [16], which facilitates the implementation of IFT algorithms, allow us to infer the reparametrised likelihood-prior-volume function in Equation (16) from the data,

$$\vec{d}_a = -\ln\left(-\ln\left(\frac{\vec{d}_L}{L_{\max}}\right)\right), \tag{21}$$

given the reconstruction prior and reconstruction likelihood model. For the inference of the likelihood-prior-volume function we first define the likelihood function a_L as a function of the Beta distributed t_i (Equation (7)),

$$a_L(z_j = \sum_{i=1}^{j} t_i) = a_0 - \int_0^{\sum_{i=1}^{j} t_i} e^{-\tau(z)} dz, \tag{22}$$

where a_0 is the likelihood for $X_0 = 1$. We perform a joint reconstruction of the function a_L and the vector $\vec{t}_d$ representing changes in prior volume according to the likelihood data $\vec{d}_a$.

$$\begin{aligned}\mathcal{P}(a, \vec{t}_d | \vec{d}_a) &\propto \mathcal{P}(\vec{d}_a | \vec{t}_d, a)\ \mathcal{P}(\vec{t}_d, a) \\ &= \mathcal{P}(\vec{d}_a | \vec{t}_d, a)\ \mathcal{P}(a | \vec{t}_d)\ \mathcal{P}(\vec{t}_d) \\ &= \delta(\vec{d}_a - a_L(\vec{t}_d))\ \mathcal{G}(\tau | T) \prod_{i=1}^{m} \mathrm{Beta}(t_i | 1, N) \\ &\approx \mathcal{G}(\vec{d}_a - a_L(\vec{t}_d), \sigma_\delta)\ \mathcal{G}(\tau | T) \prod_{i=1}^{m} \mathrm{Beta}(t_i | 1, N)\end{aligned} \tag{23}$$

Here, the Gaussian uncertainty σ_δ is supposed to be chosen small in order to approximately represent the δ-function. So far, we have managed to obtain a probabilistic model for the non-normalized reconstruction posterior. To this end, we use variational inference, in particular the geoVI algorithm supposed by [17], to get an approximate of the true reconstruction posterior distribution. In the end, this statistical approach allows us to get an estimate of the likelihood-prior-volume function and the prior volumes via the posterior mean, its uncertainty and any other quantity of interest, which can be derived from the posterior samples. The results of the here developed method are shown in Section 4.

4. Results

To test the presented method we perform a reconstruction for the simple Gaussian example discussed in Section 2 and introduced in Figure 1. The according results for the likelihood-prior-volume function are shown in Figure 3.

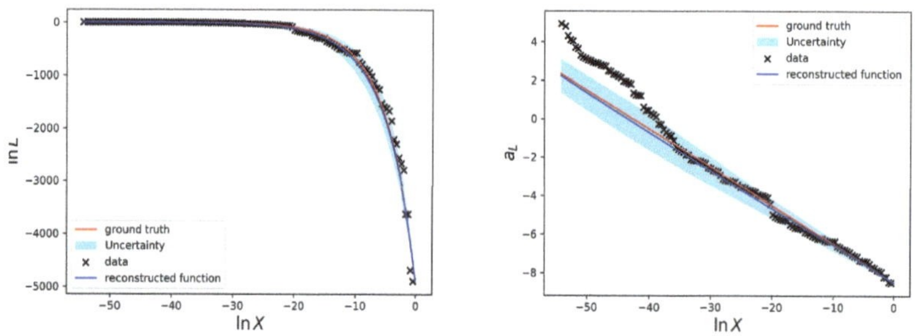

Figure 3. Reconstruction results for the likelihood-prior-volume function for the simple Gaussian example in Equation (14). The plots show the data, the ground truth and the reconstruction as well as its uncertainty. (**Left**): Reconstruction results on log-log-scale. (**Right**): Reconstruction results in reparametrised coordinates according to Equation (13).

Moreover, the posterior estimates of the logarithmic prior volumes to the according likelihood data $\vec{d}_L$ are shown in Figure 4.

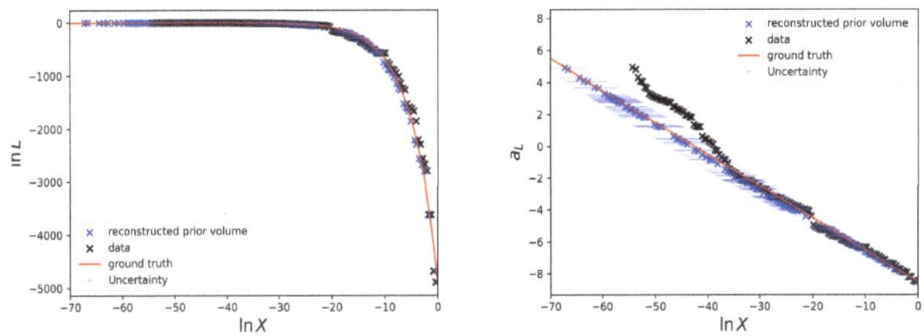

Figure 4. Reconstruction results for the prior volumes given the likelihood data $\vec{d}_L$ for the simple Gaussian example in Equation (14). The plots show the data, the ground truth and the reconstruction as well as its uncertainty. (**Left**): Reconstruction results on log-log-scale. (**Right**): Reconstruction results in reparametrised coordinates according to Equation (13).

Since the main goal of nested sampling is to compute the evidence, we want to quantify the impact of the proposed method on the evidence calculation. To do this, we use $n_{\text{samp}} = 200$ posterior samples for the prior volumes X^* and calculate the evidence given the likelihood contours $\vec{d}_L$ for each of these samples according to Equation (6),

$$Z^* = \sum_{i=1}^{m} (\vec{d}_L)_i \frac{1}{2}(X^*_{i-1} - X^*_{i+1}). \tag{24}$$

Similarly, we generate by means of anesthetic [9] $n_{\text{samp}} = 200$ samples of the prior volume via the probabilistic nested sampling approach described in Equation (7). Also for these samples we calculate the evidence according to Equation (24). A comparison of the histograms of evidences for both sample sets (classical nested sampling and reconstructed prior volumes) is shown in Figure 5.

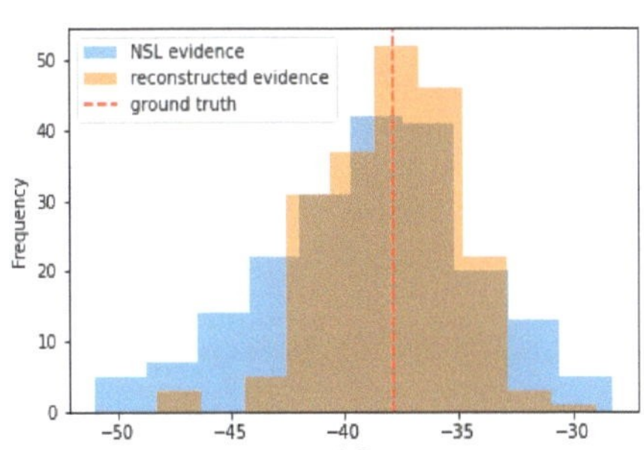

Figure 5. Comparison of histograms for logarithmic evidences for $n_{\text{samp}} = 200$ samples for the classical nested sampling (NSL) approach and the reconstructed prior volumes.

From the comparison of the histograms, one can already see that the standard deviation for the posterior sample evidences for the reconstructed prior volumes got smaller. This is also mirrored as soon as we look at numbers: The ground truth logarithmic evidence for this Gaussian case is $\ln Z_{\text{gauss}} = -37.798$. The result for the evidence for the classical nested sampling approach given $n_{\text{samp}} = 200$ is $\ln Z_d = -38.92 \pm 4.50$. And finally, the result for the evidence inferred with the here presented approach from the likelihood contours assuming smoothness and enforcing monotonicity is $\ln Z = -37.97 \pm 2.89$.

5. Conclusions & Outlook

In our search for a more accurate estimate of the evidence, we set out to reconstruct the likelihood-prior-volume function. In particular, a Bayesian method was developed to infer jointly the likelihood-prior-volume function and the vector of prior volumes from the dead-point likelihood contours given by the nested sampling algorithm. For the reconstruction we enforce monotonicity and assume smoothness. The test of the reconstruction algorithm on a Gaussian example shows a significant improvement in the accuracy of the computed logarithmic evidence.

In general, the approach presented here will only show notable improvements if the assumption of smoothness for the likelihood- prior-volume-curve holds. Fortunately, we can reasonably expect this assumption to hold in the majority of cases. Future work will apply the reconstruction algorithm to further likelihoods where the ground truth likelihood-prior-volume function is known for testing, with the ultimate goal of applying it to actual nested sampling outputs. In particular, the results on non-Gaussian likelihoods shall be tested.

Author Contributions: Conceptualization, M.W., J.R., P.F., W.H. and T.E.; methodology, M.W., J.R., P.F., W.H. and T.E.; software, M.W., J.R. and W.H.; validation, M.W., J.R., P.F., W.H. and T.E.; formal analysis, M.W., J.R., P.F., W.H. and T.E.; investigation, M.W., J.R., P.F., W.H. and T.E.; data curation, M.W., J.R. and W.H.; writing—original draft preparation, M.W.; writing—review and editing, M.W., J.R., P.F., W.H. and T.E.; visualization, M.W.; supervision, T.E.; project administration, M.W.; funding acquisition, T.E. All authors have read and agreed to the published version of the manuscript.

Funding: Margret Westerkamp acknowledges support for this research through the project Universal Bayesian Imaging Kit (UBIK, Förderkennzeichen 50OO2103) funded by the Deutsches Zentrum für Luft- und Raumfahrt e.V. (DLR). Jakob Roth acknowledges financial support by the German Federal Ministry of Education and Research (BMBF) under grant 05A20W01 (Verbundprojekt D-MeerKAT). Philipp Frank acknowledges funding through the German Federal Ministry of Education and Research for the project ErUM-IFT: Informationsfeldtheorie für Experimente an Großforschungsanlagen (Förderkennzeichen: 05D23EO1). This research was supported by the Munich Institute for Astro-, Particle and BioPhysics (MIAPbP) which is funded by the Deutsche Forschungsgemeinschaft (DFG, German Research Foundation) under Germany's Excellence Strategy- EXC-2094-390783311.

Institutional Review Board Statement: Not applicable.

Informed Consent Statement: Not applicable.

Data Availability Statement: Data sharing is not applicable to this article.

Conflicts of Interest: The authors declare no conflict of interest.

Appendix A. Gaussian Process Model in NIFTy

In NIFTy [16] we represent our reconstruction priors via generative models as described in [18]. More precisely we use the reparametrisation trick by [19] according to [20] to describe the field τ with correlation structure T in Equation (12) as a generative process,

$$\tau = A\xi, \ \mathcal{P}(\xi) = \mathcal{G}(\xi, \Xi), \ T = AA^{\dagger}. \tag{A1}$$

Under the a priori assumption of statistical homogeneity and isotropy, A becomes diagonal in Fourier space and can be fully represented by the square root of the power spectrum $p_T(|k|)$,

$$A_{kk'} = (FAF^{\dagger})_{kk'} = 2\pi\delta(k-k')\sqrt{p_T(|k|)}, \tag{A2}$$

where F is the corresponding Fourier transformation. Here, we model the logarithmic amplitude spectrum, $\sqrt{p_T(|k|)}$, as a power-law with non-parametric deviations represented by an integrated Wiener process on logarithmic coordinates $l = \ln(|k|)$ according to [14],

$$\sqrt{p_T(l)} \propto e^{\gamma(l)}, \ \frac{d^2\gamma}{dl^2} = \eta\xi_W(l), \ \mathcal{P}(\xi_W) = \mathcal{G}(\xi_W|\mathbb{1}). \tag{A3}$$

The resulting shape of the power spectrum is encoded in ξ_W, η and additional integration and normalization parameters. These additional parameters are represented by Gaussian and log-normal processes themselves and such their generative prior models are defined by hyperparameters characterising their mean and variance.

References

1. Skilling, J. Nested sampling for general Bayesian computation. *Bayesian Anal.* **2006**, *1*, 833–859. [CrossRef]
2. Handley, W.J.; Hobson, M.P.; Lasenby, A.N. polychord: Next-generation nested sampling. *Mon. Not. R. Astron. Soc.* **2015**, *453*, 4385–4399. [CrossRef]
3. Speagle, J.S. dynesty: A dynamic nested sampling package for estimating Bayesian posteriors and evidences. *Mon. Not. R. Astron. Soc.* **2020**, *493*, 3132–3158. [CrossRef]
4. Higson, E.; Handley, W.; Hobson, M.; Lasenby, A. Dynamic nested sampling: An improved algorithm for parameter estimation and evidence calculation. *Stat. Comput.* **2018**, *29*, 891–913. [CrossRef]
5. Keeton, C.R. On statistical uncertainty in nested sampling. *Mon. Not. R. Astron. Soc.* **2011**, *414*, 1418–1426. [CrossRef]
6. Walter, C. Rare Event Simulation and Splitting for Discontinuous Random Variables. *ESAIM Probab. Stat.* **2015**, *19*, 794–811.
7. Chopin, N.; Robert, C.P. Properties of nested sampling. *Biometrika* **2010**, *97*, 741–755. [CrossRef]
8. Salomone, R.; South, L.F.; Drovandi, C.C.; Kroese, D.P. Unbiased and Consistent Nested Sampling via Sequential Monte Carlo. *arXiv* **2018**, arXiv:1805.03924.
9. Handley, W. Anesthetic: Nested sampling visualisation. *J. Open Source Softw.* **2019**, *4*, 1414. [CrossRef]
10. Enßlin, T.A. Information Theory for Fields. *Ann. Phys.* **2019**, *531*, 1800127. [CrossRef]
11. Buchner, J. Nested Sampling Methods. *Stat. Surv.* **2023**, *17*, 169–215.
12. Kullback, S.; Leibler, R.A. On Information and Sufficiency. *Ann. Math. Stat.* **1951**, *22*, 79–86. [CrossRef]
13. Petrosyan, A.; Handley, W.J. SuperNest: Accelerated nested sampling applied to astrophysics and cosmology. *Phys. Sci. Forum* **2022**, *5*, 51.
14. Arras, P.; Frank, P.; Haim, P.; Knollmüller, J.; Leike, R.; Reinecke, M.; Enßlin, T. Variable structures in M87 from space, time and frequency resolved interferometry. *Nat. Astron.* **2022**, *6*, 259–269. [CrossRef]
15. Handley, W.; Lemos, P. Quantifying dimensionality: Bayesian cosmological model complexities. *Phys. Rev. D* **2019**, *100*, 023512. [CrossRef]
16. Arras, P.; Baltac, M.; Enßlin, T.A.; Frank, P.; Hutschenreuter, S.; Knollmueller, J.; Leike, R.; Newrzella, M.N.; Platz, L.; Reinecke, M.; et al. *Nifty5: Numerical Information Field Theory v5*; Astrophysics Source Code Library: College Park, MD, USA, 2019; record ascl:1903.008.
17. Frank, P.; Leike, R.; Enßlin, T.A. Geometric Variational Inference. *Entropy* **2021**, *23*, 853. [CrossRef]
18. Enßlin, T. Information Field Theory and Artificial Intelligence. *Entropy* **2022**, *24*, 374. [CrossRef]
19. Kingma, D.P.; Salimans, T.; Welling, M. Variational Dropout and the Local Reparameterization Trick. In Proceedings of the Advances in Neural Information Processing Systems 28: Annual Conference on Neural Information Processing Systems 2015, Montreal, QC, Canada, 7–12 December 2015; pp. 2575–2583.
20. Knollmüller, J.; Enßlin T.A. Metric Gaussian Variational Inference. *arXiv* **2020**, arXiv:1901.11033.

physical sciences forum

MDPI

Proceeding Paper

Knowledge-Based Image Analysis: Bayesian Evidences Enable the Comparison of Different Image Segmentation Pipelines †

Mats Leif Moskopp *, Andreas Deussen and Peter Dieterich

Institut für Physiologie, Medizinische Fakultät Carl Gustav Carus, TU Dresden, Fetscherstr. 74, 01307 Dresden, Germany; andreas.deussen@tu-dresden.de (A.D.)

* Correspondence: mats_leif.moskopp@tu-dresden.de

† Presented at the 42nd International Workshop on Bayesian Inference and Maximum Entropy Methods in Science and Engineering, Garching, Germany, 3–7 July 2023.

Abstract: The analysis and evaluation of microscopic image data is essential in life sciences. Increasing temporal and spatial digital image resolution and the size of data sets promotes the necessity of automated image analysis. Previously, our group proposed a Bayesian formalism that allows for converting the experimenter's knowledge, in the form of a manually segmented image, into machine-readable probability distributions of the parameters of an image segmentation pipeline. This approach preserved the level of detail provided by expert knowledge and interobserver variability and has proven robust to a variety of recording qualities and imaging artifacts. In the present work, Bayesian evidences were used to compare different image processing pipelines. As an illustrative example, a microscopic phase contrast image of a wound healing assay and its manual segmentation by the experimenter (ground truth) are used. Six different variations of image segmentation pipelines are introduced. The aim was to find the image segmentation pipeline that is best to automatically segment the input image given the expert knowledge with respect to the principle of Occam's razor to avoid unnecessary complexity and computation. While none of the introduced image segmentation pipelines fail completely, it is illustrated that assessing the quality of the image segmentation with the naked eye is not feasible. Bayesian evidence (and the intrinsically estimated uncertainty σ of the image segmentation) is used to choose the best image processing pipeline for the given image. This work illustrates a proof of principle and is extendable to a diverse range of image segmentation problems.

Keywords: image segmentation; Bayesian inference; nested sampling

1. Introduction

Identifying regions of interest in data sets of biological processes from microscopic imaging is of central interest in physiology and cell biology. Data sets might be recorded at high temporal and spatial resolution. Thus, the increase in data size prompts the need for automated image segmentation. However, this task is often performed manually by experimenters, who apply their expert knowledge to the data set. Knowledge-based analysis describes the concept of transferring expert knowledge into machine-readable data or a code to subsequently apply this to new data sets. Bayesian inference can be used for this purpose, where the expert knowledge is converted to posterior distributions of model parameters, as previously demonstrated by our group [1]. These parameters represent a variety of image features, such as brightness thresholds or smoothing kernel filter sizes of the applied image segmentation pipeline.

During discussions with members of the life science community on our previous work [1], the question of why certain image processing pipelines should be chosen over others arose. In the present work, we use Bayesian evidences to compare different pipelines for image segmentation. This allows for generating robust image segmentation pipelines without overfitting. The performances of these algorithms are illustrated with real-world problems, such as the analysis of the temporal cellular closure of a wound.

2. Materials and Methods

2.1. Data

Wound healing assays are commonly used to quantify collective cell motility over time. Cells are seeded in two compartments of a separating insert, resulting in two cell populations with a defined gap. After the removal of the insert, cells move into the cell-free gap area. While cell migration is recorded over time, one representative region of interest was selected on a single frame as an illustrative example in this work (see Figure 1).

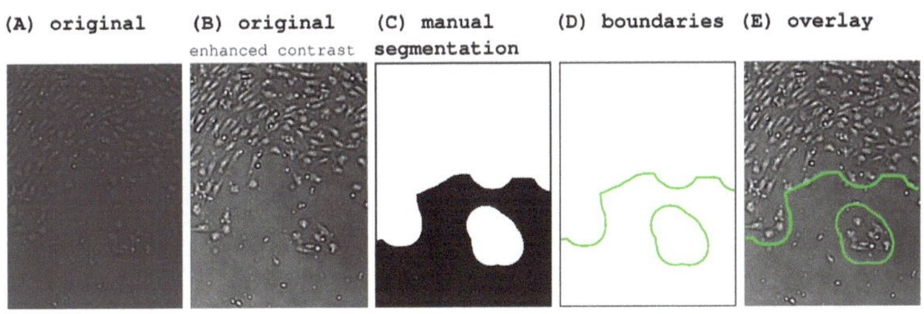

Figure 1. Image data and manual segmentation. (**A**) A typical region of interest on an image of a wound healing assay can be seen. (**B**) The same region of interest as in panel (**A**) is shown with enhanced contrast for illustrative purposes. (**C**) A manual image segmentation for a cell-free (black) and cell-covered (white) area is shown. (**D**) The boundary between the black and the white pixels of the manual segmentation is indicated by a green line. This boundary is essential for distances between manually segmented images and pipeline-segmented images (see below). (**E**) An overlay of the contrast-enhanced original image and the boundary is given for visual clarification.

2.2. Workflow

Data processing, as suggested in our previous work, [1] follows a three-step protocol:

1. **User input:** The experimenter manually marks regions with and without cells, resulting in a manually segmented image from the time series (see Figure 1).
2. **Bayesian parameter and evidence estimation:** Bayesian inference is applied to different parameter-dependent image segmentation pipelines using the manually segmented image as input data. A metric distance is applied to measure the difference between the manually segmented image and the pipeline-generated images (see below). During parameter estimation, the distance between the manually and the pipeline-segmented image will be optimized, resulting in the conversion of the expert knowledge to machine-readable posterior distributions of parameters.
3. **Application:** Based on the estimated Bayesian evidences as quality criteria of the image segmentation pipelines, one image segmentation pipeline can be chosen, and the optimized parameter set can be applied to the entire image series.

2.3. Image Segmentation Pipelines

In our previous work, we used an image segmentation pipeline consisting of a sequence of different filters [2,3]. A similar approach with manual adaptation of parameters was described previously [4]. Figure 2 illustrates the different image segmentation pipelines used throughout this work. All image segmentation pipelines were implemented in the Python programming language using scipy and opencv libraries [5,6].

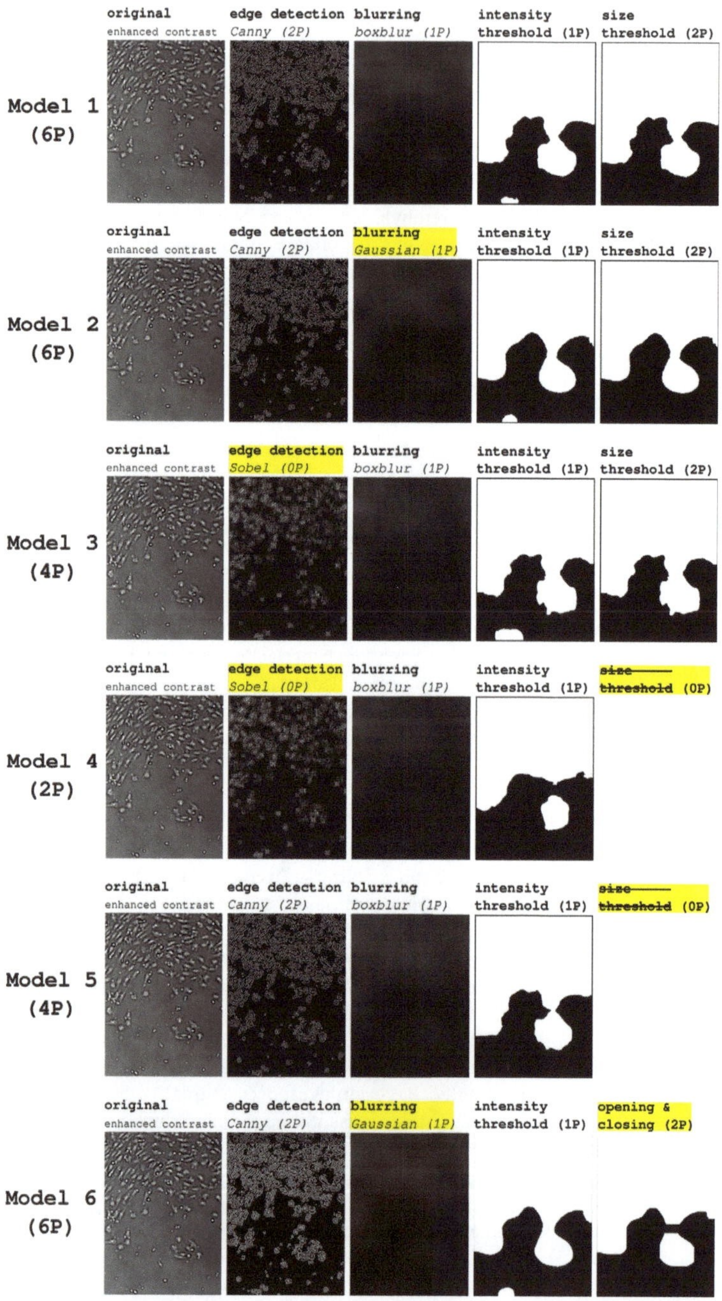

Figure 2. Image segmentation pipelines. The image segmentation pipelines (Models 1–6) consist of a sequence of image filters and algorithms that depend on one (1P) or two (2P) parameters. Further, some algorithms are applied with a fixed set of parameters (0P), so that no free parameters were used during parameter estimation. Differences of the applied filters with respect to Model 1 are highlighted in yellow. The original image is displayed with enhanced contrast for illustrative purposes only—calculations and shown results are based on the native original image (see Figure 1A).

Model 1 consists of the following elements: A canny edge detection with two free parameters is performed under the assumption that cell-covered areas have more edges than cell-free areas. Consecutively, the resulting image is blurred by using a box blur filter to merge the previously found edges into an area. An intensity threshold is used to obtain a binary image. Furthermore, on the binary image, two size thresholds for contiguous small areas of black or white pixels are applied under the assumption that, most likely, small areas represent artifacts. Model 1 has 6 free parameters.
In the following, the variations of Models 2–6 in comparison with Model 1 are highlighted.

Model 2 uses a Gaussian filter instead of the box blur. While the Gaussian filter is computationally more expensive than the box blur, it is assumed to better reproduce the natural curves of the cells in the image than the rectangular box blur. Model 2 has 6 free parameters.

Model 3 uses the Sobel filter with a fixed parameter set (of kernel size 3 pixels) instead of the canny edge detection. The Sobel filter is a less computational expensive edge detection algorithm and a more general approach. Model 3 has 4 free parameters.

Model 4 also uses a Sobel filter. Further, Model 4 does not apply size thresholds for small areas of black or white pixels. Thus, it is the least computational expensive model in this series, but does not correct for any artifacts. Model 4 has 2 free parameters.

Model 5 is the same as Model 1 without size thresholds and, therefore, does also not correct for small artifacts. Model 5 has 4 free parameters.

Model 6 is similar to Model 2 as it uses Gaussian blurring, but instead of size thresholds, it applies opening and closing image filters, which are commonly used to filter out small regions. Model 6 has 6 free parameters.

2.4. Distance Metrics

To quantify the distance between the manually segmented and the pipeline-generated images, we focused on the border between the black and the white pixels since it condenses the most important information of the binary image. For two boundaries, we performed a pixelwise distance to the closest point operation [7]. We defined that the pixelwise distance to the closest point is always measured from the longer to the shorter boundary to make the distance metric. Further, the distance from the shorter to the longer boundary would result in a convergence of the length of the pipeline-generated image towards 0.

Thus, the distance between every pixel position a_i of the longer boundary A (of length m) and the shorter boundary B (of length n) is noted as follows:

$$d_i(a_i, BP_B) = \min_{j=1\ldots n} ||b_j - a_i|| \tag{1}$$

where $||x - y||$ denotes the Euclidean distance between two points and BP_B denotes all n points of the boundary B. We call the ensemble of all distances the boundary distance BD with $\{d_1 \ldots d_m\}$.

It is noteworthy that BD is a spatial metric, which is different from other commonly used pixel-based metrics. Therefore, it not only is more intuitive for intended users, such as physiologists, but also allows for assessing the spatial uncertainty from the Bayesian inference (see below).

2.5. Bayesian Inference

We used Bayesian inference to select the optimal image segmentation pipeline and consecutively perform a parameter estimation of the parameter set $\boldsymbol{\theta}$ of the selected image segmentation pipeline. An optimal image segmentation pipeline in the Bayesian logic requires sufficient free parameters to cover the characteristics of the image, but does not overfit using too many parameters (principle of Occam's razor; see below).

Derived from the general Bayes' theorem

$$P(A|B) = \frac{P(B|A) \times P(A)}{P(B)} \quad (2)$$

we want to assess the conditional probability $P(\boldsymbol{\theta}|K)$ of the parameter set $\boldsymbol{\theta}$ given the expert knowledge K. We choose the Bayesian a priori probability $P(\boldsymbol{\theta})$ of the parameter set $\boldsymbol{\theta}$ to be constant for all parameters as they represent spatial or intensity dimensions. The term $P(K)$ corresponds to the Bayesian evidence.

$$P(\boldsymbol{\theta}|K) = \frac{P(K|\boldsymbol{\theta}) \times P(\boldsymbol{\theta})}{P(K)} \quad (3)$$

The term $P(K|\boldsymbol{\theta})$ represents the Bayesian likelihood (LH) and describes the similarity between the manually segmented image using the expert knowledge K and a pipeline-generated image using a parameter set $\boldsymbol{\theta}$. Under the assumption of a Gaussian distributed uncertainty σ of the boundary distance $BD = \{d_1 \ldots d_m\}$ between the manually segmented and the pipeline-segmented image, we denote the likelihood as

$$LH(K|\boldsymbol{\theta}) = \prod_{i=1}^{m} \frac{1}{\sqrt{2\pi\sigma^2}} \times exp\left(-\frac{1}{2}\frac{d_i^2}{\sigma^2}\right) \quad (4)$$

Model selection will be performed based on the principle of Occam's razor, which avoids unnecessary and unjustified complexity. If large areas of the parameter space have a high likelihood, the Bayesian evidence $P(K)$ has a higher value compared with parameter-likelihood spaces with large areas of low likelihood [8]. This can be paraphrased more simple as Bayesian evidence is high if the ratio of high likelihood to parameter space given by the model is justified. Thus, Bayesian evidence is taking the complexity of the models into account during model evaluation and is therefore used for model selection. For computational ease, the logarithmic evidence $ln(Z)$ will be used throughout this work.

2.6. Numerical Implementation

We used the multinested sampling algorithm [8–10] in its Python implementation [11]. This approach can be summarized in a simplified way as follows: The prior-likelihood space is scattered with a cloud of a random set of live points. For each live point, the likelihood is evaluated. Once all likelihoods of all live points in the cloud are evaluated, the one with the lowest likelihood is eliminated and replaced by a new random point with a higher likelihood. This replacement is performed iteratively. During the iteration process, the evidence can be calculated as the weighted sum of the sorted likelihoods.

We want to investigate whether the number of live points is critical to ensure that the prior likelihood space is scanned correctly (see Section 3.1).

3. Results

3.1. Determining the Necessary Number of Live Points

As explained above, the number of live points is critical for a sufficient sampling of the prior-likelihood space. Here, we applied the introduced formalism with different numbers of live points (20, 50, 100, 200, 400, 800) using Model 1. For each number of live points, Bayesian inference was applied three times independently. Consecutively, the estimated evidence, the estimated uncertainty σ, and the total number of likelihood evaluations during the iterative process were evaluated.

Results are shown in Figure 3: For 100 live points and above, the estimated logarithmic evidence $ln(Z)$ and the estimated uncertainty σ remain stable. Thus, 100 live points were chosen throughout this work as they represent the optimized balance between computational workload and stability.

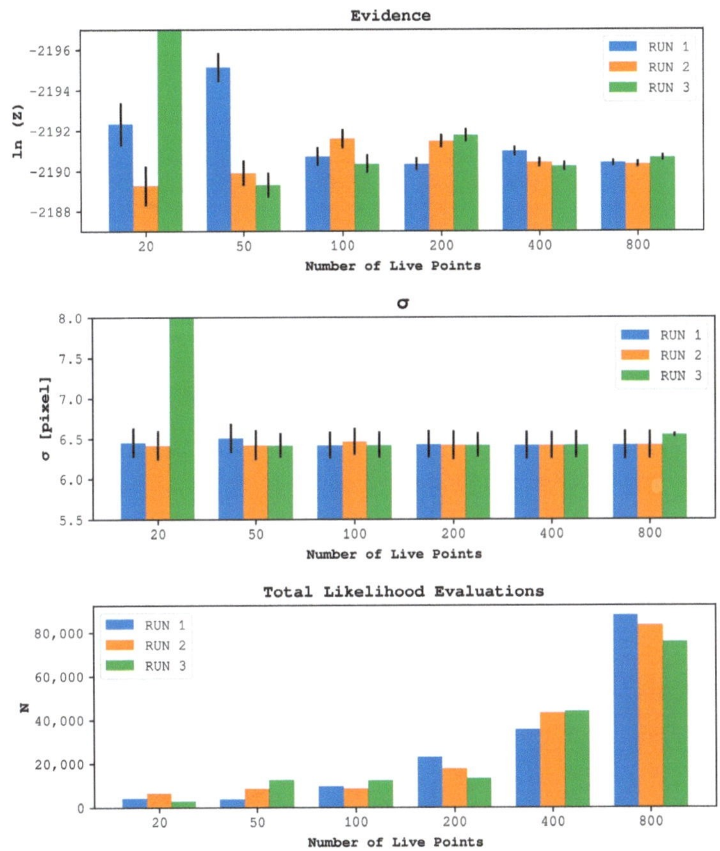

Figure 3. Evaluation of the necessary number of live points. This figure shows the results for estimated logarithmic evidence $ln(Z)$, estimated uncertainty σ, and the total number of likelihood evaluations (N) for three independent applications of the previously introduced formalism using either 20, 50, 100, 200, 400, or 800 live points. With 100 live points or more, the estimated evidences $ln(Z)$ and estimated uncertainties σ remain stable. Of the tested values with stable results, 100 live points require the least likelihood evaluations and are therefore computationally the most effective. (Data are shown as mean and Bayesian uncertainty (error bars) of the posterior distribution. Each estimation was independently run three times to evaluate reproducibility. In *RUN 3* with 20 live points, results are -3095.7 ± 1.2 for $ln(Z)$ and 40.5 ± 0.1 *pixel* for σ; both are off the charts. These extreme results indicate a failure due to very few live points. For illustrative purposes, they were not taken into consideration for the limits of the y-axes.)

3.2. Image Segmentation Using the Estimated Posterior Parameters

Using Bayesian inference, we obtained the posterior distributions for parameter sets of Models 1–6. Consecutively, these posteriors were applied to obtain pipeline-generated images (see Figure 4, red and blue). At first sight, small differences in the segmentation of Models 1–6 became obvious. However, with the naked eye, it is infeasible to tell which algorithm is closest to the manually segmented image (see Figure 4, green). Please note that, for example, Models 4 and 6 are the only two models that fully surround the small island of cells in the lower center right of the original figure—as it was suggested by the manually segmented image. Furthermore, it seems impossible to decide which algorithm to select over the others only by looking at Figure 4.

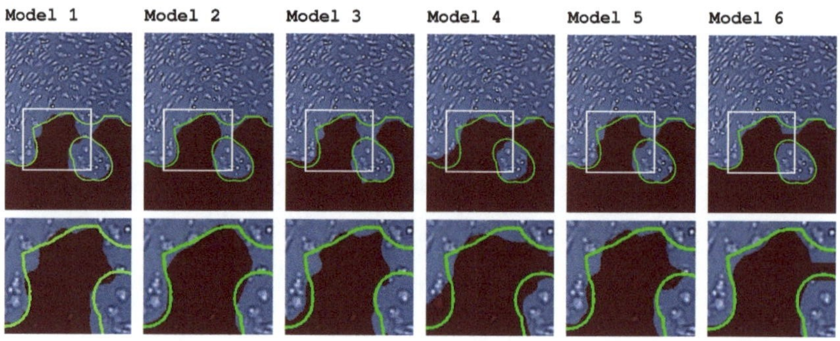

Figure 4. Image segmentation for Models 1–6. After applying the above-introduced formalism, estimated posterior parameters were used to obtain one pipeline-segmented image for Models 1–6. These images are overlays of the original input image (see Figure 1A). The pipeline-generated image is superimposed with dark red indicating a cell-free area and blue indicating a cell-covered area. The green line represents the boundary between a cell-free and a cell-covered area in the manually segmented image (see Figure 1D). The white box shows a region of interest, which is magnified in the lower part of the figure to magnify the details.

3.3. Choosing the Image Segmentation Pipeline

Bayesian inference allows for choosing an image segmentation pipeline based on the Bayesian evidence, which is the probability that the given pipeline is capable of recreating the expert knowledge. Thus, the image segmentation with the highest Bayesian evidence should be chosen. According to Table 1, Model 2 has the highest Bayesian evidence. Additionally, Model 2 shows the lowest uncertainty σ.

Table 1. Comparison of the different models based on Bayesian evidence (best results per column are marked in bold font).

	ln(Z)	σ [pxl]	Calc. Time *
Model 1	-2190.1 ± 0.5	6.40 ± 0.16	02 min 59 s
Model 2	$\mathbf{-2098.4 \pm 0.5}$	$\mathbf{5.62 \pm 0.14}$	11 min 15 s
Model 3	-2299.1 ± 0.4	7.48 ± 0.21	01 min 11 s
Model 4	-2559.5 ± 0.4	11.40 ± 0.34	**00 min 20 s**
Model 5	-2256.1 ± 0.5	7.09 ± 0.19	01 min 03 s
Model 6	-2112.2 ± 0.5	5.70 ± 0.15	27 min 23 s

* Processor: i5-2520M CPU @ 2.50 GHz.

4. Discussion

While multiple image segmentation implementations focus on Bayesian decision trees for pixel-based classification [12], our group previously demonstrated the usage of Bayesian inference to perform a parameter estimation for a pre-setup imaging pipeline based on the segmentation of an input image, which is trained with expert knowledge. A possible advantage of this approach might consist of more consistently taking into account spatial correlations such as textures within image data than pixel-based Bayesian decision trees. Contrary to pixel-based classification, this approach uses a metric distance of boundaries of conjoined regions, which is more intuitive than pixel-based uncertainties in the perception and processing. This approach enables the logically consistent handling of Bayesian uncertainties of segmentation boundaries.

Our approach enables experimenters with little knowledge in computer science to manually segment one image of an image series and then use this expert knowledge to automatically choose the best image segmentation pipeline from a collection of tested pipelines with an optimized parameter set. In this work, we were able to demonstrate that (I) the needed number of live points can be assessed in a simple straightforward approach. However, it is worth mentioning that other implementations of nested sampling have the ability to correct the number of live points on the run [13]. (II) Objectively assessing the quality of pipeline-generated images is infeasible with the naked eye. (III) Bayesian evidence logically consistently enables the systematic selection of the best image segmentation pipeline among those tested.

As a future prospect, the developed technique can be further applied to other segmentation problems, such as time lapse series or 3D volume data.

Author Contributions: Conceptualization, M.L.M. and P.D.; methodology, M.L.M. and P.D.; software, M.L.M.; validation, M.L.M., A.D., and P.D.; formal analysis, M.L.M. and P.D.; resources, A.D. and P.D.; writing—original draft preparation, M.L.M.; writing—review and editing, M.L.M. and P.D.; visualization, M.L.M.; supervision, A.D. and P.D. All authors have read and agreed to the published version of the manuscript.

Funding: This research received no external funding.

Institutional Review Board Statement: Not applicable.

Informed Consent Statement: Not applicable.

Data Availability Statement: Data sharing is not applicable to this article.

Conflicts of Interest: The authors declare no conflict of interest.

References

1. Moskopp, M.L.; Deussen, A.; Dieterich, P. Bayesian inference for the automated adjustment of an image segmentation pipeline—A modular approach applied to wound healing assays. *Knowl.-Based Syst.* **2019**, *173*, 52–61. [CrossRef]
2. Canny, J. A Computational Approach to Edge Detection. *IEEE Trans. Pattern Anal. Mach. Intell.* **1986**, *PAMI-8*, 679–698. [CrossRef]
3. Burger, W.; Burge, M.J. *Digitale Bildverarbeitung: Eine Einführung mit Java und ImageJ; mit 16 Tabellen*; eXamen.press; Springer: Berlin/Heidelberg, Germany, 2005.
4. Jonkman, J.E.N.; Cathcart, J.A.; Xu, F.; Bartolini, M.E.; Amon, J.E.; Stevens, K.M.; Colarusso, P. An introduction to the wound healing assay using live-cell microscopy. *Cell Adhes. Migr.* **2014**, *8*, 440–451. [CrossRef] [PubMed]
5. Virtanen, P.; Gommers, R.; Oliphant, T.E.; Haberland, M.; Reddy, T.; Cournapeau, D.; Burovski, E.; Peterson, P.; Weckesser, W.; Bright, J.; et al. SciPy 1.0: Fundamental algorithms for scientific computing in Python. *Nat. Methods* **2020**, *17*, 261–272. [CrossRef] [PubMed]
6. *The OpenCV Reference Manual*; Manual, Edition: 2.4.13.7.; OpenCV: Palo Alto, CA, USA, 2014.
7. Chalana, V.; Kim, Y. A methodology for evaluation of boundary detection algorithms on medical images. *IEEE Trans. Med. Imaging* **1997**, *16*, 642–652. [CrossRef] [PubMed]
8. Feroz, F.; Hobson, M.P. Multimodal nested sampling: An efficient and robust alternative to Markov Chain Monte Carlo methods for astronomical data analyses: Multimodal nested sampling. *Mon. Not. R. Astron. Soc.* **2008**, *384*, 449–463. [CrossRef]
9. Skilling, J. Nested sampling for general Bayesian computation. *Bayesian Anal.* **2006**, *1*, 833–859. [CrossRef]
10. Von Toussaint, U. Bayesian inference in physics. *Rev. Mod. Phys.* **2011**, *83*, 943–999. [CrossRef]
11. Buchner, J.; Georgakakis, A.; Nandra, K.; Hsu, L.; Rangel, C.; Brightman, M.; Merloni, A.; Salvato, M.; Donley, J.; Kocevski, D. X-ray spectral modelling of the AGN obscuring region in the CDFS: Bayesian model selection and catalogue. *Astron. Astrophys.* **2014**, *564*, A125. [CrossRef]
12. Arganda-Carreras, I.; Kaynig, V.; Rueden, C.; Eliceiri, K.W.; Schindelin, J.; Cardona, A.; Sebastian Seung, H. Trainable Weka Segmentation: A machine learning tool for microscopy pixel classification. *Bioinformatics* **2017**, *33*, 2424–2426. [CrossRef] [PubMed]
13. Buchner, J. UltraNest—A robust, general purpose Bayesian inference engine. *J. Open Source Softw.* **2021**, *6*, 3001. [CrossRef]

physical sciences forum

MDPI

Proceeding Paper

Flow Annealed Kalman Inversion for Gradient-Free Inference in Bayesian Inverse Problems †

Richard D. P. Grumitt [1,*], Minas Karamanis [2,3] and Uroš Seljak [2,3]

1 Department of Astronomy, Tsinghua University, Beijing 100084, China
2 Berkeley Center for Cosmological Physics and Department of Physics, University of California, Berkeley, CA 94720, USA; mkaramanis@berkeley.edu (M.K.); useljak@berkeley.edu (U.S.)
3 Physics Department, Lawrence Berkeley National Laboratory, Cyclotron Rd, Berkeley, CA 94720, USA
* Correspondence: rgrumitt@mail.tsinghua.edu.cn

† Presented at the 42nd International Workshop on Bayesian Inference and Maximum Entropy Methods in Science and Engineering, Garching, Germany, 3–7 July 2023.

Abstract: For many scientific inverse problems, we are required to evaluate an expensive forward model. Moreover, the model is often given in such a form that it is unrealistic to access its gradients. In such a scenario, standard Markov Chain Monte Carlo algorithms quickly become impractical, requiring a large number of serial model evaluations to converge on the target distribution. In this paper, we introduce Flow Annealed Kalman Inversion (FAKI). This is a generalization of Ensemble Kalman Inversion (EKI) where we embed the Kalman filter updates in a temperature annealing scheme and use normalizing flows (NFs) to map the intermediate measures corresponding to each temperature level to the standard Gaussian. Thus, we relax the Gaussian ansatz for the intermediate measures used in standard EKI, allowing us to achieve higher-fidelity approximations to non-Gaussian targets. We demonstrate the performance of FAKI on two numerical benchmarks, showing dramatic improvements over standard EKI in terms of accuracy whilst accelerating its already rapid convergence properties (typically in $\mathcal{O}(10)$ steps).

Keywords: inverse problems; Bayesian inference; normalizing flows; Ensemble Kalman Inversion; gradient-free

1. Introduction

Many scientific inference tasks are concerned with inverse problems of the form

$$y = \mathcal{G}(x) + \eta, \tag{1}$$

where $y \in \mathbb{R}^{d_y}$ are the data, $x \in \mathbb{R}^d$ are the model parameters, $\mathcal{G}$ is the forward map, and η is the observation noise. Throughout this work, we will assume that we do not have access to gradients of $\mathcal{G}$ with respect to the parameters, and that $\eta \sim \mathcal{N}(0, \Gamma)$, where Γ is a fixed noise covariance. The assumption of additive Gaussian noise is the standard setting for Ensemble Kalman Inversion (EKI) [1–8], and whilst we are restricted to problems with Gaussian likelihoods, this covers a large family of scientific inverse problems. The goal of the Bayesian inverse problem is then to recover the posterior distribution over the model parameters given our observations, $p(x|y)$.

Typical gradient-free inference methods often involve some variant on Markov Chain Monte Carlo (MCMC) algorithms, e.g., random walk Metropolis [9–11] or Sequential Monte Carlo (SMC) [12]. However, these methods typically require $\gtrsim 10^3$ serial model evaluations to achieve convergence, making them intractable for problems with expensive forward models. In contrast, EKI utilizes embarrassingly parallel model evaluations to update parameter estimates, typically converging to an approximate solution in $\mathcal{O}(10)$ iterations [2,6–8].

EKI leverages ideas originally developed in the context of Ensemble Kalman Filtering (EKF) for data assimilation [13]. Since its development, EKI has seen applications across a range of disciplines, including studies of fluid flow [14], climate models [15], and machine learning tasks [16]. EKI can be understood in the context of annealing, where we seek to move from the prior to the posterior through a sequence of intermediate measures. In standard EKI, this involves constructing a sequence of Gaussian approximations to the intermediate measures. In the regime where we have a Gaussian prior $\pi_0(x) = \mathcal{N}(m_0, C_0)$ and a linear forward model $\mathcal{G}(x) = Gx$, the particle distribution obtained via EKI converges to the true posterior in the limit where the ensemble size $J \to \infty$. However, outside this linear Gaussian regime, EKI is an uncontrolled approximation to the posterior that is constructed on the basis of matching first and second moments of the target distribution. Nonetheless, EKI has been shown to perform well on problems with non-linear forward models and slightly non-Gaussian targets [1,2,6].

In this work, we propose the application of normalizing flows (NFs) [17–20] to relax the Gaussian ansatz made by standard EKI for the intermediate measures. Instead of assuming a Gaussian particle distribution at each iteration, the NF is used to fit the empirical particle distribution and map it to a Gaussian latent space, where the EKI updates are performed. Thus, we are better able to capture non-Gaussian target geometries. The structure of this paper is as follows: in Section 2 we describe the Flow Annealed Kalman Inversion (FAKI) algorithm; in Section 3, we demonstrate the performance of the method on two Bayesian inference tasks with non-Gaussian target geometries; and we summarize our work in Section 4.

2. Methods

2.1. Regularized Ensemble Kalman Inversion

A number of versions of EKI have been proposed in the literature. Of interest here is the regularized, perturbed observation form of EKI [6]. Starting with an ensemble of particles drawn from the prior, $\{x_0^j\}_{j=1}^J$, the particles are updated at each iteration according to

$$x_{n+1}^j = x_n^j + C_n^{x\mathcal{G}}(C_n^{\mathcal{G}\mathcal{G}} + \alpha_n\Gamma)^{-1}(y - \mathcal{G}(x_n^j) + \sqrt{\alpha_n}\xi_n^j). \tag{2}$$

The empirical covariances $C_n^{x\mathcal{G}}$ and $C_n^{\mathcal{G}\mathcal{G}}$ are given by

$$C_n^{x\mathcal{G}} = \frac{1}{J-1}\sum_{j=1}^{J}(x_n^j - \langle x_n\rangle) \otimes (\mathcal{G}(x_n^j) - \langle \mathcal{G}_n\rangle), \tag{3}$$

$$C_n^{\mathcal{G}\mathcal{G}} = \frac{1}{J-1}\sum_{j=1}^{J}(\mathcal{G}(x_n^j) - \langle \mathcal{G}_n\rangle) \otimes (\mathcal{G}(x_n^j) - \langle \mathcal{G}_n\rangle). \tag{4}$$

At each iteration, we perturb the forward model evaluations with the Gaussian observation noise $\xi_n^j \sim \mathcal{N}(0, \Gamma)$. The parameter α_n is a Tikhonov regularization parameter, which can be viewed as an inverse step size in the Bayesian annealing context. In particular, given a set of annealing parameters $\beta_0 \equiv 0 < \beta_1 < \ldots < \beta_N < \beta_{N+1} \equiv 1$, we have the corresponding set of target distributions

$$\pi_n(x) \propto \pi_0(x)\exp\left(-\frac{\beta_n}{2}\left\|\Gamma^{-1/2}(y - \mathcal{G}(x))\right\|^2\right), \tag{5}$$

with

$$\alpha_n = \beta_{n+1} - \beta_n. \tag{6}$$

EKI proceeds by constructing a sequence of ensemble approximations to Gaussian distributions that approximate the intermediate targets.

The choice of the regularization parameter, α_n, controls the transition from the prior to the posterior. Previous proposals for an adaptive choice have taken inspiration from SMC by using a threshold on the effective sample size (ESS) of the particles at each temperature level [21,22]. In this work, we adopt the same approach, calculating pseudo-importance weights at each temperature given by

$$w_n^j = \exp\left(-\frac{1}{2}(\beta_{n+1} - \beta_n)\left\|\Gamma^{-1/2}(y - \mathcal{G}(x_n^j))\right\|^2\right). \tag{7}$$

The next temperature level can then be selected by solving

$$\left(\sum_{j=1}^{J} w_n^j(\beta_{n+1})^2\right)^{-1}\left(\sum_{j=1}^{J} w_n^j(\beta_{n+1})\right)^2 = \tau J, \tag{8}$$

using the bisection method, where $0 < \tau < 1$ is the target fractional ESS threshold. Throughout our work we set $\tau = 0.5$. The full pseudo-code for EKI is given in Algorithm 1.

Algorithm 1 Ensemble Kalman Inversion

Input: J prior samples $\{x_0^j \sim \pi_0(x)\}_{j=1}^J$, data y, observation error covariance Γ and fractional ESS target threshold τ
Initialize inverse temperature $\beta_0 = 0$, iteration counter $n = 0$
while $\beta < 1$ **do**
  Evaluate $\mathcal{G}_j = \mathcal{G}(x_n^j)$, $j \in \{1, \ldots, J\}$
  Solve for β_{n+1} using the bisection method with Equation (8)
  $\alpha_n \leftarrow \beta_{n+1} - \beta_n$
  Update particles using Equation (2) to obtain $\{x_{n+1}^j\}_{j=1}^J$
  $n \leftarrow n + 1$
end while
Output: Converged particle ensemble $\{x_N^j\}_{j=1}^J$

2.2. Normalizing Flows

As discussed above, standard EKI proceeds by constructing a sequence of ensemble approximations to Gaussian distributions. The procedure works well in the situation where the target and all the intermediate measures are close to Gaussian. However, when any of these measures are far from Gaussian, EKI can dramatically fail to capture the final target geometry.

To address this shortcoming, we propose the use of NFs to approximate each intermediate target, instead of using the Gaussian ansatz of standard EKI. NFs are powerful generative models that can be used for flexible density estimation and sampling [17–20]. An NF model maps from the original space $x \in \mathbb{R}^d$ to a latent space $z \in \mathbb{R}^d$ through a sequence of invertible transformations $f = f_1 \circ f_2 \circ \ldots \circ f_L$, such that we have a bijective mapping $z = f(x)$. The mapping is such that the latent variables are mapped to some simple base distribution, typically chosen to be the standard Normal distribution, giving $z \sim p_z(z) = \mathcal{N}(0, I)$.

The NF density can be evaluated through the change of variables formula,

$$q(x) = p_z(f(x))|\det Df(x)| = p_z(f(x))\prod_{l=1}^{L}|\det Df_l(x)|, \tag{9}$$

where $Df(x) = \partial f(x)/\partial x$ denotes the Jacobian of f. The efficient evaluation of this density requires the Jacobian of the transformation to be easy to evaluate, and efficient sampling requires the inverse of the mapping f to be easy to calculate. In this work, we use Masked Autoregressive Flows (MAFs) [18], which have previously been found to perform well in the context of preconditioned MCMC sampling within SMC without the need for expensive hyper-parameter searches during sampling [23].

2.3. Flow Annealed Kalman Inversion

Given particles distributed as $\pi_n(x)$, the subsequent target can be written as

$$\pi_{n+1}(x) \propto \pi_n(x) \exp\left(-\frac{1}{2\alpha_n}\left\|\Gamma^{-1/2}(y - \mathcal{G}(x))\right\|^2\right). \tag{10}$$

We may therefore view $\pi_n(x)$, i.e., the posterior at the temperature level β_n, as an effective prior for $\pi_{n+1}(x)$, with a data likelihood annealed by α_n^{-1}. By fitting an NF to the particles $\{x_n^j\}_{j=1}^J$, we obtain an approximate map from the intermediate target $\pi_n(x)$ to $\mathcal{N}(z|0, I)$. The latent space target is then given by the change of variables formula as

$$\pi_{n+1}(z) = \pi_{n+1}(x = f_n^{-1}(z))\left|\det Df_n^{-1}(z)\right|. \tag{11}$$

By controlling the choice of α_n, we control the distance between the Gaussianized effective prior and this latent space target density. For FAKI, we therefore perform the EKI updates in the NF latent space at each temperature level, allowing us to relax the Gaussian ansatz of standard EKI by constructing an approximate map from each $\pi_n(x)$ to a Gaussian latent space. It is worth noting that, whilst this method relaxes the Gaussianity assumptions of standard EKI, it does not address the linearity assumptions used in deriving EKI.

The FAKI update for the latent space particle locations is given by

$$z_{n+1}^j = z_n^j + C_n^{z\mathcal{G}}(C_n^{\mathcal{G}\mathcal{G}} + \alpha_n\Gamma)^{-1}(y - \mathcal{G}(f_n^{-1}(z_n^j)) + \sqrt{\alpha_n}\xi_n^j), \tag{12}$$

where the latent space empirical covariances are given by

$$C_n^{z\mathcal{G}} = \frac{1}{J-1}\sum_{j=1}^{J}(z_n^j - \langle z_n\rangle) \otimes (\mathcal{G}(f_n^{-1}(z_n^j)) - \langle \mathcal{G}_n\rangle), \tag{13}$$

$$C_n^{\mathcal{G}\mathcal{G}} = \frac{1}{J-1}\sum_{j=1}^{J}(\mathcal{G}(f_n^{-1}(z_n^j)) - \langle \mathcal{G}_n\rangle) \otimes (\mathcal{G}(f_n^{-1}(z_n^j)) - \langle \mathcal{G}_n\rangle). \tag{14}$$

The full pseudocode for FAKI is given in Algorithm 2.

Algorithm 2 Flow Annealed Kalman Inversion

Input: J prior samples $\{x_0^j \sim \pi_0(x)\}_{j=1}^J$, data y, observation error covariance Γ and fractional ESS target threshold τ
Initialize inverse temperature $\beta_0 = 0$, iteration counter $n = 0$
while $\beta < 1$ **do**
  Evaluate $\mathcal{G}_j = \mathcal{G}(x_n^j), j \in \{1, \ldots, J\}$
  Solve for β_{n+1} using the bisection method with Equation (8)
  $\alpha_n \leftarrow \beta_{n+1} - \beta_n$
  Fit NF map f_n to current samples $\{x_n^j\}_{j=1}^J$
  Map particles to latent space $z_n^j = f_n(x_n^j), j \in \{1, \ldots, J\}$
  Update particles using Equation (12) to obtain $\{z_{n+1}^j\}_{j=1}^J$
  Map back to the data space $x_{n+1}^j = f_n^{-1}(z_{n+1}^j), j \in \{1, \ldots, J\}$
  $n \leftarrow n + 1$
end while
Output: Converged particle ensemble $\{x_N^j\}_{j=1}^J$

3. Results

In this section, we demonstrate the performance of FAKI compared to standard EKI on two numerical benchmarks, a two-dimensional Rosenbrock distribution and a stochastic

Lorenz system [24,25]. Both models display significant non-Gaussianity at some point during the transition from prior to posterior, severely frustrating the performance of EKI. This manifests in both the reduced fidelity of the final ensemble approximations to the posterior and a larger number of iterations being required for convergence following the ESS-based annealing scheme described in Section 2.1.

In Table 1, we provide statistics summarizing the performance of EKI and FAKI on our numerical benchmarks. We measured the quality of the posterior approximations by computing the 1-Wasserstein distance, W_1 [26,27], between the samples obtained through FAKI and EKI against reference posterior samples obtained via long runs of the Hamiltonian Monte Carlo (HMC) algorithm [28,29]. These reference samples were thinned to be approximately independent when computing the 1-Wasserstein distances (we use the Python Wasserstein library: https://github.com/pkomiske/Wasserstein/ (accessed on 30 June 2023)). The 1-Wasserstein distance may be interpreted as the cost involved in rearranging one probability measure to look like another, with lower values indicating the two probability measures are closer to one another. In addition to this assessment of the approximation quality, we report the number of iterations, N_{iter}, required by FAKI and EKI for convergence. For both quantities, we report the median and median absolute deviation (MAD) estimated over 10 independent runs using different random seeds.

Table 1. Median and MAD values of N_{iter} and 1-Wasserstein distances for each model and algorithm combination, calculated over 10 independent runs using different random seeds. For both numerical benchmarks, we see that FAKI resulted in a reduced number of iterations for convergence and a lower value of the 1-Wasserstein distance between the converged samples and the ground truth.

Model	Algorithm	Median (N_{iter})	MAD (N_{iter})	Median (W_1)	MAD (W_1)
Rosenbrock	EKI	100	7.0	0.72	0.05
Rosenbrock	FAKI	34.0	7.0	0.43	0.14
Lorenz	EKI	10.0	0.0	69.8	1.08
Lorenz	FAKI	8.0	0.0	5.65	0.86

3.1. $d = 2$ Rosenbrock

In our first numerical experiment, we considered the two-dimensional Rosenbrock distribution. This toy model allowed us to clearly see the impact of non-Gaussianity on the performance of EKI and how FAKI is able to alleviate these issues. For the Rosenbrock model, we assumed a Gaussian prior over the parameters $x \in \mathbb{R}^2$,

$$x \sim \mathcal{N}(0, 10^2 I). \tag{15}$$

The data, $y \in \mathbb{R}^2$, were distributed according to the likelihood

$$y \sim \mathcal{N}(\mathcal{G}(x) = (x_1 - x_0^2, x_0)^\top, \Gamma = \text{diag}(0.01^2, 1^2)). \tag{16}$$

To generate simulated data, we evaluated $y = G((1,1)^\top) + \eta$, where $\eta \sim \mathcal{N}(0, \text{diag}(0.01^2, 1^2))$. The large difference in noise scales resulted in a highly non-Gaussian posterior geometry that posed a significant challenge for EKI. For each run of EKI and FAKI, we used 100 particles.

In Figure 1, we show pair plots comparing the final particle distributions obtained with EKI and FAKI against samples obtained through a long run of HMC. The NF mapping meant that the ensemble approximation obtained by FAKI was able to capture the highly non-linear target geometry. In comparison, EKI struggled to fill the tails of the Rosenbrock target. Moreover, whilst FAKI converged within ∼34 iterations, EKI required a median number of ∼100 iterations to converge using the ESS-based annealing scheme.

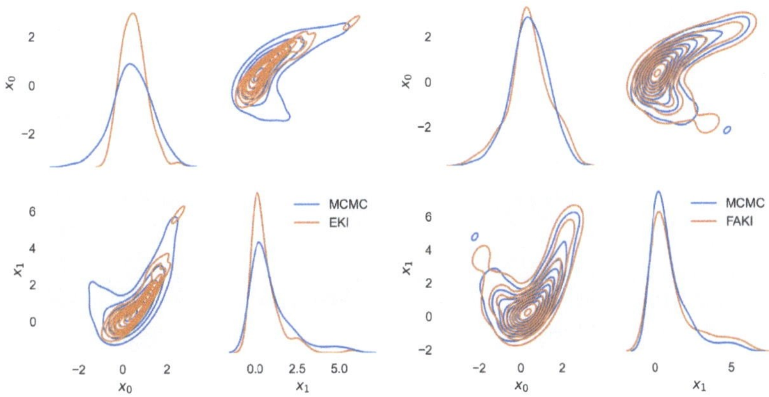

(**a**) EKI–MCMC comparison (**b**) FAKI–MCMC comparison

Figure 1. Pair plots for the Rosenbrock target. Panel (**a**): pair-plot comparison of samples from EKI and a long HMC run. Panel (**b**): pair-plot comparison of samples from FAKI and a long HMC run. Samples from FAKI were able to correctly capture the highly non-linear target geometry. Standard EKI struggled to fill the tails of the target and required ∼100 iterations to converge, compared to ∼34 iterations for FAKI.

3.2. Stochastic Lorenz System

The Lorenz equations are a set of coupled differential equations used as a simple model of atmospheric convection. Notably, for certain parameter values, the Lorenz equations are known to exhibit chaotic behavior [24]. In this work, we followed [25] and considered the stochastic Lorenz system

$$\mathrm{d}X_t = 10(Y_t - X_t)\mathrm{d}t + \mathrm{d}W_t^x, \tag{17}$$
$$\mathrm{d}Y_t = X_t(28 - Z_t)\mathrm{d}t - Y_t\mathrm{d}t + \mathrm{d}W_t^y, \tag{18}$$
$$\mathrm{d}Z_t = X_tY_t\mathrm{d}t - \frac{8}{3}Z_t\mathrm{d}t + \mathrm{d}W_t^z, \tag{19}$$

where W_t^x, W_t^y, and W_t^z are Gaussian white noise processes with standard deviation $\sigma_0 = 0.1$. To generate simulated, data we integrated these equations using a Euler–Maruyama scheme with $\mathrm{d}t = 0.02$ for 30 steps, with initial conditions $X_0, Y_0, Z_0 \sim \mathcal{N}(0, 1^2)$. The observations were then taken to be the X_t values over these 30 time steps, with Gaussian observational noise $\eta_t \sim \mathcal{N}(0, \sigma^2 = 1^2)$.

The goal of our inference here was to recover the initial conditions, the trajectories (X_t, Y_t, Z_t) and the innovation noise scale σ_0, giving a parameter space of $d = 94$ dimensions. We assigned priors over these parameters as

$$\log \sigma_0 \sim \mathcal{N}(-1, 1^2), \tag{20}$$
$$X_0, Y_0, Z_0 \sim \mathcal{N}(0, 1^2), \tag{21}$$
$$X_t \sim \mathcal{N}(X_{t-1} + f_X(X_{t-1}, Y_{t-1}, Z_{t-1}, t-1)\mathrm{d}t, \sigma_0^2\mathrm{d}t), t \in \{1, \ldots, 30\}, \tag{22}$$
$$Y_t \sim \mathcal{N}(Y_{t-1} + f_Y(X_{t-1}, Y_{t-1}, Z_{t-1}, t-1)\mathrm{d}t, \sigma_0^2\mathrm{d}t), t \in \{1, \ldots, 30\}, \tag{23}$$
$$Z_t \sim \mathcal{N}(Z_{t-1} + f_Z(X_{t-1}, Y_{t-1}, Z_{t-1}, t-1)\mathrm{d}t, \sigma_0^2\mathrm{d}t), t \in \{1, \ldots, 30\}, \tag{24}$$

where f_X, f_Y, and f_Z are the transition functions corresponding to Equations (17)–(19), respectively. The Gaussian likelihood has the form

$$\hat{X}_t \sim \mathcal{N}(X_t, \sigma^2), t \in \{1, \ldots, 30\}, \tag{25}$$

where $\hat{X}_t$ are the observations of the X_t trajectory. The chaotic dynamics of the Lorenz system resulted in a highly non-Gaussian prior distribution, with the inversion having to proceed through a sequence of highly non-Gaussian intermediate measures towards the posterior. This severely frustrated the performance of EKI, with the Gaussian ansatz failing to describe the geometry of the intermediate measures. For each run of EKI and FAKI, we used 940 particles.

In Figure 2, we show the ensemble estimates for the mean and standard deviation along each dimension obtained by EKI and FAKI, compared to reference estimates obtained through long runs of HMC. FAKI was able to obtain accurate mean estimates along each dimension, whereas EKI was unable to obtain the correct means for much of the Z_t trajectory. EKI severely overestimated the marginal standard deviations along many dimensions. This situation was alleviated by the NF mappings learned by FAKI. The greater fidelity of the FAKI posterior approximations is reflected in the median estimates for the 1-Wasserstein distances, with a value of 5.65 for FAKI and 69.8 for EKI.

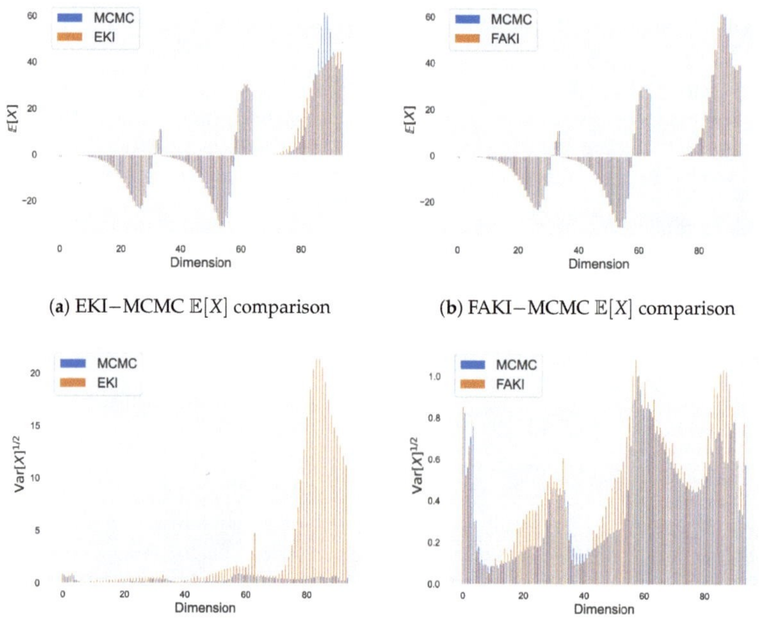

(**a**) EKI−MCMC $\mathbb{E}[X]$ comparison (**b**) FAKI−MCMC $\mathbb{E}[X]$ comparison

(**c**) EKI−MCMC $\mathrm{Var}[X]^{1/2}$ comparison (**d**) FAKI−MCMC $\mathrm{Var}[X]^{1/2}$ comparison

Figure 2. Comparison of first and second moment estimates along each dimension for the stochastic Lorenz system. Panel (**a**): comparison between the mean estimates from EKI and a long HMC run. Panel (**b**): comparison between the mean estimates from FAKI and a long HMC run. Panel (**c**): comparison between the standard deviation estimates from EKI and a long HMC run. Panel (**d**): comparison between the standard deviation estimates from FAKI and a long HMC run. Blue bars indicate the moment estimates obtained via HMC along each dimension, with the adjacent orange bars showing the estimates obtained through EKI/FAKI. EKI was unable to obtain accurate mean estimates for much of the Z_t trajectory, whilst FAKI was able to obtain accurate mean estimates for each dimension. FAKI outperformed EKI in its estimates of the marginal standard deviations, with EKI drastically overestimating the standard deviations along many dimensions.

4. Conclusions

In this work, we introduced Flow Annealed Kalman Inversion (FAKI), a gradient-free inference algorithm for Bayesian inverse problems with expensive forward models. This is a generalization of Ensemble Kalman Inversion (EKI) where we utilize normalizing flows (NFs) to replace the Gaussian ansatz made in EKI. Instead of constructing a sequence of ensemble approximations to Gaussian measures that approximate a sequence of intermediate measures, as we move from the prior to the posterior, we learn an NF mapping at each iteration to a Gaussian latent space. Provided the transition between temperature levels is controlled, we can perform Kalman inversion updates in the NF latent space. In the NF latent space, the Gaussianity assumptions of EKI are more closely satisfied, resulting in a more stable inversion at each temperature level.

We demonstrated the performance of FAKI on two numerical benchmarks, a $d = 2$ Rosenbrock distribution and a $d = 94$ stochastic Lorenz system. Both examples exhibited significant non-Gaussianity in the transition from prior to posterior that frustrated standard EKI. In the presence of strong non-Gaussianity, we found that FAKI produced higher-fidelity posterior approximations compared to EKI, as measured by the 1-Wasserstein distance between FAKI/EKI samples and reference HMC samples. In addition to the improved fidelity of the posterior approximations, we found that FAKI tended to reduce the number of iterations required for convergence.

Whilst the application of NFs is able to relax the Gaussian ansatz of EKI, it does not address the linearity assumptions used in deriving EKI. As such, FAKI is still not exact for general forward models. In future work, it will be interesting to explore methods to address this: for example, the combination of FAKI with unbiased MCMC or importance sampling methods. It would also be interesting to consider generalizations of FAKI that are able to accommodate non-Gaussian likelihoods and/or parameter-dependent noise covariances. The use of NFs means that we typically require ensemble sizes $J \gtrsim 10d$ to learn accurate NF maps with the MAF architecture employed in this work. It would be useful to explore alternative NF architectures and regularization schemes that are able to learn accurate NF maps with smaller ensemble sizes, in order to enable FAKI to scale to higher dimensions. In this work, we found that the MAF architecture is able to capture a wide range of target geometries without the need for expensive NF hyper-parameter searches. However, it may be possible to exploit NF architectures with inductive biases that are particularly suited to common target geometries, e.g., the non-linear correlations that often appear in hierarchical models.

Author Contributions: Conceptualization, U.S.; methodology, R.D.P.G. and U.S.; software, R.D.P.G. and M.K.; validation, R.D.P.G.; formal analysis, R.D.P.G.; investigation, R.D.P.G. and M.K.; resources, U.S.; data curation, R.D.P.G. and M.K.; writing—original draft preparation, R.D.P.G.; writing—review and editing, R.D.P.G., M.K. and U.S.; visualization, R.D.P.G.; supervision, U.S.; project administration, U.S.; funding acquisition, R.D.P.G. and U.S. All authors have read and agreed to the published version of the manuscript.

Funding: This research was funded by NSFC (grant No. 12250410240) and the U.S. Department of Energy, Office of Science, Office of Advanced Scientific Computing Research under Contract No. DE-AC02-05CH11231 at Lawrence Berkeley National Laboratory to enable research for Data-intensive Machine Learning and Analysis. RDPG was supported by a Tsinghua Shui Mu Fellowship.

Institutional Review Board Statement: Not applicable.

Informed Consent Statement: Not applicable.

Data Availability Statement: The code and data underlying this work can be made available upon contacting the corresponding author.

Acknowledgments: The authors thank Qijia Jiang and David Nabergoj for helpful discussions.

Conflicts of Interest: The authors declare no conflicts of interest.

References

1. Iglesias, M.A.; Law, K.J.; Stuart, A.M. Ensemble Kalman methods for inverse problems. *Inverse Probl.* **2013**, *29*, 045001. [CrossRef]
2. Iglesias, M.A. A regularizing iterative ensemble Kalman method for PDE-constrained inverse problems. *Inverse Probl.* **2016**, *32*, 025002. [CrossRef]

3. Schillings, C.; Stuart, A.M. Analysis of the ensemble Kalman filter for inverse problems. *SIAM J. Numer. Anal.* **2017**, *55*, 1264–129 [CrossRef]
4. Chada, N.K.; Iglesias, M.A.; Roininen, L.; Stuart, A.M. Parameterizations for ensemble Kalman inversion. *Inverse Probl.* **201** *34*, 055009. [CrossRef]
5. Schillings, C.; Stuart, A.M. Convergence analysis of ensemble Kalman inversion: The linear, noisy case. *Appl. Anal.* **201** *97*, 107–123. [CrossRef]
6. Iglesias, M.; Yang, Y. Adaptive regularisation for ensemble Kalman inversion. *Inverse Probl.* **2021**, *37*, 025008. [CrossRef]
7. Huang, D.Z.; Schneider, T.; Stuart, A.M. Iterated Kalman methodology for inverse problems. *J. Comput. Phys.* **2022**, *463*, 11126 [CrossRef]
8. Huang, D.Z.; Huang, J.; Reich, S.; Stuart, A.M. Efficient derivative-free Bayesian inference for large-scale inverse problem *Inverse Probl.* **2022**, *38*, 125006. [CrossRef]
9. Geyer, C.J. Practical Markov Chain Monte Carlo. *Stat. Sci.* **1992**, *7*, 473–483. [CrossRef]
10. Gelman, A.; Gilks, W.R.; Roberts, G.O. Weak convergence and optimal scaling of random walk Metropolis algorithms. *Ann. App Probab.* **1997**, *7*, 110–120. [CrossRef]
11. Cotter, S.L.; Roberts, G.O.; Stuart, A.M.; White, D. MCMC Methods for Functions: Modifying Old Algorithms to Make Then Faster. *Stat. Sci.* **2013**, *28*, 424–446. [CrossRef]
12. Del Moral, P.; Doucet, A.; Jasra, A. Sequential Monte Carlo Samplers. *J. R. Stat. Soc. Ser. B Stat. Methodol.* **2006**, *68*, 411–43 [CrossRef]
13. Evensen, G. Sequential data assimilation with a nonlinear quasi-geostrophic model using Monte Carlo methods to forecast erro statistics. *J. Geophys. Res. Ocean.* **1994**, *99*, 10143–10162. [CrossRef]
14. Xiao, H.; Wu, J.L.; Wang, J.X.; Sun, R.; Roy, C. Quantifying and reducing model-form uncertainties in Reynolds-average Navier–Stokes simulations: A data-driven, physics-informed Bayesian approach. *J. Comput. Phys.* **2016**, *324*, 115–136. [CrossRef
15. Schneider, T.; Lan, S.; Stuart, A.; Teixeira, J. Earth system modeling 2.0: A blueprint for models that learn from observations an targeted high-resolution simulations. *Geophys. Res. Lett.* **2017**, *44*, 12–396. [CrossRef]
16. Kovachki, N.B.; Stuart, A.M. Ensemble Kalman inversion: A derivative-free technique for machine learning tasks. *Inverse Prob* **2019**, *35*, 095005. [CrossRef]
17. Dinh, L.; Sohl-Dickstein, J.; Bengio, S. Density estimation using Real NVP. In Proceedings of the 5th International Conferenc on Learning Representations, ICLR 2017, Toulon, France, 24–26 April 2017; Conference Track Proceedings. Available online https://openreview.net/ (accessed on 30 June 2023).
18. Papamakarios, G.; Murray, I.; Pavlakou, T. Masked Autoregressive Flow for Density Estimation. In Proceedings of the Advance in Neural Information Processing Systems 30: Annual Conference on Neural Information Processing Systems 2017, Long Beac CA, USA, 4–9 December 2017; Guyon, I., von Luxburg, U., Bengio, S., Wallach, H.M., Fergus, R., Vishwanathan, S.V.N., Garnet R., Eds.; Neural Information Processing Systems Foundation, Inc. (NeurIPS): La Jolla, CA, USA, 2017; pp. 2338–2347.
19. Kingma, D.P.; Dhariwal, P. Glow: Generative Flow with Invertible 1x1 Convolutions. In Proceedings of the Advances in Neura Information Processing Systems 31: Annual Conference on Neural Information Processing Systems 2018, NeurIPS 2018, Montréa QC, Canada, 3–8 December 2018; Bengio, S., Wallach, H.M., Larochelle, H., Grauman, K., Cesa-Bianchi, N., Garnett, R., Eds NeurIPS: La Jolla, CA, USA, 2018; pp. 10236–10245.
20. Dai, B.; Seljak, U. Sliced Iterative Normalizing Flows. In Proceedings of the 38th International Conference on Machine Learning ICML 2021, Virtual Event, 18–24 July 2021; Meila, M., Zhang, T., Eds.; Proceedings of Machine Learning Research; PMLR New York, NY, USA, 2021; Volume 139, pp. 2352–2364.
21. De Simon, L.; Iglesias, M.; Jones, B.; Wood, C. Quantifying uncertainty in thermophysical properties of walls by means o Bayesian inversion. *Energy Build.* **2018**, *177*, 220–245. [CrossRef]
22. Iglesias, M.; Park, M.; Tretyakov, M. Bayesian inversion in resin transfer molding. *Inverse Probl.* **2018**, *34*, 105002. [CrossRef]
23. Karamanis, M.; Beutler, F.; Peacock, J.A.; Nabergoj, D.; Seljak, U. Accelerating astronomical and cosmological inference wit preconditioned Monte Carlo. *Mon. Not. R. Astron. Soc.* **2022**, *516*, 1644–1653. [CrossRef]
24. Lorenz, E.N. Deterministic Nonperiodic Flow. *J. Atmos. Sci.* **1963**, *20*, 130–148. [CrossRef]
25. Ambrogioni, L.; Lin, K.; Fertig, E.; Vikram, S.; Hinne, M.; Moore, D.; van Gerven, M. Automatic structured variational inference In Proceedings of the International Conference on Artificial Intelligence and Statistics, Virtual, 13–15 April 2021; PMLR: New York NY, USA, 2021; pp. 676–684.
26. Villani, C. *Optimal Transport—Old and New*; Springer: Berlin/Heidelberg, Germany, 2008; Volume 338, pp. xxii+973.
27. Zhang, L.; Carpenter, B.; Gelman, A.; Vehtari, A. Pathfinder: Parallel quasi-Newton variational inference. *J. Mach. Learn. Res* **2022**, *23*, 13802–13850.
28. Neal, R.M. MCMC using Hamiltonian dynamics. *Handb. Markov Chain. Monte Carlo* **2011**, *2*, 2.
29. Hoffman, M.D.; Gelman, A. The No-U-Turn sampler: Adaptively setting path lengths in Hamiltonian Monte Carlo. *J. Mach Learn. Res.* **2014**, *15*, 1593–1623.

physical sciences forum

MDPI

roceeding Paper

Nested Sampling—The Idea †

ohn Skilling

Maximum Entropy Data Consultants, Killaha East, V93 H7VW Kenmare, Ireland; john@skilling.co.uk

† Presented at the 42nd International Workshop on Bayesian Inference and Maximum Entropy Methods in Science and Engineering, Garching, Germany, 3–7 July 2023.

Abstract: We seek to add up $Q = \int f \, dX$ over unit volume in arbitrary dimension. Nested sampling locates the bulk of Q by geometrical compression, using a Monte Carlo ensemble constrained within a progressively more restrictive lower limit $f \leq f^*$. This domain is divided into a core $f > f^*$ and a shell $f = f^*$, with the core kept adequately populated.

Keywords: nested sampling

1. The Idea

Quantification means counting, which can be performed either outwards by construction, or inwards by peeling items away until there are none left. Quantification extends to volumes, which underpin integration so numerical estimation of volume is fundamental.

Nested sampling [1,2] counts inwards, estimating volumes statistically from random scatterings of points $\mathbf{x}$, ranked by some quality function $F(\mathbf{x})$ appropriate to the current application (Figure 1). The procedure delves arbitrarily deep by recursively taking proportions of arbitrarily big spaces.

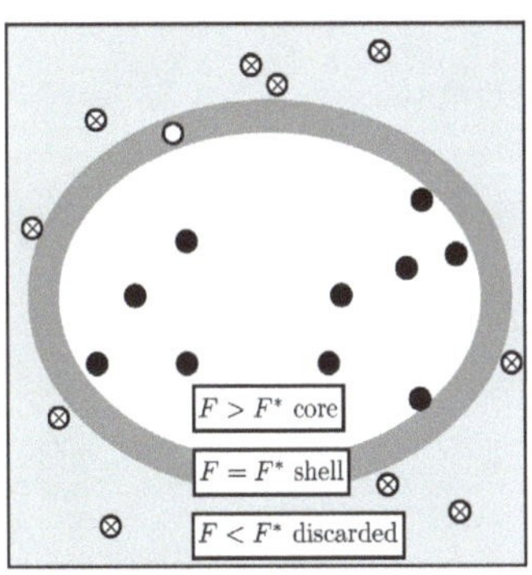

Figure 1. Sample ranked r out of n encloses about r/n of the volume.

Although suggested by intuition trained in the three dimensions of physical space, geometrical estimation of volume becomes unhelpful in high dimension, where directions tend to be mostly orthogonal, volumes collect around outer boundaries, and spanning the space requires at least as many points as dimensions. Counting is immune to all that.

2. Quantity and Shape

A central use of volume is estimation of the mass (or "quantity") Q and shape p

$$Q = \int F(\mathbf{x}) \, dV \,, \qquad p(\mathbf{x}) = F(\mathbf{x})/Q \tag{1}$$

of a non-negative density function $F(\mathbf{x})$ defined over a known volume, often thought of as a unit prior measure $V = 1$. The information

$$H = \int p(\mathbf{x}) \log p(\mathbf{x}) \, dV \tag{2}$$

about $\mathbf{x}$ carried by F, measured in nats (logs base e) or bits (base 2), quantifies the corresponding shape. Nested sampling is general but intended for applications where $H \gg 1$ so that the bulk of Q occupies some tiny $O(e^{-H})$ fraction of the original volume (Figure 2 left).

According to fable, mathematicians find a needle in a haystack by iteratively halving the haystack, repeatedly discarding the half that does not contain the needle. Such compression proceeds exponentially, so is linear in H (one bit per step). This neatly outclasses simple point-by-point search, which is proportional to the volume ratio $\exp(H)$. Nested sampling operates similarly, with its locations ranked by value F. Volumetric compression controlled by F suggests replacing Q in (1) by the equivalent ($\int y \, dx = \int x \, dy$) Lebesgue integral [3]

$$Q = \int_0^\infty X(f) \, df \tag{3}$$

where $X(f) = \int_{F(\mathbf{x}) \geqslant f} dV$ is the volume enclosed by the contour $F(\mathbf{x}) = f$ (Figure 2 right). (I am indebted to Ning Xiang in private communication buttressed by [4] for pointing out the Lebesgue connection, which to my embarrassment I had as inventor failed to notice).

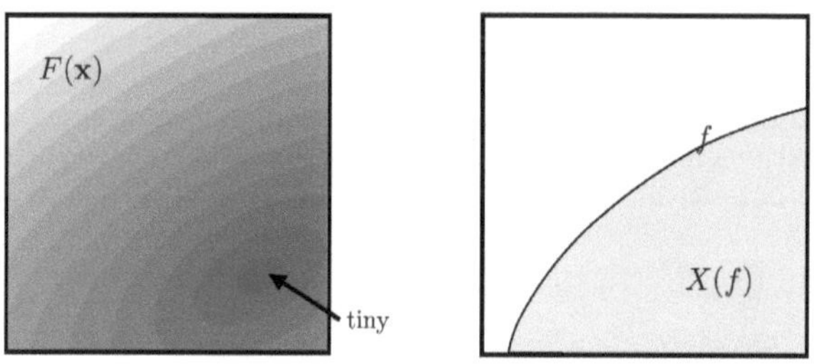

Figure 2. (**left**) unit volume V modulated by F; (**right**) volume $X(f)$ covering $F \geqslant f$.

X parameterizes a one-dimensional decomposition of quantity Q according to value F. Nested sampling accumulates the quantity and discovers the shape by tracking inwards and upwards, compressing X from the original 1 (complete) towards 0 (empty).

3. Nested Sampling

By hypothesis, F is a function accessible pointwise through evaluation at specified locations $\mathbf{x}$, whereas related volumes X have no such easy access. Hence, estimates of Q can realistically only be built from evaluations of F. A priori, we have no knowledge of where good (high value of F) locations $\mathbf{x}$ might be, so we start with a Monte Carlo ensemble of random locations. In any ensemble, there will be one or more locations with the worst (lowest) value f. This outer "*shell*" $F = f$ surrounds the inner "*core*" $F > f$ of other locations with better values. For example, the four-object ensemble in Figure 3 (top) has $c = 3$ objects in the core and $s = 1$ in the shell.

Compression is achieved by discarding the shell while retaining the core. Actually, any of the ensemble values could be used to divide inner from outer, and the mathematician of the fable would have used the median value. However, compression is smoother, and results are more precise if only the outer shell is discarded. To avoid eroding the ensemble, it is rebuilt with more locations randomly chosen within the current constraint $F \geqslant f$ until the core $F > f$ contains enough objects that survive. For example, in Figure 3 (bottom), three extra random locations extended the original ensemble until the core built up to the original four objects, with three in the shell that is about to be discarded. For continuous applications, coincident values of F would be vanishingly rare, so a shell would never hold more than one object.

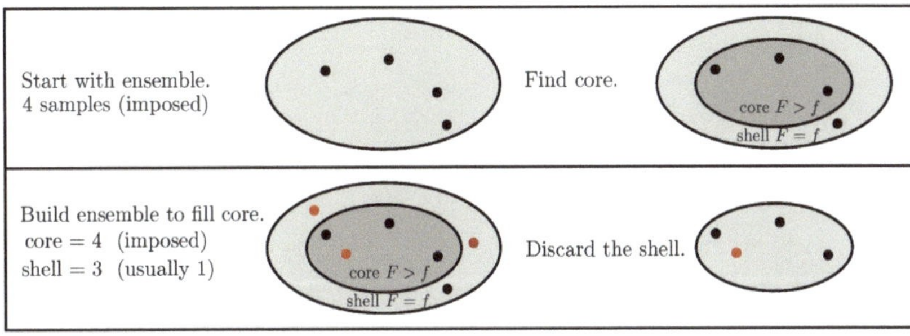

Figure 3. Nested sampling iterate with ensemble size 4.

Discarding s (=3) shell objects to leave just the core c (=4) reflects a volume compression ratio around $\frac{c}{c+s}$ (=4/7). This estimate is obtained without any reference to triangulation or geometry or even topology. After discarding the shell, the outermost (lowest) value in the core will define an updated and increased lower limit f, which can seed a subsequent iterate.

With all points distributed uniformly and equivalently, the compression ratio γ (< 1) will have been beta-distributed

$$\gamma \sim \texttt{Beta}(c, s), \quad \text{explicitly} \quad \Pr(\gamma) \propto \gamma^{c-1}(1-\gamma)^{s-1}. \tag{4}$$

So, when we wish to estimate what γ actually was numerically, we can do no better than sample $\texttt{Beta}(c, s)$, either just once or (preferably) many times. Successive compressions $\gamma_1, \gamma_2, \gamma_3, \ldots$ starting from the initial volume $X_0 = 1$ lead to core volumes (Figure 4 left)

$$X_0 = 1 \xrightarrow{\times\gamma_1} X_1 = \gamma_1 \xrightarrow{\times\gamma_2} X_2 = \gamma_1\gamma_2 \xrightarrow{\times\gamma_3} X_3 = \gamma_1\gamma_2\gamma_3 \xrightarrow{\times\gamma_4} \ldots \tag{5}$$

It is best to keep the shells as thin as possible to maximise the overall compression per sample, which is why we choose to discard just one outermost value at a time. It does not matter if sampling omits intermediate values of F because $\texttt{Beta}(c, 0)$ for an invisibly empty shell ($s = 0$) would give $\gamma = 1$ so would not contribute to compression.

These multiplicative factors are better accumulated additively as logarithms, for which the mean and standard deviation

$$\log \gamma = -\left(\sum_{j=c}^{c+s-1} \frac{1}{j}\right) \pm \left(\sum_{j=c}^{c+s-1} \frac{1}{j^2}\right)^{1/2} \tag{6}$$

imply well-behaved moments of logarithmic compression $\log X_k$. Conversely, the moments of raw compression X_k rapidly become misleading and unusable. Just as in statistical mechanics where the variable of interest is not raw degeneracy Ω but the entropy $S = \log \Omega$, here it is $\log Q$ that matters.

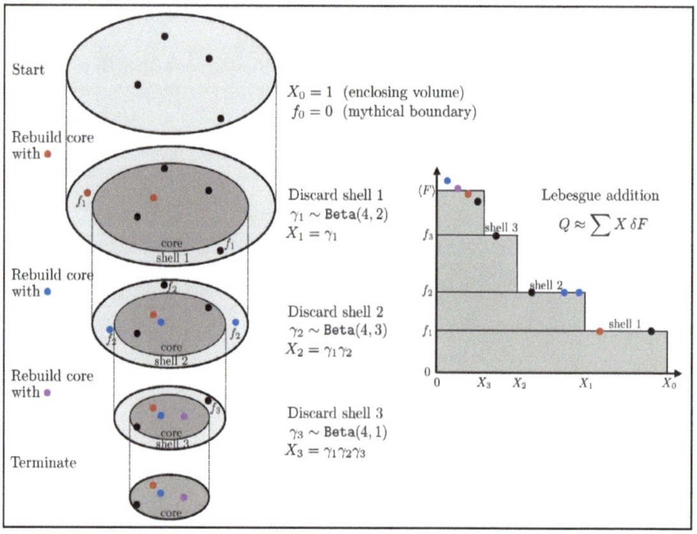

Figure 4. Nested sampling trajectory.

4. Quantity

From the γ's, Q is estimated by Lebesgue (3) as

$$Q \approx \underbrace{(f_1 - 0)X_0 + (f_2 - f_1)X_1 + \cdots + (f_k - f_{k-1})X_{k-1}}_{\text{nested sampling trajectory}} + \underbrace{(\langle F \rangle - f_k)X_k}_{\text{termination}} \tag{7}$$

where $\langle F \rangle$ is the terminating mean value (Figure 4 right). The result is a distribution $\Pr(Q)$—better represented in view of the large dynamic ranges as $\Pr(\log Q)$.

5. Approximation

There is no universally valid "best" single-value representative of $\Pr(\log Q)$, which need not even be unimodal. Neither is there any "unbiassed estimator" for Q or $\log Q$. Users who seek a single value may instead plausibly fix each γ at its logarithmically mean value (6).

Typically in large applications ($H \gg 1$), the terms in Q rise to a maximum (as increasing F overcomes diminishing volumes X) and then decay (as diminishing volume overcomes limited values of F). Correspondingly, many iterations are needed to scan the volume range. For continuous applications, s will always be 1 so that each γ is distributed as $\Pr(\gamma) = c\gamma^{c-1}$ with logarithmic mean and standard deviation

$$\log \gamma = -\frac{1}{c} \pm \frac{1}{c} \tag{8}$$

Compression by a factor of e^H to the bulk of Q will take about $cH \pm \sqrt{cH}$ such iterates, after which the volume will be estimated with uncertainty $\delta \log X \approx \sqrt{H/c}$. The uncertainty will be reflected in Q

$$\delta \log Q \approx \sqrt{H/c} \tag{9}$$

which (under limited computer resources) is minimised by keeping the retained core size c constant.

Of course, that is merely what is anticipated for a typical application. Particular cases may behave worse, and $\delta \log Q$ should be estimated statistically if an application has risk of that.

6. Shape

As shells of volume $X_{k-1}-X_k$ are peeled away (Figure 4) to form the nested sampling trajectory, the corresponding value f_k gives a Riemann weight $w_k = (X_{k-1}-X_k)f_k$ (Figure 5 left). That weight can then be randomly assigned among the shell objects to provide a decomposition of quantity according to volume.

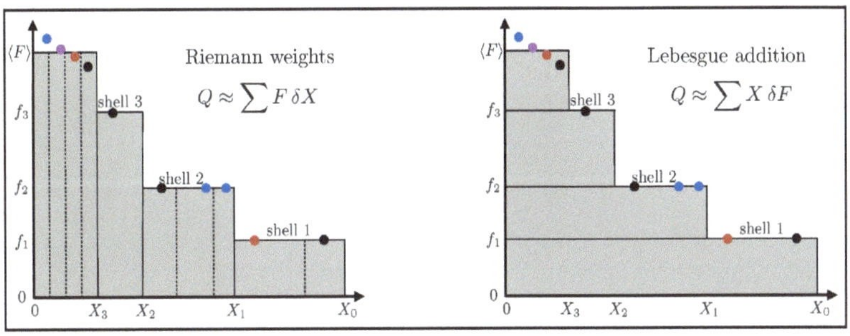

Figure 5. Riemann and Lebesgue.

Objects drawn randomly in proportion to w then give a usefully compact representation of shape $p(\mathbf{x})$ as a set of equally weighted locations. These introductory locations can be used to seed standard Metropolis MCMC exploration if more samples are wanted.

7. Programming

It may be assumed that the user has, through some such method as importance weighting, already extracted from the original problem whatever structure is analytically available so that the remaining numerical task is reduced as far as reasonably possible. Nested sampling is then to be employed for compressing through the remaining information H.

The following is the minimal skeleton program for nested sampling:

INITIALISE (Figure 6 left)	
Set N and allocate $(\mathbf{x}, F)$ for N objects	Stored ensemble $0, 1, \ldots, N-1$
$X = 1$	Initialise volume
$f = 0$	Initialise lower bound
$Q = 0$	Initialise quantity Q
for $i = 0, 1, \ldots, N-1$	Initialise ensemble …
$\quad F_i = F(\mathbf{x}_i \text{ uniform})$	… with random locations $\mathbf{x}_i$
ITERATE (Figure 6 right, $\Delta\xi \equiv \xi - \xi_{\text{previous}}$)	
until(terminate)	Iterate until termination
$\quad f = \mathbf{min}(F_0, F_1, \ldots, F_{N-1})$	Update lower bound f
$\quad \Delta Q = X \times \Delta f$	Update Q (Figure 5 right, lower)
$\quad M = N$	Initial membership = retained storage
$\quad$ **for** $i = 0, 1, \ldots, N-1$ randomly	For each stored object, …
$\qquad$ **while**(F_i equals f)	… keep trying again until out of shell
	OUTPUT $(\mathbf{x}_i, F(\mathbf{x}_i))$ and M
$\qquad\quad F_i = F(\mathbf{x}_i \text{ uniform in } F \geqslant f)$	Replace $(\mathbf{x}_i, F(\mathbf{x}_i))$ in $F \geqslant f$
$\qquad\quad \Delta \log X = \log \gamma(M)$	Compress.
$\qquad\quad \Delta M = 1$	Increment core+shell membership
TERMINATE	
	OUTPUT each $(\mathbf{x}_i, F(\mathbf{x}_i))$ and $\varnothing$
$f = \mathbf{mean}(F_0, F_1, \ldots, F_{N-1})$	Final ensemble.
$\Delta Q = X \times \Delta f$	Update Q (Figure 5 right, top)

The output forms the nested sampling trajectory, both pre- and post-termination, with each location $\mathbf{x}$ representing its shell of volume ΔX having values at or around $F(\mathbf{x})$, and compression defined by M with core size N. The compression ratio $\gamma(M)$ in line 14 can be either the probabilistic estimate $\mathtt{Beta}(M, 1)$ (4) or the approximating logarithmic mean value $e^{-1/M}$ (8).

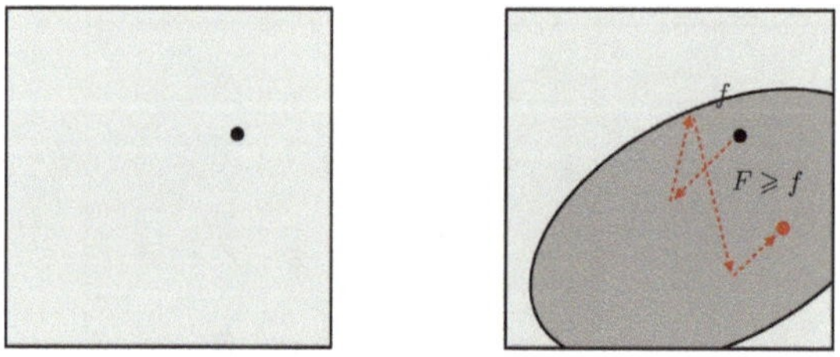

Figure 6. (**left**) set prior object by MC; (**right**) generate new object by MCMC.

The program's overt purpose is to accumulate the quantity Q according to Lebesgue summation (7). This accumulation is parasitic upon nested exploration which is driven only by ranking of quality, so that arbitrary monotonic quantities $\Theta = \int X(f)\, d\theta(f)$ could be accumulated consistently at the same time.

Apart from the procedure for evaluating $F(\mathbf{x})$, the program requires three inputs from the user.

A: A Monte Carlo procedure for a random $\mathbf{x}$ uniformly distributed over the original unit volume (Figure 6 left), used in line 6.

B: A procedure for generating a new $\mathbf{x}$ uniformly distributed within the current constraint $F(\mathbf{x}) \geqslant f$, used in line 13. In practice, this will be a Markov chain (MCMC) procedure seeded at a random member of the current ensemble and exploring the constrained volume with moves obeying detailed balance (Figure 6 right). Note that these moves are *not* modulated by F, except that destinations below f are prohibited. Geometrical properties of an application may assist construction of the MCMC procedure, as when ellipsoidal domains are constructed around the walkers in suitably smooth applications in suitably restricted dimension ([5] etc.), but nested sampling itself is not dependent on geometry or topology or continuity or differentiability or convenient shapes. It is for the user to program a suitable procedure for the application in hand. Or fail in the attempt.

C: A termination criterion, needed in line 7. There is no universally valid criterion because numerical experimentation alone can never exclude the possibility of high values in unreached locations which could render termination premature. Your author's default criterion is to terminate when H, which can be accumulated through $\Delta(QH + Q\log Q) = X\,\Delta(f\log f)$, appears to have stopped increasing significantly, indicating that most of the relevant structure has likely been found.

Lines 1–6 of the program are straightforward initialisation of a random N-object ensemble.

Lines 7–15 are the iterative loop. On entry, the ensemble has N core objects. The bounding (lowest) value f is appropriately increased (line 8) and Q updated (line 9). At least one object then lies on the new boundary with $F = f$ (the new shell), with the others in the diminished core $F > f$. The aim is to add new randomly located objects until the core has again built up to N, following which the shell can be discarded (or, more usefully, output as the trajectory).

Whenever a shell object is discovered within the N-object stored ensemble (Lines 11 and 12), it is written out to the trajectory to make room for a new random location. When that new location is generated (Line 13), it is included in the extended membership M (line 15). Meanwhile, the corresponding 1-in-M contribution to compression is incorporated in Line 14.

This loop (Lines 11 to 15) ends when the new location falls into the now-more-populated core. The iterate ends when every stored object is in the core, with every shell object having been written to the trajectory ($c = N, s = M-N$).

Compression could be programmed using a single ratio $\gamma = \texttt{Beta}(c,s)$ (Equation (4)) just before Line 9 to discard the whole shell in one go. However, it is better to use the individual steps shown in Line 13. That is equivalent to grouped compression, and $\texttt{Beta}(M,1)$ is quick to compute.

$$\texttt{Beta}(c,s) = \prod_{k=c}^{c+s-1} \texttt{Beta}(k,1), \qquad \texttt{Beta}(k,1) = \left(\texttt{Uniform}(0,1)\right)^{1/k}. \tag{10}$$

Lastly, if the program has been properly terminated with the bulk of the structure found, the contribution of the final ensemble (Lines 16 to 17) will be negligible and could be omitted.

Because of the dynamic range inherent in large problems, the skeleton program as written is susceptible to computer over/underflow. Therefore any useful implementation should store X, F, Q, w as logarithms.

A professional refinement is to track several chains of Beta-distributed compressions in line 14 to obtain the distribution of Q. Production of a nested sampling trajectory $\{\mathbf{x}, F\}$ was statistical, so its interpretation ought also to be statistical.

8. Convergence

There has been a view in the community [6,7] that interest lies in, and convergence should be proved for, Q rather than $\log Q$. That view is a relic of bygone concentration on small problems—misleading nowadays because the extremely heavy-tailed distribution of Q in applications of appreciable size requires excessive resources to decrease δQ below Q. As mentioned below (6), it is variation in $\log Q$ that matters, even if that residual uncertainty allows orders of magnitude of uncertainty in Q.

In most practical applications, a run can be continued until compression has scanned through most of the posterior distribution, as indicated by flattening off of H as it rises toward its presumed final value. In such cases, rms convergence of $\log Q$ proceeds as the usual statistical inverse-square-root $O(n^{-1/2})$, where $n \gtrsim cH$ is the number of iterative steps as in (9). That fact can be demonstrated by observing that the number of iterates n required to compress volume from 1 to X is a Poisson distribution with mean $-c \log X$, which consequently has inverse-square-root uncertainty. The compression $-\log X$ inferred from n is then exponential with mean n/c, with that same inverse-square-root uncertainty. The overall effect is then inverse square root as stated.

Incidentally, the behaviour of nested sampling depends on the prior-to-posterior information H, which can be unrelated to dimension. Dimension, which is a geometrical construction, is not part of nested sampling. Applications need not even have a dimension.

Exceptionally, there may be a localised but important quality peak hiding beyond termination and yielding substantial or even dominant termination error. It was shown in [8] that termination error is controllable if H can be bounded above. Resources can then be adjusted to yield convergence as $O(n^{-1/4})$. Of course, the information H is always bounded in any practical application: the question would be how to define a convincing upper limit.

Those analyses assumed that the ensemble size was held constant, with implicitly unit shells $s = 1$ surrounding a constant core size c. But, in discrete applications, plateaus of constant quality F may necessitate larger shells, which require extra resources to traverse. Indeed, Mother Nature can supply arbitrarily challenging problems, to which we have no general answer.

Ultimately, finding a 1-in-e^{-H} posterior domain may require e^{H} exploratory samples, thus defeating the enterprise. Numerical exploration will never be able to find a flagpole in the Atlantic Ocean.

Funding: This research received no external funding.

Institutional Review Board Statement: Not applicable.

Informed Consent Statement: Not applicable.

Data Availability Statement: No new data were created or analyzed in this study. Data sharing is not applicable to this article.

Conflicts of Interest: The author declares no conflicts of interest.

References

1. Skilling, J. Nested Sampling. In *Bayesian Inference and Maximum Entropy Methods in Science and Engineering, Garching, Germany 200*[illegible] Fischer, R., Dose, V., Preuss, R., von Toussaint, U., Eds.; AIP Publishing LLC: New York, NY, USA, 2004; Volume 735, pp. 395–40[illegible] [CrossRef]
2. Skilling, J. Nested sampling for general Bayesian computation. *Bayesian Anal.* **2006**, *1*, 833–859. [CrossRef]
3. Lebesgue, H. *Leçons Sur l'intégration et la Recherche des Fonctions Primitive*; Gauthier-Villars: Paris, France, 1904.
4. Jasa, T.; Xiang, N. Using Nested Sampling in the Analysis of Multi-Rate Sound Energy Decay in Acoustically Coupled Rooms. I[illegible] *Bayesian Inference and Maximum Entropy Methods in Science and Engineering*; Knuth, K.H., Abbas, A.E., Morris, R.D., Castle, J.F[illegible] Eds.; AIP Publishing LLC: New York, NY, USA, 2005; Volume 803, pp. 189–196. [CrossRef]
5. Mukherjee, P.; Parkinson, D.; Liddle, A.R. A nested sampling algorithm for cosmological model selection. *Astrophys. J. Lett.* **200**[illegible] *638*, L51. [CrossRef]
6. Evans, M.J. Discussion of "Nested Sampling for Bayesian Computations". In *Bayesian Statistics 8*; Bernardo, J.M., Bayarri, M.J[illegible] Berger, J.O., Dawid, A.P., Heckerman, D., Smith, A.F.M., West, M., Eds.; Oxford University Press: Oxford, UK, 2007; pp. 507–51[illegible]
7. Chopin, N.; Robert, C.P. Properties of nested sampling. *Biometrika* **2010**, *97*, 741–755. [CrossRef]
8. Skilling, J. Nested sampling's convergence. In Proceedings of the Bayesian Inference and Maximum Entropy Methods in Scienc[illegible] and Engineering, Oxford, MS, USA, 5–10 July 2009; Goggans, P.M., Chan, C.Y., Eds.; AIP Publishing LLC: New York, NY, USA[illegible] 2009; Volume 1193, pp. 277–291.

physical sciences forum

MDPI

Proceeding Paper

Preconditioned Monte Carlo for Gradient-Free Bayesian Inference in the Physical Sciences †

Minas Karamanis [1,2,*] and Uroš Seljak [1,2]

1 Berkeley Center for Cosmological Physics, University of California, Berkeley, CA 94720, USA; useljak@berkeley.edu
2 Lawrence Berkeley National Laboratory, 1 Cyclotron Road, Berkeley, CA 94720, USA
* Correspondence: mkaramanis@berkeley.edu
† Presented at the 42nd International Workshop on Bayesian Inference and Maximum Entropy Methods in Science and Engineering, Garching, Germany, 3–7 July 2023.

Abstract: We present preconditioned Monte Carlo (PMC), a novel Monte Carlo method for Bayesian inference in complex probability distributions. PMC incorporates a normalizing flow (NF) and an adaptive Sequential Monte Carlo (SMC) scheme, along with a novel past resampling scheme to boost the number of propagated particles without extra computational costs. Additionally, we utilize preconditioned Crank–Nicolson updates, enabling PMC to scale to higher dimensions without the gradient of target distribution. The efficacy of PMC in producing samples, estimating model evidence, and executing robust inference is showcased through two challenging case studies, highlighting its superior performance compared to conventional methods.

Keywords: Bayesian inference; sequential Monte Carlo; normalizing flows; gradient free

1. Introduction

Bayesian inference (BI) is a crucial tool in the physical sciences, offering a mathematical framework for quantifying uncertainty [1–5]. BI rests on three core elements: prior and posterior distributions, and the likelihood function. The prior distribution encapsulates our initial understanding of a problem, representing our beliefs before any data are observed. The likelihood function signifies how likely it is our observed data are under varying model parameters. Posterior distributions emerge from Bayes' theorem, combining the prior and likelihood to update our beliefs after data observation.

However, most BI analyses are analytically intractable, demanding numerical methods such as Markov chain Monte Carlo (MCMC). These techniques approximate the posterior, generating samples from it. State-of-the-art MCMC methods hinge on the knowledge of the gradient of the target probability density, which is often intractable in the physical sciences. Moreover, MCMC faces other challenges: its inherently serial nature makes parallelization difficult, it struggles with multimodal or highly correlated distributions, and it is generally unable to compute the marginal likelihood, a critical component for model comparison. These limitations underscore the need for more flexible and efficient computational tools.

Sequential Monte Carlo (SMC) is a Monte Carlo method designed to address these challenges [6–8]. SMC distinguishes itself with straightforward parallelization, an ability to handle multimodal posteriors, and the provision of a marginal likelihood estimate. In SMC, MCMC acts as a tool for moving the particles, typically utilizing gradient-free MCMC variants, such as random-walk metropolis (RWM) [9] and slice sampling (SS) [10]. Despite its advantages, the SMC based on these MCMC methods still struggles to scale to high-dimensional problems and handle highly correlated and multimodal posteriors.

To overcome these challenges, we introduce preconditioned Monte Carlo (PMC), a refined variant of SMC. PMC employs a normalizing flow (NF) [11] to Gaussianize the target distribution, simplifying MCMC sampling [12–14]. In contrast to canonical

SMC, PMC includes a novel resampling scheme termed past resampling (PR), enhancing the number of propagated particles without incurring additional computational costs. Furthermore, PMC departs from the reliance on RWM or SS for sampling, instead choosing preconditioned Crank–Nicolson (pCN) updates [15,16]. As such, PMC can scale to higher dimensions without requiring the gradient of the target distribution, enhancing the overall sampling performance.

2. Methods

2.1. Sequential Monte Carlo

2.1.1. Background

Sequential Monte Carlo (SMC) propagates a set of particles, drawn from a known distribution $\rho(\theta)$, through multiple intermediary distributions towards a target distribution $p(\theta)$. The transition speed depends on the number of intermediary distributions. Like annealed importance sampling (AIS), SMC uses multiple MCMC steps per iteration to balance particles towards each stage's equilibrium distribution. SMC differentiates from AIS by employing resampling to avoid weight degeneracy, where a few particles hold high-importance weights. The resampling equalizes the weights to manage their high variance.

2.1.2. Bridging the Prior and Posterior

To develop a series of intermediate distributions that help the transition from a known initial distribution $\rho(\theta)$ to the desired posterior distribution $p(\theta)$, a method of interpolation is typically employed as follows:

$$p_t(\theta) \propto \rho^{1-\beta_t}(\theta)p^{\beta_t}(\theta), \quad t = 1, \ldots, m \tag{1}$$

where β_t refers to a sequence of annealing temperatures, arranged such that

$$0 = \beta_1 < \beta_2 < \cdots < \beta_m = 1 \tag{2}$$

In a Bayesian setting, the prior is naturally chosen as the auxiliary density $\rho(\theta) = p(\theta|\mathcal{M})$, and the posterior as the target density $p(\theta) = p(\theta|d, \mathcal{M})$. In this case, Equation (1) simplifies to:

$$p_t(\theta) \propto p^{\beta_t}(d|\theta, \mathcal{M})p(\theta|\mathcal{M}) \tag{3}$$

2.1.3. Key Steps

The process of transitioning from one distribution to the next, $p_{t-1}(\theta)$ to $p_t(\theta)$, is performed in three primary steps: correction, selection, and mutation. These steps are encapsulated in a single SMC iteration, which is repeated until the target distribution is adequately approximated.

- Correction—The importance weights w_t^i of the particles θ_{t-1}^i are calculated as:

$$w_t^i = \frac{p_t(\theta_{t-1}^i)}{p_{t-1}(\theta_{t-1}^i)} \tag{4}$$

  and then they are used to estimate the normalizing constant (i.e., marginal likelihood) of $p_t(\theta)$:

$$\mathcal{Z}_t = \mathcal{Z}_{t-1} \times \frac{1}{N}\sum_{i=1}^{N} w_t^i \tag{5}$$

  Initially, all weights are equal such that $w_1^i = 1/N \, \forall i \in \{1, 2, \ldots, N\}$ with $\mathcal{Z}_1 = 0$.
- Selection—the particles θ_{t-1}^i are resampled according to their importance weights. The goal is to retain those particles that are highly represented in the new distribution $p_t(\theta)$, while eliminating those that are less represented. Particles with high weights are more likely to be selected, thus forming the resampled set θ_t^i.

- Mutation—The resampled particles θ_t^i are perturbed to generate diversity and explore more of the parameter space. This step uses a transition kernel $K_t(\theta_t^i, \theta')$. In practice, K_t takes the form of an MCMC kernel (e.g., several steps of an Metropolis–Hastings (MH) transition). The careful design of the transition kernel is required to ensure that the SMC sampler maintains good mixing properties and computational efficiency.

2.1.4. Adaptive Temperature Schedule

The effective sample size (ESS) measures the number of statistically independent samples in the weighted ensemble. It provides an estimate of the "quality" of the samples and is given by:

$$ESS = \frac{1}{\sum_{i=1}^{N}(w_t^i)^2}, \tag{6}$$

where the weights w_t^i are normalized to sum to one. In practice, a constant ESS is maintained throughout the run by determining each β_t adaptively by setting the next temperature level as:

$$\beta_{t+1} = \arg\min_{\beta}\{\beta : ESS_\beta \geq ESS_{target}\}, \tag{7}$$

where ESS_β represents the effective sample size at a temperature of β, N is the total number of samples, and ESS_{target} is the target ESS. The algorithm proceeds with the new temperature until the final target distribution (i.e., the posterior) is achieved. This adaptive temperature selection scheme, guided by ESS, is a common feature of SMC. This enables the efficient bridging of the gap between the prior and posterior while ensuring a diverse set of samples.

2.2. *Past Resampling*

In the canonical formulation of SMC, the new particles are resampled from the previous generation of particles with probabilities proportional to their importance weights given by Equation (4). However, this scheme exhibits several limitations. First, the number of particles, N, of a generation needs to be large enough to effectively capture the characteristics of the tempered distribution of the Equation (1). This effect is further exaggerated in higher dimensions. Secondly, the ESS_{target} needs to amount to a significant fraction of N (i.e., 80–99%) to suppress the high variance of the importance weights. Third, even when the N and ESS_{target} are large, the resampled particles will include many copies of the same particles. This requires running MCMC longer during the mutation stage to diversify these particles.

To address these challenges without increasing the computational cost of the method, we propose to resample particles from all past generations, instead of just the last one. We refer to this modification of SMC as past resampling (PR). As the number of resampled particles is substantially lower than that of the particles of past generations, N', the former are effectively independent and do not include copies of the same particle. This addresses the third challenges of the canonical approach. Furthermore, the target ESS, which determines the convergence rate of the algorithm, can be significantly higher than in the canonical formulation, addressing the first and second issues.

PR requires the computation of importance weights for the particles of all (or a subset of) past iterations. In the t_{th} iteration, the weights of the particles of the t'_{th} iteration can be written as:

$$w_{tt'}^i = \Lambda_{tt'}\tilde{W}_{tt'}^i, \tag{8}$$

where $\tilde{W}_{tt'}^i$ are the normalized importance weights:

$$\tilde{W}_{tt'}^i = \frac{\tilde{w}_{tt'}^i}{\sum_{i=1}^{N}\tilde{w}_{tt'}^i}, \tag{9}$$

where:

$$\tilde{w}^i_{tt'} = \frac{p_t(\theta_{t'})}{p_{t'}(\theta_{t'})}. \tag{10}$$

The coefficients $\Lambda_{tt'}$ determine the influence of past iterations on the current one. Generally, the further in the past t' is, the less significant its influence. The estimation of $\Lambda_{tt'}$ can be formulated as the variational calculus problem of maximizing the total ESS. The solution of this problem leads to:

$$\Lambda_{tt'} = \frac{\lambda_{tt'}}{\sum_{t'=1}^{t} \lambda_{tt'}}, \tag{11}$$

where:

$$\lambda_{tt'} = \frac{1}{\sum_{i=1}^{N} \left(\tilde{W}^i_{tt'}\right)^2}. \tag{12}$$

In other words, the importance weights of the particles of the t'_{th} iteration with respect to the current t_{th} iteration are adjusted by the normalized ESS given by Equation (11). Equations (8)–(12) are related to the recycling scheme proposed by [17,18] and applied at the end of the run. The main difference is that PR is applied to every iteration of SMC, instead of just the last one. In practice, only the iterations with $\lambda_{tt'} > 10$ are used to avoid noisy contributions.

The weights of Equation (8) can be employed to design a lower-variance estimate of the normalizing constant of $p_t(\theta)$, which generalizes Equation (5):

$$\mathcal{Z}_t = \sum_{t'=1}^{t-1} \left(\Lambda_{tt'} \mathcal{Z}_{t'} \frac{1}{N} \sum_{i=1}^{N} \tilde{w}^i_{tt'} \right) \tag{13}$$

2.3. Preconditioning

2.3.1. Background

The mutation step of SMC typically involves the use of the MH algorithm. MH comprises a general class of MCMC methods that generate samples, in the form of a Markov chain, from the target distribution $p_\theta(\theta)$ by iterating a proposal and acceptance step. In the proposal step, a new state θ' is sampled from the proposal distribution $q_\theta(\theta'|\theta)$, that is, conditioned on the current state θ in the Markov chain. In the acceptance step, the new state is added to the Markov chain with a probability given by:

$$\alpha = \min\left\{1, \frac{p_\theta(\theta')q_\theta(\theta|\theta')}{p_\theta(\theta)q_\theta(\theta'|\theta)}\right\} \tag{14}$$

If the new state is rejected, then the current state θ is added to the Markov chain instead and the process is repeated until a sufficient number of samples have been collected. The form of the acceptance probability of Equation (14) is designed to ensure detailed balance, a sufficient condition for the Markov chain to have $p_\theta(\theta)$ as its invariant distribution. In the case of SMC, the MH algorithm is employed during the mutation step targeting the current tempered distribution p_t.

The choice of a proposal distribution $q_\theta(\theta'|\theta)$ is crucial as it determines the overall sampling efficiency of the method. Generally, a good proposal distribution needs to balance exploration and exploitation, by generating states that are sufficiently different from the current one while maintaining a high acceptance rate. A common choice for $q_\theta(\theta'|\theta)$ is the normal distribution $N(\theta'|\theta, C)$ leading to the RWM algorithm. The efficiency of RWM can be substantially improved by setting $C = 2.38^2/D \times \Sigma$, where D is the number of dimensions and Σ is the covariance matrix describing the target distribution $p_\theta(\theta)$.

However, if the target distribution $p_\theta(\theta)$ is significantly non-Gaussian, then the simple choice of $q_\theta(\theta'|\theta) = N(\theta'|\theta, C)$ is insufficient and a more sophisticated proposal distribution is required. Designing advanced proposal distributions is an active field of research within the statistics community. The problem is further exaggerated in higher dimensions where the $2.38^2/D$ factor leads to vanishingly small steps in parameter space. We argue that the method of preconditioning coupled with the appropriate proposal distribution has the potential to address both of these issues.

The basic idea behind preconditioning is to transform the target distribution $p_\theta(\theta)$ into the simpler distribution $p_z(z)$ using a bijective transformation $z = f(\theta)$. The two probability densities are linked through the following transformation:

$$p_z(z) = p_\theta(\theta)\left|\det\frac{\partial f^{-1}(z)}{\partial z}\right| \tag{15}$$

Sampling from $p_z(z)$ is generally easier than $p_\theta(\theta)$ due to its higher symmetry and simpler geometry. For instance, one may choose $z = f(\theta)$ such that $p_z(z)$ is approximately a standard normal distribution $\mathcal{N}(z|0,1)$. The MH algorithm is trivially generalized to the latent space of the preconditioning transformation with a proposal distribution $q_z(z'|z)$ and an acceptance probability of:

$$\alpha = \min\left\{1, \frac{p_z(z')q_z(z|z')}{p_z(z)q_z(z'|z)}\right\} \tag{16}$$

2.3.2. Normalizing Flows

Normalizing flows (NFs) are a natural choice for the preconditioning transformation $z = f(\theta)$. NFs are bijective transformations typically parameterized using a neural network (NN) architecture. Generally, the forms of NN and NF have to be chosen carefully to maintain the reversibility of the transformation and the tractability of the Jacobian determinant in Equation (15). Recent developments have led to flexible NFs that are easily trained using samples from $p_\theta(\theta)$. Although not limited to this, NFs typically aim to map $p_\theta(\theta)$ to a standard normal distribution $\mathcal{N}(\theta|0,\mathbb{1}_D)$. The NF can be trained to approximate a certain probability density by minimizing the forward Kullback–Leibler (KL) divergence as the loss function:

$$L = -\frac{1}{N}\sum w_i \log p(\theta_i) \tag{17}$$

2.3.3. Preconditioned Crank–Nicolson

In principle, any MH method can be applied in the latent space of z targeting $p_z(z)$. Refs. [12–14] utilized RWM, whereas [19] employed the No U-Turn Sampler (NUTS). Despite their performance, those methods do not take full advantage of the symmetries of $p_z(z)$.

Preconditioned Crank–Nicolson (pCN) provides a more suitable alternative [15,16]. Assuming that the target density can be decomposed as:

$$p(\theta) \propto \mathcal{L}(\theta)\mathcal{N}(\theta|0,\Sigma) \tag{18}$$

where $\mathcal{L}(\theta)$ is the likelihood function and $\mathcal{N}(\theta|0,\Sigma)$ is the normal prior density, a new state is proposed as follows:

$$\theta' = \sqrt{1-\epsilon^2}\theta + \epsilon v \tag{19}$$

where $v \sim \mathcal{N}(\theta|0,\Sigma)$ and $0 < \epsilon < 1$. In the limit that $\epsilon \to 1$, θ' is a sample from the prior, whereas when $\epsilon \to 0$, $\theta' = \theta$. For values between 0 and 1, the proposed states shift towards zero. For a weak likelihood function, θ' is shifted along the direction of the gradient, leading to higher acceptance rates. The acceptance criterion takes the form:

$$\alpha = \min\left\{1, \frac{\mathcal{L}(\theta')}{\mathcal{L}(\theta)}\right\}, \tag{20}$$

where the term corresponding to the prior density is missing, as the latter is already included in the proposal.

The main limitations of pCN are the assumptions of a normal prior and of a weak likelihood function. For pCN to be efficient, the prior density needs to dominate any contributions from the likelihood, thus maintaining the symmetry of the normal distribution. Preconditioned pCN (p^2CN) allows us to bypass these difficulties. Given the approximately standard normal target density $p_z(z)$, the p^2CN proposal is:

$$z' = \sqrt{1-\epsilon^2} z + \epsilon v, \tag{21}$$

where $v \sim \mathcal{N}(v|0,1)$. The acceptance probability of the new state is:

$$\alpha = \min\left\{1, \frac{\tilde{\mathcal{L}}(z')}{\tilde{\mathcal{L}}(z)}\right\}, \tag{22}$$

where $\tilde{\mathcal{L}}(z) = p_z(z)/\mathcal{N}(z|0,1)$ is the pseudo-likelihood function, which reduces to the proper target density $p_z(z)$ in latent space when multiplied with the pseudo-prior $\mathcal{N}(v|0,1)$.

Unlike RWM, p^2CN can scale to high dimensions. In fact, for a sufficiently preconditioned target distribution, p^2CN exhibits a non-zero acceptance rate for $\epsilon > 0$, even in a very high D. The efficiency of the method solely relies on the capacity of the preconditioning transformation to map the target distribution to a standard normal density and the value of ϵ.

2.4. Preconditioned Monte Carlo

PMC seamlessly integrates the SMC framework, PR, and p^2CN into a comprehensive algorithm, as illustrated in Algorithm 1. A noteworthy feature is the use of an NF as a preconditioner. PMC is primarily characterized by three key hyperparameters. The first is the ESS, which regulates the algorithm's convergence rate, and is implicitly tied to the number of beta levels. The second is the number of resampled particles, a value that should ideally be lower than the ESS. The algorithm's performance shows a relatively low sensitivity to variations in this parameter, provided that the number of resampled particles remains below half the ESS value. Lastly, the number of p^2CN steps per iteration is set to:

$$M = \frac{D}{2} \times \min\left(1, \frac{2.38/\sqrt{D}}{\epsilon}\right)^{3/2} \tag{23}$$

where ϵ was adjusted at the beginning of each iteration to yield an acceptance rate of 40%.

Algorithm 1 Preconditioned Monte Carlo

input Number of particles N and desired ESS
$t \leftarrow 1$, $\beta_1 \leftarrow 0$, $\mathcal{Z}_1 \leftarrow 1$, $\{\theta_1^k\}_{k=1}^N \sim \pi(\theta)$
while $\beta_t \neq 1$ **do**
  $t \leftarrow t+1$
  $\beta_t \leftarrow$ solution to Equation (7)
  compute weights $\left\{\{w_{tt'}^i\}_{t'=1}^t\right\}_{i=1}^N$ using Equation (8)
  compute evidence $\mathcal{Z}_t$ using Equation (13)
  train $\theta = f(u)$ on $\left\{\{\theta_{t'}^i, w_{tt'}^i\}_{t'=1}^t\right\}_{i=1}^N$ using Equation (17)
  $\{\tilde{\theta}^i\}_{i=1}^N \leftarrow$ resample N particles from $\left\{\{\theta_{t'}^i, w_{tt'}^i\}_{t'=1}^t\right\}_{i=1}^N$
  $\{\theta_t^k\}_{k=1}^N \leftarrow$ propagate $\{\tilde{\theta}^i\}_{i=1}^N$ according to $\mathcal{K}_t(\{\theta_t^i\}_{i=1}^N \leftarrow \{\tilde{\theta}_{t-1}^i\}_{i=1}^N; f)$
end while
return weighted samples $\left\{\{\theta_{t'}^i, w_{tt'}^i\}_{t'=1}^t\right\}_{i=1}^N$ and estimate of the marginal likelihood $\mathcal{Z}_t$

3. Results

To verify the sampling performance of our approach, we perform an ablation study comparing PMC with PR and p^{2}CN (PMC-PR-p^{2}CN) to the following SMC variants: SMC with PR and RWM updates (SMC-PR-RWM), SMC with no PR and RWM updates (SMC-RWM), PMC with PR and RWM (PMC-PR-RWM), PMC with no PR and RWM (PMC-RWM), and PMC with no PR and p^{2}CN (PMC-p^{2}CN). The combinations of SMC with pCN were not included due to the low performance of pCN in the non-preconditioned setting. The total cost of each method in terms of likelihood evaluations along with the estimate of the logarithm of the marginal likelihood, $\log \mathcal{Z}$, subtracted from the exact value, are shown in Table 1. In all cases, the variance of $\log \mathcal{Z}$ was computed based on 20 independent runs. The number of MCMC steps per iteration was determined using Equation (23) for p^{2}CN. In RWM, the proposal covariance $\lambda^2\Sigma$ was tuned to yield an acceptance rate of 23.4% and the number of MCMC steps was set to $D/2 \times [(2.38/\sqrt{D})/\lambda]^2$.

For methods employing PR, we set the ESS target to 1500 and the number of resampled particles to 500. When PR was not employed, the total number of particles was 2000 and the ESS was 1500. A masked autoregressive flow (MAF) with six blocks of transformations featuring 3×128 hidden units was employed as the NF preconditioner [20]. The flow was trained using the Adam optimizer with a learning rate of 10^{-3}, until no further improvement was observed in the validation loss function for at least 50 training iterations [21]. The training batch size was 1000 and the validation fraction was 30%. The same NF was trained in all iterations without resetting its weights.

Table 1. Computational cost in terms of the number of likelihood evaluations and marginal likelihood estimates.

Algorithm	Rosenbrock-10 (Calls $\times 10^6$)	Sonar-61 (Calls $\times 10^6$)	Rosenbrock-10 ($\Delta \log \mathcal{Z}$)	Sonar-61 ($\Delta \log \mathcal{Z}$)
PMC-PR-p^{2}CN	**0.11**	**0.32**	$\mathbf{0.09 \pm 0.21}$	$\mathbf{-0.08 \pm 0.12}$
PMC-PR-RWM	0.23	2.31	0.18 ± 0.27	0.36 ± 0.69
PMC-p^{2}CN	0.21	1.21	-0.33 ± 0.66	0.27 ± 0.38
PMC-RWM	0.48	6.41	-0.39 ± 0.72	0.30 ± 0.58
SMC-PR-RWM	0.62	4.20	0.15 ± 0.33	-0.18 ± 0.63
SMC-RWM	1.12	9.81	-0.28 ± 0.41	-0.32 ± 0.93

3.1. Rosenbrock Distribution in 10-D

To evaluate the performance of PMC in sampling strongly correlated target distributions, we utilized the Rosenbrock distribution in 10-D, with a log-likelihood function given by:

$$\log \mathcal{L}(\theta) = -\sum_{i=1}^{D/2} \left[10(\theta_{2i-1}^2 - \theta_{2i})^2 + (\theta_{2i-1}^2 - 1)^2\right], \tag{24}$$

where $D = 10$, and a $\mathcal{N}(\theta|0, 3^2 \times \mathbb{1}_D)$ is the prior density. The value of the marginal likelihood for this example is $\log \mathcal{Z} = -21.39$, as computed using an extensive run and verified using all methods. Figure 1 depicts the 2D marginal posterior sampled using PMC-PR-p^{2}CN and SMC-RWM. Although the other methods produced an indistinguishable contour, they are not shown for the sake of clarity. As shown in Table 1, PMC-PR-p^{2}CN is the least computationally expensive algorithm, requiring an order of magnitude with fewer likelihood calls compared to the other SMC/PMC variants. The values for the number of likelihood calls per method shown in Table 1 demonstrate that both PR and p^{2}CN substantially improve the sampling efficiency. In terms of the estimate of the marginal likelihood, PMC-PR-p^{2}CN exhibits the highest accuracy and precision. Although less significant than in the case of computational cost evaluation, both PR and p^{2}CN aid in reducing the uncertainty of the marginal likelihood estimate.

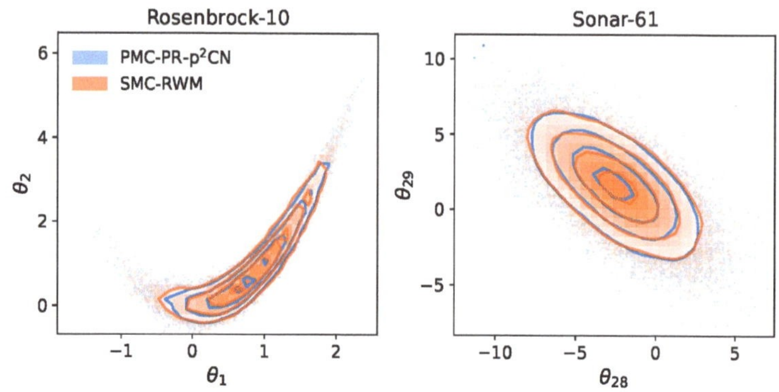

Figure 1. Two-dimensional marginal posteriors of 10-D Rosenbrock (**left**) and 61-D logistic regression with sonar data (**right**) as obtained using PMC-PR-p^{2}CN (blue) and SMC-RWM (orange).

3.2. The 61-D Logistic Regression with Sonar Data

To demonstrate the favorable scaling of PMC with the number of dimensions, as opposed to SMC/PMC variants employing RWM proposals, we used the 61-D problem of logistic regression with sonar data [22–24]. In this case, the likelihood function is:

$$\mathcal{L}(\theta) = \prod_{i=1}^{N} \sigma\left[y_i\left(\theta_1 + x_i^T \theta_{2:D}\right)\right], \tag{25}$$

where $\sigma(t) = (1 + e^{-t})^{-1}$ is the sigmoid function, N is the number of datapoints, x_i are the predictor variables, and $y_i \in [-1, +1]$ are the labels. Following standard practice, each predictor is re-scaled to have a mean of zero and a standard deviation of 0.5. The prior distribution is $\pi(\theta_1) = \mathcal{N}(\theta_1|0, 20^2)$ for the intercept and $\pi(\theta_{2:D}) = \mathcal{N}(\theta_{2:D}|0, 5^2 \times \mathbb{1}_{D-1})$ for all other parameters. The value of the marginal likelihood for this example is $\log \mathcal{Z} = -125.46$. Figure 1 depicts the 2D marginal posterior sampled using PMC-PR-p^{2}CN and SMC-RWM. Table 1 highlights that PMC-PR-p^{2}CN is the most efficient algorithm from a computational perspective, requiring significantly fewer likelihood calls than its SMC/PMC counterparts. Table 1 also underscores the key role of both PR and p^{2}CN in enhancing the sampling efficacy, as evidenced by the number of likelihood calls per method. Furthermore, when evaluating the marginal likelihood, PMC-PR-p^{2}CN displays the highest accuracy and precision. Similarly to the Rosenbrock example, PR and p^{2}CN contribute to reducing the estimate's uncertainty of the marginal likelihood.

4. Discussion

In this study, we developed and thoroughly tested PMC, a novel variant of SMC. Our key innovation is the incorporation of a PR scheme and p^{2}CN updates in the PMC methodology. The resulting method displayed remarkable efficiency, outperforming other variants of PMC and conventional SMC in terms of total computational cost, measured by the total number of likelihood evaluations. The use of p^{2}CN enabled superior scaling with the number of dimensions compared to RWM, as shown in the logistic regression example. Furthermore, PMC demonstrated improved precision and accuracy in estimating the marginal likelihood, a critical component of Bayesian model comparison.

The performance of PMC represents a promising stride forward in making Bayesian computation more feasible in the physical sciences, particularly in scenarios where the models are not easily differentiable and the gradient of the target posterior is intractable. Our work lays the groundwork for more sophisticated computational tools for tackling problems in the physical sciences. Looking to the future, we envision further refining PMC by developing more advanced tuning procedures, such as determining the optimal number of MCMC steps per iteration. Additionally, we plan to undertake a comprehensive comparison of PMC with nested sampling, a method widely adopted in astronomy and cosmology. This will solidify the position of PMC within the Bayesian computational toolkit and explore its potential to bring new insights into the physical world. PR, p^2CN, and new developments will be integrated in the `pocoMC` Python package version 1.0.0 [13] available in https://pocomc.readthedocs.io, accessed on 16 July 2023.

Author Contributions: Conceptualization, M.K. and U.S.; methodology, M.K.; software, M.K.; validation, M.K.; formal analysis, M.K.; investigation, M.K.; resources, U.S.; data curation, M.K.; writing—original draft preparation, M.K.; writing—review and editing, M.K. and U.S.; visualization, M.K.; supervision, U.S.; project administration, U.S.; funding acquisition, U.S. All authors have read and agreed to the published version of the manuscript.

Funding: This research was funded by the U.S. Department of Energy, Office of Science, Office of Advanced Scientific Computing Research under Contract No. DE-AC02-05CH11231 at Lawrence Berkeley National Laboratory to enable research for data-intensive machine learning and analysis.

Institutional Review Board Statement: Not applicable.

Informed Consent Statement: Not applicable.

Data Availability Statement: The sonar data used in the logistic regression example are available at the UCI machine learning repository [22].

Acknowledgments: The authors thank David Nabergoj for helpful discussions.

Conflicts of Interest: The authors declare no conflict of interest. The funders had no role in the design of the study; in the collection, analyses, or interpretation of data; in the writing of the manuscript; or in the decision to publish the results.

References

1. Jaynes, E.T. *Probability Theory: The Logic of Science*; Cambridge University Press: Cambridge, UK, 2003.
2. Gregory, P. *Bayesian Logical Data Analysis for the Physical Sciences: A Comparative Approach with Mathematica® Support*; Cambridge University Press: Cambridge, UK, 2005.
3. MacKay, D.J. *Information Theory, Inference and Learning Algorithms*; Cambridge University Press: Cambridge, UK, 2003.
4. Trotta, R. Bayesian methods in cosmology. *arXiv* **2017**, arXiv:1701.01467.
5. Sharma, S. Markov chain Monte Carlo methods for Bayesian data analysis in astronomy. *Annu. Rev. Astron. Astrophys.* **2017**, *55*, 213–259. [CrossRef]
6. Del Moral, P.; Doucet, A.; Jasra, A. Sequential monte carlo samplers. *J. R. Stat. Soc. Ser. B (Stat. Methodol.)* **2006**, *68*, 411–436. [CrossRef]
7. Chopin, N.; Papaspiliopoulos, O. *An Introduction to Sequential Monte Carlo*; Springer: Berlin/Heidelberg, Germany, 2020; Volume 4.
8. Naesseth, C.A.; Lindsten, F.; Schön, T.B. Elements of sequential monte carlo. *arXiv* **2019**, arXiv:1903.04797.
9. Hastings, W.K. Monte Carlo Sampling Methods using Markov Chains and their Applications. *Biometrika* **1970**, *57*, 97–109. [CrossRef]
10. Neal, R.M. Slice sampling. *Ann. Stat.* **2003**, *31*, 705–767. [CrossRef]
11. Papamakarios, G.; Nalisnick, E.; Rezende, D.J.; Mohamed, S.; Lakshminarayanan, B. Normalizing flows for probabilistic modeling and inference. *J. Mach. Learn. Res.* **2021**, *22*, 1–64.
12. Karamanis, M.; Beutler, F.; Peacock, J.A.; Nabergoj, D.; Seljak, U. Accelerating astronomical and cosmological inference with preconditioned Monte Carlo. *Mon. Not. R. Astron. Soc.* **2022**, *516*, 1644–1653. [CrossRef]
13. Karamanis, M.; Nabergoj, D.; Beutler, F.; Peacock, J.A.; Seljak, U. pocoMC: A Python package for accelerated Bayesian inference in astronomy and cosmology. *arXiv* **2022**, arXiv:2207.05660.
14. Moss, A. Accelerated Bayesian inference using deep learning. *Mon. Not. R. Astron. Soc.* **2020**, *496*, 328–338. [CrossRef]
15. Beskos, A.; Roberts, G.; Stuart, A.; Voss, J. MCMC methods for diffusion bridges. *Stochastics Dyn.* **2008**, *8*, 319–350. [CrossRef]
16. Cotter, S.L.; Roberts, G.O.; Stuart, A.M.; White, D. MCMC methods for functions: Modifying old algorithms to make them faster. *Stat. Sci.* **2013**, *28*, 424–446. [CrossRef]
17. Le Thu Nguyen, T.; Septier, F.; Peters, G.W.; Delignon, Y. Improving SMC sampler estimate by recycling all past simulated particles. In Proceedings of the 2014 IEEE Workshop on Statistical Signal Processing (SSP), Gold Coast, QLD, Australia, 29 June–2 July 2014; pp. 117–120.
18. Gramacy, R.; Samworth, R.; King, R. Importance tempering. *Stat. Comput.* **2010**, *20*, 1–7. [CrossRef]
19. Hoffman, M.; Sountsov, P.; Dillon, J.V.; Langmore, I.; Tran, D.; Vasudevan, S. Neutra-lizing bad geometry in hamiltonian monte carlo using neural transport. *arXiv* **2019**, arXiv:1903.03704.

20. Papamakarios, G.; Pavlakou, T.; Murray, I. Masked autoregressive flow for density estimation. In Proceedings of the 31 Conference on Neural Information Processing Systems (NIPS 2017), Long Beach, CA, USA, 4–9 December 2017; Curran Associat Inc.: Red Hook, NY, USA, 2017; pp. 2335–2344.
21. Kingma, D.P.; Ba, J. Adam: A method for stochastic optimization. *arXiv* **2014**, arXiv:1412.6980.
22. Sejnowski, T.; Gorman, R. Connectionist Bench (Sonar, Mines vs. Rocks). UCI Machine Learning Repository. Available onlin https://archive.ics.uci.edu (accessed on 16 July 2023).
23. Chopin, N.; Ridgway, J. Leave Pima Indians alone: Binary regression as a benchmark for Bayesian computation. *Stat. Sci.* **201** *32*, 64–87. [CrossRef]
24. Dau, H.D.; Chopin, N. Waste-free sequential monte carlo. *J. R. Stat. Soc. Ser. B Stat. Methodol.* **2022**, *84*, 114–148. [CrossRef]

physical sciences forum

MDPI

Proceeding Paper

Analysis of Ecological Networks: Linear Inverse Modeling and Information Theory Tools †

Valérie Girardin [1], Théo Grente [1,2,*], Nathalie Niquil [3] and Philippe Regnault [4]

1 Laboratoire de Mathématiques Nicolas Oresme, CNRS 6139, Normandie Université, UNICAEN, 14000 Caen, France; valerie.girardin@unicaen.fr
2 France Énergies Marines, 29280 Plouzané, France
3 Laboratoire BOREA MNHN, CNRS 8067, IRD 207, UNICAEN, 14000 Caen, France; nathalie.niquil@unicaen.fr
4 Laboratoire de Mathématiques de Reims, CNRS 9008, Université de Reims Champagne Ardenne, 51097 Reims, France; philippe.regnault@univ-reims.fr
* Correspondence: theo.grente@gmail.com

† Presented at the 42nd International Workshop on Bayesian Inference and Maximum Entropy Methods in Science and Engineering, Garching, Germany, 3–7 July 2023.

Abstract: In marine ecology, the most studied interactions are trophic and are in networks called food webs. Trophic modeling is mainly based on weighted networks, where each weighted edge corresponds to a flow of organic matter between two trophic compartments, containing individuals of similar feeding behaviors and metabolisms and with the same predators. To take into account the unknown flow values within food webs, a class of methods called Linear Inverse Modeling was developed. The total linear constraints, equations and inequations defines a multidimensional convex-bounded polyhedron, called a polytope, within which lie all realistic solutions to the problem. To describe this polytope, a possible method is to calculate a representative sample of solutions by using the Monte Carlo Markov Chain approach. In order to extract a unique solution from the simulated sample, several goal (cost) functions—also called Ecological Network Analysis indices—have been introduced in the literature as criteria of fitness to the ecosystems. These tools are all related to information theory. Here we introduce new functions that potentially provide a better fit of the estimated model to the ecosystem.

Keywords: divergence; entropy; linear inverse modeling; trophic systems

1. Introduction

The description of ecosystems is often based on networks of interactions of different types. For terrestrial ecosystems, recent developments concern different types of interactions, sometimes gathered into a common model called multiplex [1]. In marine ecology, the most studied interactions are trophic, i.e., interactions between predators. These highly complex food webs have been described by numerous models widely used to describe the impact of human activities on marine ecosystems; see [2,3]. They are also important tools for the sustainable management of marine and coastal environments [4].

There are many ways to model food webs. Flows from prey to predators can be represented by nonquantified interactions, and then the model is a graph (web) with vertices (the species) and edges (the interactions). Trophic modeling for the ecosystem-based management of fisheries and other management questions is mainly based on weighted networks; see, for example, the Ecopath–Ecosim–Ecospace model [5]. Each link corresponds to a flow of organic matter between two trophic compartments, collecting individuals of similar feeding behaviors and metabolisms and with the same predators.

In ecology, field studies are often associated with laboratory experiments to estimate some flows within food webs, but many flow values remain unknown, which the Linear Inverse Modeling (LIM) takes into account; see [6]. LIM relies on the principle of a steady state for the biomass of all compartments, i.e., the sum of the inflows and outflows through the components of the system equals the rate of change in their standing stocks, most often considered negligible or null. This yields a set of linear equations, the Mass-Balance Equations (MBE), describing the steady state or mass balance. Then, constraints are added from field measurements of mass transfers like local estimations of primary production, respiration or diet contents, etc. Additional constraints come from experiments or the study of other ecosystems. All these constraints constitute a set of linear inequalities that impose linear combinations of flows to be between certain values. The total of the MBE and inequalities defines a bounded multidimensional polyhedron, called a polytope, within which lie all realistic solutions to the problem.

The first idea to describe the polytope was to select a single solution inside this space of possible solutions, assuming a least square of the flows; see [7]. Then, methods were developed to describe the solution space by calculating a representative sample of all possible solutions by using the Monte Carlo approach, see [8], and more efficiently by using the Monte Carlo Markov Chain (MCMC) approach, see [9,10]. Further, Linear Inverse Modeling Monte Carlo Markov Chain (LIM-MCMC) methods involve mass balanced models (the biomass of each species is assumed to be constant) that consider the uncertainty in the estimations and link them to the variability of the living; see [6,10]. As such, a range of possible results is provided that makes LIM-MCMC advantageous by allowing one to include uncertainties in the data; see [11].

Still, the question remains to extract a unique solution from this simulated sample. In [12], we proposed a numerical optimization of some pertinent goal functions that directly yield a unique functioning point of the ecosystem with a low computing time. This straightforward approach allowed us to infer properties on the stability of the selected state and hence on the evolution of the ecosystem. Precisely, the assumption is that the biomasses are the stationary state of a dynamical system determined by the constraints on the flows. Determining the stability of this fixed point informs us on how the biomasses will vary. In this sense, this approach is data-driven by the information on both the biomasses and the biological constraints.

Such an optimization may also be inferred in a probabilistic framework where the flows are all seen as random variables. MCMC thus amounts to simulating samples drawn from the probability density functions (pdfs) of the flows, while the optimization can be processed in the space of pdfs satisfying the biological constraints.

In ecology, Ecological Network Analysis (ENA) indices have been introduced in the literature as criteria of fitness to the ecosystems; see [9,12–14]. They are all related to Shannon information theory. Some are based on Shannon entropy while others are based on Kullback–Leibler divergence [15] or Shannon mutual information. In this paper, we present several of these classical goal (cost) functions. Then, we consider goal functions that take into account reference distributions that may not belong to the polytope but are of interest; for instance, those resulting from previous studies of the same ecosystem. We also consider goal functions based on Rényi entropy and divergence that may fit better to particular food systems by adjusting the associated parameter. Their potential optimization opens the way to a better fit of the chosen unique solution among the MCMC sample.

2. The Constraints, the Polytope of Solutions and the Probabilistic Framework

2.1. The Constraints

We will denote by V the set of all vertices of the trophic network (directed graph), and E is the set of all oriented edges—denoted by ij when going from vertice i to vertice j, with $|E| = n$.

In many situations, all biomasses B_i of the species can be measured with accuracy. On the other hand, the flows F_{ij} between the species (or compartments) are much more difficult to evaluate. Therefore, we will adopt here the standard viewpoint that the biomasses are given and the flows between nodes are unknown.

Mathematically, the flow is a vector $F = (F_{ij})_{ij \in E}$ of dimension $n = |E|$ with non-negative components, also called flows, satisfying the Kirchhoff law on all vertices:

$$\sum_{j \in N^+(i)} F_{ij} - \sum_{j \in N^-(i)} F_{ji} = 0, \quad i \in V, \tag{1}$$

where the successors and predecessors of a vertex i are $N^+(i) = \{j \in V : ij \in E\}$ and $N^-(i) = \{j \in V : ji \in E\}$.

The flows F_{ij} satisfy biological constraints. For example, a fish cannot eat more than a certain percentage of its biomass. These constraints are numerous thanks to the important knowledge that has been acquired over the years on the functioning of species ecosystems and can be summarized as a system of linear equations and inequations over the flow space.

2.2. The Polytope of Solutions

The set $\mathcal{S}$ of solutions F to the constraints system is a convex subset of $\mathbb{R}^n$, where n is the number of edges of the graph. Precisely, for any such ecosystem,

$$\mathcal{S} = \{F = (F_{ij})_{ij \in E} \in \mathbb{R}^n : F_{ij} \geq 0, \ AF = b, \ GF \leq h\}, \tag{2}$$

where A and G are matrices of sizes $m \times n$ and $k \times n$, respectively, and b and h vectors of sizes m and k, respectively, where m and k are, respectively, the number of equality and inequality constraints (with $m < n$ so that the system $AF = b$ is underdetermined).

The defined domain is the intersection of hyperplanes and half spaces and hence is a convex polyhedron. In addition, biological constraints encompass bounds on flows, so $\mathcal{S}$ is bounded. The same is true for most alike ecosystems. Such bounded polyhedrons are called polytopes. Apart from the MBE which involve many flows in the same equation, most constraints involve a small number of flows. Therefore, matrices A and G are mainly composed of 0 and a few 1.

2.3. The Probabilistic Framework

The problem can also be presented in terms of pdfs or proportions. Let $f_{ij} = F_{ij}/F_{..}$: denote the proportion of flows from vertex (i.e., compartment) i to vertex j with respect to the total flow in the system, where $F_{..} = \sum_{ij \in E} F_{ij}$. The marginal distributions $(f_{i.})_{i \in V}$ and $(f_{.j})_{j \in V}$, where $f_{i.} = \sum_{j \in N^+(i)} f_{ij}$ and $f_{.j} = \sum_{i \in N^-(j)} f_{ij}$, represent the proportions of, respectively, the outflows and inflows of trophic compartments. The polytope of admissible flows defined by (2) is identified as

$$\mathcal{S} = \{f = (f_{ij})_{ij \in E} \in \mathbb{R}^n : 0 \leq f_{ij} \leq 1, \ Af = a, \ Gf \leq g\}, \tag{3}$$

where $a = b/F_{..}$ and $g = h/F_{..}$.

Actually, $\mathcal{S}$ can be considered, and hence MCMC can be applied in a smaller space. Let d be the rank of A, that is, the dimension of the subspace of $\mathbb{R}^n$ defined by the system $Af = b$. One way to present this is to write $f = f_p + \widetilde{A}x$, where f_p is a particular solution in $\mathcal{S}$ and $\widetilde{A}$ is the orthonormal matrix obtained from A, e.g., through the well-known QR decomposition. Indeed, in this subspace, only inequality constraints remain, say $\widetilde{G}x \leq \widetilde{h}$ with $\widetilde{G} = G\widetilde{A}$ and $\widetilde{h} = Gf_p - h$, which define the volume delimited by the intersection of hyperplanes. MCMC methods are then applied to the latter system that is in a smaller space.

3. The Goal Functions

In ecological studies, goal functions appear either as indices of the behavior of the system or as a means of comparison between estimated solutions. Actually, they are potential tools for determining one solution from the numerous MCMC-simulated ones.

Many goal functions can be considered in relation to ecosystems. They are either statistical tools, information theory quantities or purely ecological indexes. The most classical are the mean, the quadratic energy and the so-called ENA, all quantities derived from entropy. An important characteristics is whether the goal function includes information from the constraints or not. Further, only convex (or concave) functions yield a unique optimum corresponding to a unique functioning state of the system.

The natural setting is to give all these functions in terms of the pdfs f. Indeed, the advantage of considering pdfs for ENA indices, computed in different situations, is that different ecosystems can be compared. This makes it possible to compare different stations or different time periods. For example, Le Guen et al. [16] compares seven spatial subsections of the Seine estuary at two time periods; they were thus able to describe the change in the functioning of the estuary's food web, linked to the construction of the Le Havre port extension called Port2000, but also to the change in the precipitation regime that occurred over the same period. Such comparisons are essential in the search for indicators of marine ecosystem health.

3.1. Classical ENA

The most usual function is the empirical mean of all fluxes,

$$E(f) = \sum_{ij \in E} f_{ij}. \tag{4}$$

The quadratic energy, giving rise to the least squares method, is also classical, with

$$Q(f) = \sum_{ij \in E} f_{ij}^2. \tag{5}$$

Various functions involving the concept of the entropy of distributions have been considered in the literature on ecosystems under various specific names. See [17] and the references therein for a synthesis of these Shannon entropic indices and [18] for details and more on information theory tools. Shannon's definition of entropy $\mathbb{S}$ gave birth to information theory; see [18,19]. It was introduced in the field of ecological systems by [20], through the so-called Mc Arthur index C,

$$\begin{aligned} C(f) &= \mathbb{S}(f) && (6) \\ &= -\sum_{ij \in E} f_{ij} \log(f_{ij}) = -\sum_{ij \in E} \frac{F_{ij}}{F_{..}} \log\left(\frac{F_{ij}}{F_{..}}\right). && (7) \end{aligned}$$

Note that the basis of the logarithm is non pertinent since only comparisons or optimization are to be considered.

Mutual information is also a classical tool in information theory. It was introduced in ecology by [21], ans is called ascendency, defined as

$$A(f) = \sum_{ij \in E} f_{ij} \log\left(\frac{f_{ij}}{f_{i.}f_{.j}}\right) = \sum_{ij \in E} \frac{F_{ij}}{F_{..}} \log\left(\frac{F_{ij}F_{..}}{F_{i.}F_{.j}}\right). \tag{8}$$

The ascendency can be written as $K(f|(f_{i.}) \otimes (f_{.j}))$; that is, the Kullback–Leibler divergence with respect to the product distribution $(f_{i.}) \otimes (f_{.j}) = (f_{i.}f_{.j})$. This function is neither convex nor concave, which hinders it from being used as an optimization function; see [12].

While A/C also serves to measure the degree of constraint of the system, in order to keep a balance between A and C, Fath [22] prefers

$$G_{A/C}(f) = \frac{A(f)}{C(f)} \log\left[\frac{A(f)}{C(f)}\right], \tag{9}$$

for which we fail to find an interpretation in terms of pdfs.

The symmetrized conditional entropy becomes, in ecology, the system redundancy (overhead), introduced in [23] as the negative quantity $\Phi = C - A$, that is to say

$$\Phi(f) = -\sum_{ij \in E} f_{ij} \log\left(\frac{f_{ij}\, f_{ij}}{f_{i.}\, f_{.j}}\right) = -\sum_{ij \in E} \frac{F_{ij}}{F_{..}} \log\left(\frac{F_{ij}^2}{F_{i.}F_{.j}}\right). \tag{10}$$

Several indices drawn from overhead and ascendency have been proposed as health indicators based on marine ecosystem functioning; see [24,25].

Note that in the literature, these ENA indices have sometimes been multiplied by $F_{..}$, to make them depend on the total mass of the system. Then, they can only be used as tools when analyzing the system and neither for determining a flow solution to the constraints nor for comparing two systems.

3.2. Goal Function Involving a Reference

An added value for determining a pertinent solution in $\mathcal{S}$ is to incorporate information on the system that is not given in terms of the set of the MBE and inequalities. With this aim, the following goal functions involve some reference pdf, say f^*, known a priori to be informative on the behavior of the ecosystem. In particular, a solution to the problem obtained in a previous study, or a previous year, may not be a solution to the present but can be considered as a reasonable reference. Another natural way is to consider the middle of the constraint intervals, in particular to take into account the regular feeding habits of most species; see [12].

The most classical tool in information theory is the Kullback–Leibler divergence

$$\mathbb{K}(f|f^*) = \sum_{ij \in E} f_{ij} \log\left(\frac{f_{ij}}{f_{ij}^*}\right), \tag{11}$$

where $f^* = (f_{ij}^*)$ is some reference pdf of flows that makes sense in terms of the ecosystem. Note that [17] considers it as a "structural term".

The divergence is not a mathematical distance because it is not symmetric in f and f^*. Still, it is non-negative and null only if $f = f^*$, and minimizing K determines the projection in terms of the divergence of the reference f^* on the set $\mathcal{S}$ of solutions f to the constraints; see [18,26].

It is worth remembering that the entropy $\mathbb{S}$ in Equation (6) is, up to constants, the Kullback–Leibler divergence with respect to the uniform distribution on the flows, with $f_{ij}^* = 1/n$. This uniform distribution is well known to maximize the entropy when no information is a priori available.

A simple generalization of the quadratic function Q is the well-known χ^2-distance:

$$\chi^2(f) = \sum_{ij \in E} \frac{(f_{ij} - f_{ij}^*)^2}{(f_{ij}^*)^2};$$

see [12] for an application, where for both $\mathbb{K}$ and χ^2, we have chosen f^* for each flow as the middle of the interval defined by the inequality constraints.

3.3. Rényi Goal Functions

Shannon-type entropy quantities have been generalized to many others that—to the best of our knowledge—have not yet been used in ecological networks. For a better fit of the goal function to the problem through the choice of a parameter s, we propose considering the family of Rényi entropy, introduced in [27], and defined for positive $s \neq 1$ by

$$R_s(f) = \frac{1}{1-s} \log\left[\sum_{ij \in E} (f_{ij})^s\right], \tag{12}$$

with associated divergence:

$$R_s(f|f^*) = \frac{1}{1-s}\log\left[\sum_{ij\in E}\frac{(f_{ij})^s}{(f^*_{ij})^{s-1}}\right]. \tag{13}$$

The value of the parameter s has to be chosen. In a first stage, $s = 1/2$ and $s = 2$ can illustrate the role of the regions separated by the threshold $s = 1$. Note that Shannon entropy appears as the limit of Rényi entropy when s tends to one; see [18].

Similarly to the Shannon case, Rényi mutual information is Rényi divergence with respect to the product distribution $(f_{i.}f_{.j})$:

$$A_s(f) = R_s(P|(f_{i.}) \otimes (f_{.j})) = \frac{1}{1-s}\log\left[\sum_{ij\in E}\frac{(f_{ij})^s}{(f_{i.}f_{.j})^{s-1}}\right]. \tag{14}$$

The associated quantities $A_s(f)/R_s(f)$ and $R_s(f) - A_s(f)$ may also be of use in further studies.

4. Conclusions

Classically in ecology, ENA indices are used to bring more information to an estimated model or as a tool for comparing two of them. Considering them as goal (cost) functions in [12] opens the field to full mathematical optimization methods as soon as the convexity of the function is assessed. Adding to the collection divergence-like functions that take into account reference pdfs should lead to a better fit to a priori information on the ecosystem. Tools from extended information theory, such as Rényi's, may also yield an optimized solution that is a better fit to certain complex systems by a pertinent choice of the parameter.

A fully determined food web model (one in which all the flows are known, mainly from the mean of the observations) is presented in [28], which describes the eight habitats of the Sylt-Romo Bight. These food webs are composed of 56 living compartments and three nonliving compartments. In [29], the known flows are replaced with inequalities at four decreasing levels of information block per block. Then, a unique solution is chosen within the LIM-MCMC sample by optimizing a variety of goal functions. Some technical issues of computation time regarding the R package `limsolve` used in this paper made the randomization difficult. This will be corrected by using a new version of `limsolve` updated in C++, which will appear soon; see [30]. Then, a meaningful comparison of the goal functions on the ecosystems of [28] at decreasing levels of information will be conducted by the authors.

Author Contributions: All authors contributed equally to the writing of the original draft, the writing of the final version, the review and the editing of the present paper. All authors have read and agreed to the published version of the manuscript.

Funding: This research received no external funding.

Institutional Review Board Statement: Not applicable.

Informed Consent Statement: Not applicable.

Data Availability Statement: Data sharing is not applicable to this article.

Conflicts of Interest: The authors declare no conflicts of interest.

Abbreviations

The following abbreviations are used in this manuscript:

MBE	Mass-Balance Equations
pdf	Probabilistic density function
ENA	Ecological Network Analysis
LIM	Linear Inverse Modeling
MCMC	Monte Carlo Markov Chain

References

1. Fontaine, C.; Guimarães, P.R., Jr.; Kéfi, S.; Loeuille, N.; Memmott, J.; van Der Putten, W.H.; Van Veen, F.J.; Thébault, E. The ecological and evolutionary implications of merging different types of networks. *Ecol. Lett.* **2011**, *14*, 1170–1181. [CrossRef]
2. Belgrano, A.; Scharler, U.M.; Dunne, J.; Ulanowicz, R.E. (Eds.) *Aquatic Food Webs: An Ecosystem Approach*; Oxford University Press: Oxford, UK, 2005.
3. Jørgensen, S.E.; Fath, B.D. *Fundamentals of Ecological Modelling: Applications in Environmental Management and Research*; Elsevier: Amsterdam, The Netherlands, 2011.
4. Langlet, D.; Rayfuse, R. *The Ecosystem Approach in Ocean Planning and Governance: Perspectives from Europe and Beyond*; Brill: Leiden, The Netherlands, 2018; p. 492.
5. Christensen, V.; Walters, C.J. Ecopath with Ecosim: Methods, capabilities and limitations. *Ecol. Model.* **2004**, *172*, 109–139. [CrossRef]
6. Niquil, N.; Saint-Béat, B.; Johnson, G.A.; Soetaert, K.; Van Oevelen, D.; Bacher, C.; Vézina, A.F. Inverse modeling in modern ecology and application to coastal ecosystems. In *Treatise on Estuarine and Coastal Science*; Academic Press: Waltham, MA, USA, 2011; pp. 115–133.
7. Vézina, A.F.; Platt, T. Food web dynamics in the ocean. 1. Best-estimates of flow networks using inverse methods. *Mar. Ecol. Prog. Ser.* **1988**, *42*, 269–287. [CrossRef]
8. Kones, J.K.; Soetaert, K.; van Oevelen, D.; Owino, J.O.; Mavuti, K. Gaining insight into food webs reconstructed by the inverse method. *J. Mar. Syst.* **2006**, *60*, 153–166. [CrossRef]
9. Kones, J.K.; Soetaert, K.; van Oevelen, D.; Owino, J.O. Are network indices robust indicators of food web functioning? A Monte Carlo approach. *Ecol. Model.* **2009**, *220*, 370–382. [CrossRef]
10. Van Oevelen, D.; Van den Meersche, K.; Meysman, F.J.; Soetaert, K.; Middelburg, J.J.; Vézina, A.F. Quantifying food web flows using linear inverse models. *Ecosystems* **2010**, *13*, 32–45. [CrossRef]
11. Grami, B.; Rasconi, S.; Niquil, N.; Jobard, M.; Saint-Béat, B.; Sime-Ngando, T. Functional effects of parasites on food web properties during the spring diatom bloom in Lake Pavin: A linear inverse modeling analysis. *PLoS ONE* **2011**, *6*, e23273. [CrossRef]
12. Caputo, J.G.; Girardin, V.; Knippel, A.; Nguyen, M.H.; Niquil, N.; Noguès, Q. Analysis of trophic networks: An optimisation approach. *J. Math. Biol.* **2021**, *83*, 1–29. [CrossRef] [PubMed]
13. Johnson, G.A.; Niquil, N.; Asmus, H.; Bacher, C.; Asmus, R.; Baird, D. The effects of aggregation on the performance of the inverse method and indicators of network analysis. *Ecol. Model.* **2009**, *220*, 3448–3464. [CrossRef]
14. Ulanowicz, R.E. Biodiversity, functional redundancy and system stability: Subtle connections. *J. R. Soc. Interface* **2018**, *15*, 20180367. [CrossRef]
15. Kullback, S.; Leibler, R.A. On information and sufficiency. *Ann. Math. Stat.* **1951**, *22*, 79–86. [CrossRef]
16. Le Guen, C.; Tecchio, S.; Dauvin, J.C.; De Roton, G.; Lobry, J.; Lepage, M.; ; Morin, J.; Lassalle, G.; Raoux, A.; Niquil, N. Assessing the ecological status of an estuarine ecosystem: Linking biodiversity and food-web indicators. *Estuar. Coast. Shelf Sci.* **2019**, *228*, 106339. [CrossRef]
17. Christensen, V. Ecosystem maturity—Towards quantification. *Ecol. Model.* **1995**, *77*, 3–32. [CrossRef]
18. Cover, T.M.; Thomas, J.A. *Elements of Information Theory*, 2nd ed.; Wiley Online Books: Hoboken, NJ, USA, 2006.
19. Shannon, C.E.; Weaver, W. *The Mathematical Theory of Communication*; University of Illinois Press: Urbana, IL, USA, 1949.
20. MacArthur, R. Fluctuations of animal populations and a measure of community stability. *Ecology* **1955**, *36*, 533–536. [CrossRef]
21. Hirata, H.; Ulanowicz, R.E. Large-scale systems perspective on ecological modelling and analysis. *Ecol. Model.* **1986**, *31*, 79–104. [CrossRef]
22. Fath, B.D. Quantifying economic and ecological sustainability. *Ocean. Coast. Manag.* **2015**, *108*, 13–19. [CrossRef]
23. Ulanowicz, R.E.; Norden, J.S. Symmetrical overhead in flow networks. *Int. J. Syst. Sci.* **1990**, *21*, 429–437. [CrossRef]
24. Safi, G.; Giebels, D.; Arroyo, N.L.; Heymans, J.J.; Preciado, I.; Raoux, A.; ; Schückel, U.; Tecchio, S.; de Jonge, V.N.; Niquil, N. Vitamine ENA: A framework for the development of ecosystem-based indicators for decision makers. *Ocean. Coast. Manag.* **2019**, *174*, 116–130. [CrossRef]
25. Fath, B.D.; Asmus, H.; Asmus, R.; Baird, D.; Borrett, S.R.; de Jonge, V.N.; Ludovisi, A.; Niquil, N.; Scharler, U.M.; Schückel, U.; et al. Ecological network analysis metrics: The need for an entire ecosystem approach in management and policy. *Ocean. Coast. Manag.* **2019**, *174*, 1–14. [CrossRef]
26. Csiszár, I. I-divergence geometry of probability distributions and minimization problems. *Ann. Probab.* **1975**, *3*, 146–158. [CrossRef]
27. Rényi, A. On the dimension and entropy of probability distributions. *Acta Math. Acad. Sci. Hung.* **1959**, *10*, 193–215. [CrossRef]
28. Baird, D.; Asmus, H.; Asmus, R. Trophic dynamics of eight intertidal communities of the Sylt-Rømø Bight ecosystem, northern Wadden Sea. *Mar. Ecol. Prog. Ser.* **2007**, *351*, 25–41. [CrossRef]
29. Saint-Béat, B.; Vézina, A.F.; Asmus, R.; Asmus, H.; Niquil, N. The mean function provides robustness to linear inverse modelling flow estimation in food webs: A comparison of functions derived from statistics and ecological theories. *Ecol. Model.* **2013**, *258*, 53–64. [CrossRef]
30. Girardin, V.; Grente, T.; Niquil, N.; Regnault, P. Comparing and Updating R Packages using MCMC Algorithms for Linear Inverse Modeling of Metabolic Networks. *HAL Preprint* **2024**, hal-04455831.

physical sciences forum

MDPI

Proceeding Paper

Manifold-Based Geometric Exploration of Optimization Solutions †

Guillaume Lebonvallet *, Faicel Hnaien and Hichem Snoussi

Laboratory of Computer Science and Digital Society (LIST3N), Université de Technologie de Troyes, 12 rue Marie Curie, 10300 Troyes, France; faicel.hnaien@utt.fr (F.H.); hichem.snoussi@utt.fr (H.S.)

* Correspondence: guillaume.lebonvallet@utt.fr; Tel.: +33-6-04-49-20-58

† Presented at the 42nd International Workshop on Bayesian Inference and Maximum Entropy Methods in Science and Engineering, Garching, Germany, 3–7 July 2023.

Abstract: This work introduces a new method for the exploration of solutions space in complex problems. This method consists of the build of a latent space which gives a new encoding of the solution space. We map the objective function on the latent space using a manifold, i.e., a mathematical object defined by an equations system. The latent space is built with some knowledge of the objective function to make the mapping of the manifold easier. In this work, we introduce a new encoding for the Travelling Salesman Problem (TSP) and we give a new method for finding the optimal round.

Keywords: manifold; optimization; Traveling Salesman Problem (TSP)

1. Introduction

Manifolds are mathematical objects which can represent high-dimension points intrinsically linked to a low-dimension space [1]. Manifolds are used in dimension reduction for the classification of complex objects like facial expressions or different views from the same scene. We want to achieve the resolution of complex problems like the Travelling Salesman Problem (TSP) [2,3] by using manifolds to map the objective function to the solutions space. We can then use the manifold to explore the solutions space and find the optimum solution. However, the solutions space is often too complex and various dimensionsto allow this. Our solution is to describe the solutions space with a latent space of lower dimensions, which is simpler to explore. The manifold can then be defined with simple functions which make the findings of the optimum solution easier. We will apply this protocol on the TSP and give an example.

2. Materials and Methods

The goal of this paper is to find a latent space which is linked to the objective function. This latent space will be mapped by a tree formed by the distance matrix. In fact, we normally define the objective function of TSP problems by the Equation (1) [4–6].

$$Z = \sum_{i \neq j} d_{ij} x_{ij} \tag{1}$$

The link between the solution represented by x and the objective value Z is the distance matrix d. We want to include a part of the knowledge from this distance matrix in the encoding of the solution.

To achieve this, we will use a tree to sort the different solutions based on the links used in the travel. We assume that travels with short links have lower lengths than travels which do not use them. We can then repeatedly split the set of solutions in respect of the use of some links until all solutions have been allocated to a leaf node of the tree. The new encoding is then obtained by numbering the solutions by the use of the best links.

The Algorithms 1 and 2 shows the recursive building of the encoding tree.

Algorithm 1 Calculate the encoding tree

```
S ← set of all solutions
L ← list of links sorted accordingly to the distance matrix
tree ← buildNode(S, L)
```

Algorithm 2 Function buildNode(Solution Set S, Links List L)

```
if S has more than one solution then
  l ← first link of L (the best link)
  while S has no solution with l or S has no solution without l do
    Remove l from L
    l ← first link of L (the new best link)
  end while
  S1 ← the set of solutions of S with the link l
  S2 ← the set of solutions of S without the link l
  node ← new intermediate node with l
  Remove l from L
  node.leftChild ← buildNode(S1, L)
  node.rightChild ← buildNode(S2, L)
  return node
else
  return new leaf node with the unique solution of S
end if
```

3. Results

For example, the Equation (2) gives the distance matrix of a TSP with 4 nodes.

$$d = \begin{pmatrix} 0 & 8 & 39 & 37 & 50 \\ 8 & 0 & 45 & 47 & 49 \\ 39 & 45 & 0 & 9 & 21 \\ 37 & 47 & 9 & 0 & 15 \\ 50 & 49 & 21 & 15 & 0 \end{pmatrix} \tag{2}$$

The list of the ten links of this problem is sorted in the Table 1.

Table 1. List of the links for the TSP example sorted by the distance.

Distance	Link
8	(0, 1)
9	(2, 3)
15	(3, 4)
21	(2, 4)
37	(0, 3)
39	(0, 2)
45	(1, 2)
47	(1, 3)
49	(1, 4)
50	(0, 4)

With this list, we can build the encoding tree (Figure 1). On each node, we choose the first link in the list allowing the split of the solutions space. The left child of this node takes the solutions using that link and the right child the others which do not use the link. By continually splitting the solutions space, each solution is assigned to a leaf node of the tree. We can now number the solutions from left to right and we obtain the new encoding in the Table 2.

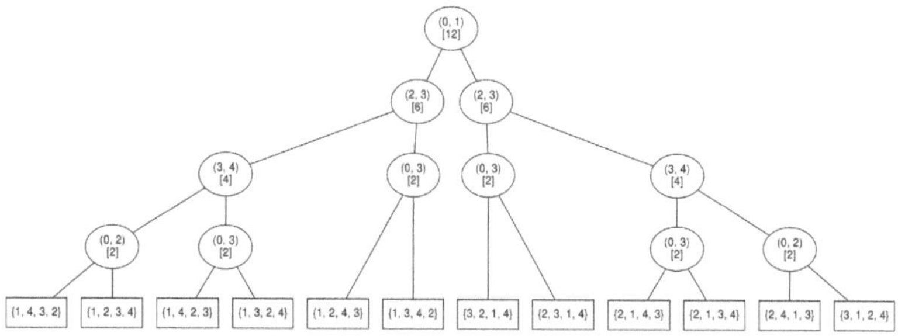

Figure 1. Encoding tree: each intermediate node (round shape) represents a splitting of the solution space based on the use of a link (the left branch keeps the solutions using the link, the right branch the solutions without the link), the number in square brackets indicates the number of solutions; the leaf nodes (rectangle shape) represents the solutions.

Table 2. The new encoding of the solutions of the TSP example.

Encoding	Solution
0	{1, 4, 3, 2}
1	{1, 2, 3, 4}
2	{1, 4, 2, 3}
3	{1, 3, 2, 4}
4	{1, 2, 4, 3}
5	{1, 3, 4, 2}
6	{3, 2, 1, 4}
7	{2, 3, 1, 4}
8	{2, 1, 4, 3}
9	{2, 1, 3, 4}
10	{2, 4, 1, 3}
11	{3, 1, 2, 4}

Figure 2 compares this new encoding with the real rank of the solution, i.e., a sort of the solutions from the best (numbered 0) to the worst. Despite being imperfect, the new encoding shows a tendency between the new encoding and the objective value represented by the rank. We can also see the different branches of the encoding tree.

To demonstrate the drawbacks of the method, the new encoding has been calculated on several simulated TSPs with seven nodes. The results are displayed in Figure 3. The plots show the distribution of the objective value according to the encoding. Three cases are represented with two examples in each:

- The plots (a) and (b) represent a good resolution of the TSP. The optimum is the left most point, encoded with the value 0. The plots clearly show some branches of the encoding tree;
- The plots (c) and (d) display two examples where the optimum is not encoded with 0. It indicates that the tree incorrectly splits the solutions space at some point in the tree. Furthermore, the plot (d) shows that the optimum is not seen clearly in the first branch, which indicates that the wrong split is situated in the tree;
- The plots (e) and (f) show some examples where the encoding fails. The reason could be the low range of distances used for these TSPs. On the plot (e), the branches are difficult to see.

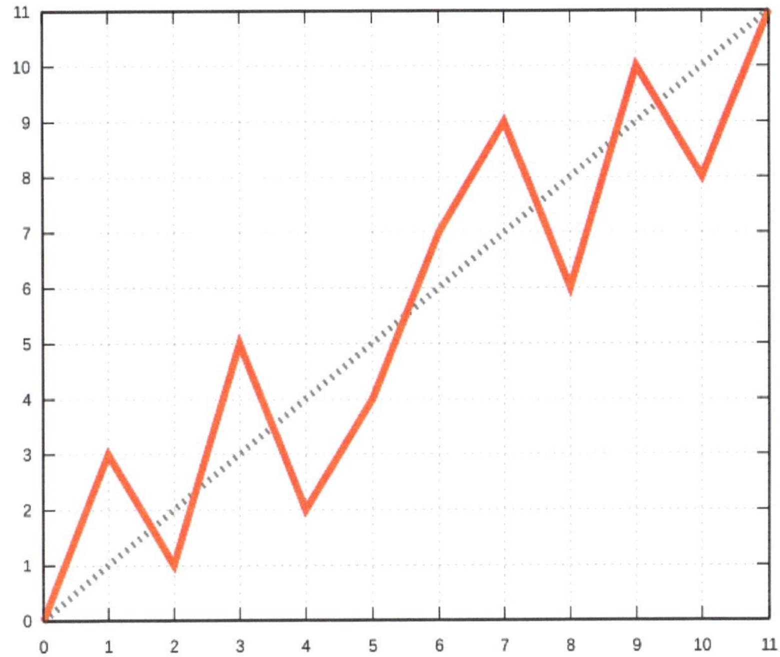

Figure 2. Graph of the new encoding: the horizontal axis represents the rank of a solution (0 is the optimum) and the vertical axis the encoding of that solution.

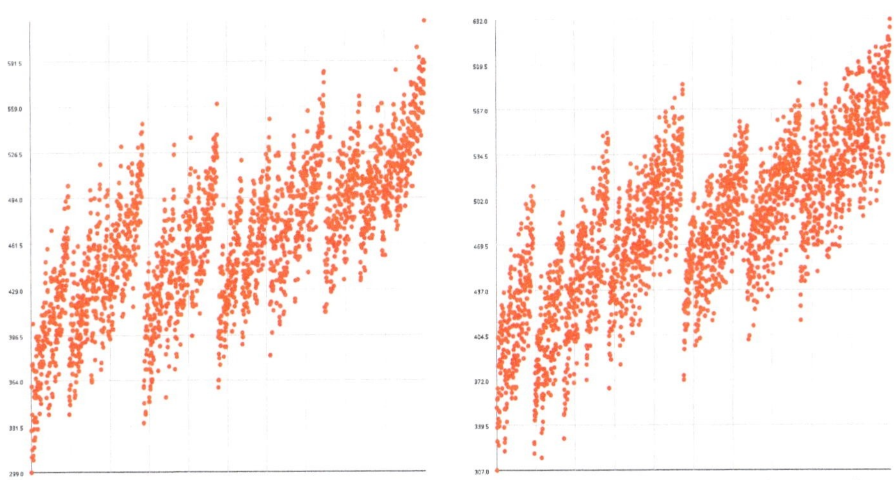

(a) (b)

Figure 3. *Cont.*

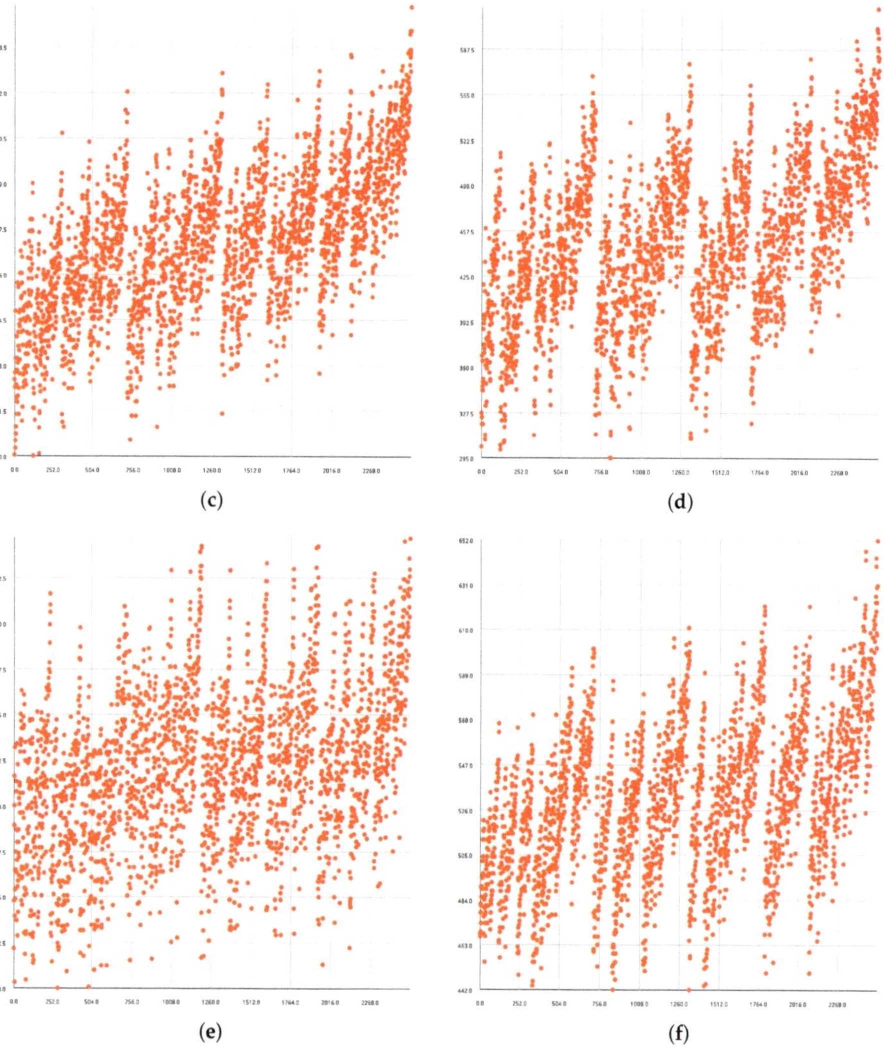

Figure 3. Simulations of the new encoding on several TSP problems with 7 nodes.

4. Discussion

We have an encoding which links the solution to the distance matrix; we need to link this encoding to the objective value. As the encoding is partially aware of the objective value, the neighbourhood of each solution is defined on the objective function. A simple function can be used to link this encoding to the objective value, e.g., a polynomial function. The fitting of this polynomial function to the different points representing the solutions can be made with a simple least-square optimization. Finding the minimum of the polynomial in the encoding space is also simple but can lead to multiple stationary points. As the fitting is not perfect, we need to explore a neighbourhood of each minimum to expect finding the optimum. As the encoding is scalar, it is fairly easy. This neighbourhood will be defined on the latent space of low dimension, which is easier to explore. In fact, the parametrization creates a new encoding of a solution which allows a guided exploration of the solutions space according to the objective, which can simplify the search of the optimal value.

Our new method for finding the optimum is describe as follows:

1. Build a portion of the encoding tree: the stopping criteria can be based on an upper bound of the optimum value [7];
2. For each leaf node which has multiple solutions, choose the solution which uses the better links: the encoding is found by counting the solutions on the branches placed on the left side;
3. Map a manifold of the objective value on the solutions space with the subset of solutions chosen:
4. Find the minimum/minima of this manifold: as the function used is very simple, it should not be too difficult to differentiate and search for the stationary points;
5. The optimum should be found on a neighbourhood of the minimum or one of the neighbourhoods of the minimum/minima if there are multiple local minimums on the solutions space.

Author Contributions: Conceptualization, methodology, software, and writing, G.L.; validation and review, F.H. and H.S. All authors have read and agreed to the published version of the manuscript.

Funding: This research received no extra funding.

Institutional Review Board Statement: Not applicable.

Informed Consent Statement: Not applicable.

Data Availability Statement: Data is contained within the article and/or generated as described.

Conflicts of Interest: The authors declare no conflict of interest.

References

1. Hu, J.; Liu, X.; Wen, Z.W.; Yuan, Y.X. A Brief Introduction to Manifold Optimization. *J. Oper. Res. Soc. China* **2020**, *8*, 199–248. [CrossRef]
2. Pop, P.C.; Cosma, O.; Sabo, C.; Sitar, C.P. A comprehensive survey on the generalized traveling salesman problem. *Eur. J. Oper. Res.* 2023, *in press*. [CrossRef]
3. Bektas, T. The multiple traveling salesman problem: An overview of formulations and solution procedures. *Omega* **2006**, *34*, 209–219. [CrossRef]
4. Miller, C.E.; Tucker, A.W.; Zemlin, R.A. Integer Programming Formulation of Traveling Salesman Problems. *J. ACM* **1960**, *7*, 326–329. [CrossRef]
5. Dantzig, G.; Fulkerson, R.; Johnson, S. Solution of a Large-Scale Traveling-Salesman Problem. *J. Oper. Res. Soc. Am.* **1954**, *2*, 393–410. [CrossRef]
6. Papadimitriou, C.H. The Euclidean travelling salesman problem is NP-complete. *Theor. Comput. Sci.* **1977**, *4*, 237–244. [CrossRef]
7. Fischetti, M.; Salazar González, J.J.; Toth, P. A Branch-and-Cut Algorithm for the Symmetric Generalized Traveling Salesman Problem. *Oper. Res.* **1997**, *45*, 378–394. [CrossRef]

physical sciences forum

MDPI

Proceeding Paper

Nested Sampling for Detection and Localization of Sound Sources Using a Spherical Microphone Array †

Ning Xiang [1,*] and Tomislav Jasa [2]

1 Graduate Program in Architectural Acoustics, Rensselaer Polytechnic Institute, Troy, NY 12180, USA
2 Thalgorithm Research, San Jose, CA 95134, USA; tomislav.jasa@gmail.com
* Correspondence: xiangn@rpi.edu
† Presented at the 42nd International Workshop on Bayesian Inference and Maximum Entropy Methods in Science and Engineering, Garching, Germany, 3–7 July 2023.

Abstract: Since its inception in 2004, nested sampling has been used in acoustics applications. This work applies nested sampling within a Bayesian framework to the detection and localization of sound sources using a spherical microphone array. Beyond an existing work, this source localization task relies on spherical harmonics to establish parametric models that distinguish the background sound environment from the presence of sound sources. Upon a positive detection, the parametric models are also involved to estimate an unknown number of potentially multiple sound sources. For the purpose of source detection, a no-source scenario needs to be considered in addition to the presence of at least one sound source. Specifically, the spherical microphone array senses the sound environment. The acoustic data are analyzed via spherical *Fourier* transforms using a Bayesian model comparison of two different models accounting for the absence and presence of sound sources for the source detection. Upon a positive detection, potentially multiple source models are involved to analyze direction of arrivals (DoAs) using Bayesian model selection and parameter estimation for the sound source enumeration and localization. These are two levels (enumeration and localization) of inferential estimations necessary to correctly localize potentially multiple sound sources. This paper discusses an efficient implementation of the nested sampling algorithm applied to the sound source detection and localization within the Bayesian framework.

Keywords: nested sampling; Bayesian model comparison; Bayesian model selection; parameter estimation; sound source detection; sound source localization

1. Introduction

Nested sampling (NS) was introduced by Skilling [1] as a numerical method for efficient Bayesian calculations. Soon afterward, this method was applied to acoustics problems [2], where Jasa and Xiang explored using *Lebesgue* integral as the mathematical foundation of the NS algorithm. Since then, that effort has resulted in a series of publications in acoustic applications [3–5]. This paper showcases that the NS algorithm has recently been applied in sound source detection and localization within a Bayesian framework. To detect and localize sound sources, the sound environment is sensed by a spherical microphone array whose signals are processed using a spherical *Fourier* transform. Spherical harmonics are exploited to process the acoustic data and to formulate the signal models. This paper emphasizes that source detection represents a model comparison problem, source enumeration represents a model selection problem, and source localization represents a parameter estimation problem, all of which can be efficiently accomplished within the Bayesian framework using the NS algorithm.

This paper presents a further development from the previous work [6] in that a background model for a no-source scenario needs to established. The source detection problem is critically based on the model comparison between the no-source and one-source models. Special attention is given to the spherical harmonics when dealing with the background model and is separately dealt with in Section 2.2. In addition to these model improvements, higher-order (fourth) spherical components have been achieved due to the further development of a 32-channel microphone array as illustrated in Figure 1.

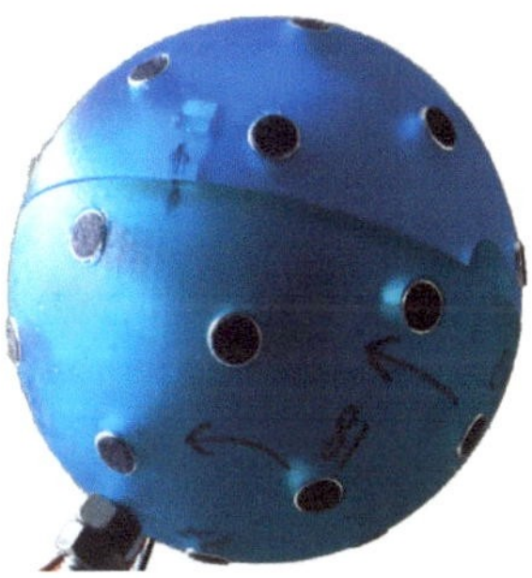

Figure 1. Spherical microphone array of radius $a = 3.5$ cm. Altogether, 32 microphones are nearly uniformly flush-mounted over the rigid spherical surface.

2. Spherical Microphone Data and Models

This section briefly introduces the data processing and the prediction models used for the sound source detection, enumeration, and direction of arrival estimations.

2.1. Microphone Array Data

When Q microphones are arranged flush on a rigid sphere of radius a nearly equidistantly, the sound pressure signals $\underline{P}_{mic}$ are processed by

$$D(\Theta) = \frac{4\pi}{Q} \sum_{n=0}^{N} \sum_{m=-n}^{n} \frac{\underline{Y}_n^m(\Theta)}{B(k\,a)} \sum_{q=1}^{Q} \underline{P}_{\text{mic}}(\Theta_q)\, \underline{Y}_n^m(\Theta_q)^*, \tag{1}$$

with the third sum over $q = 1, \ldots, Q$ being a spherical *Fourier* transform of Q microphone signals $\underline{P}_{\text{mic}}(\Theta_q)$ at angular positions Θ_q, and symbol $*$ standing for a complex conjugate. Function $B(k\,a)$ is a modal strength of the rigid sphere of radius a,

$$B(k\,a) = \mathrm{j}k\left[j_n(k\,a) - \frac{j_n'(k\,a)}{h_n'(k\,a)}\, h_n(k\,a)\right], \tag{2}$$

where $\mathrm{j} = \sqrt{-1}$, $k = \omega/c$ is the propagation coefficient of sound waves. Functions $j_n(\cdot)$ and $h_n(\cdot)$ are spherical Bessel and Hankel functions, and $j_n'(\cdot)$ and $h_n'(\cdot)$ are their derivatives, respectively. $\Theta = \{\theta, \phi\}$, collectively represents elevation and azimuth angles, while Θ_q specifies Q microphone locations flush-mounted on the spherical surface of the rigid sphere of radius a. Figure 1 shows a spherical microphone array of $Q = 32$ channels developed for this research. The spherical array is built upon a rigid sphere of radius $a = 3.5$ cm. In the following, we denote $\mathbf{D} = \{D(\Theta)\}$ as a two-dimensional matrix (vectors) representing the *experimental data* in the context of Bayesian inferential inversion.

2.2. Prediction Models

In Equation (1), $\underline{Y}_n^m(\Theta)$ is so-called spherical harmonics of order n and degree m, it is orthonormal and complete in a sense, and

$$g(\Theta_s, \Theta) = 2\sqrt{\pi} \sum_{n=0}^{N} \sum_{m=-n}^{n} \underline{Y}_n^m(\Theta_s)^* \underline{Y}_n^m(\Theta) \to \delta(\cos\theta - \cos\theta_s)\delta(\phi - \phi_s), \tag{3}$$

when $N \to \infty$. Θ_s is the source angle in the form of the direction of arrival (DoA). $\delta(\cdot)$ represents the *Kronecker* delta function. Using Equation (3), we establish a predictive model of spherical beamforming as

$$M_S(\boldsymbol{\Theta}_S, \boldsymbol{\Theta}) = \sum_{s=1}^{S} \frac{A_s\, g^2(\Theta_s, \Theta)}{\max[\, g^2(\Theta_s, \Theta)]}, \tag{4}$$

where A_s counts for different source energy strengths of individual sound sources. Note that $\boldsymbol{\Theta}_S = \{A_1, A_2, \ldots, A_S, \Theta_1, \Theta_2, \ldots, \Theta_S\}$ collectively denotes both strength vector A_S and angular directions (vectors) of S sound sources, and each angular vector contains one pair of elevation and azimuth angles $\Theta_s = \{\theta_s, \phi_s\}$. Variable $\boldsymbol{\Theta}$ represents the angular range for possible sound sources to be localized. For $S \geq 1$, the kernel function $g(\Theta_s, \Theta)$ in Equation (3) is processed for the upper order $(N+1)^2 \leq Q$. This means that the integer-valued order N of the spherical harmonics is limited by the number of microphone channels instead of infinity. Figure 2 illustrates a superposition of two simultaneous sources of equal amplitudes predicted by the model kernel in Equation (3) for $N = 4$ before squaring the operation to build source energy. The finite upper order N is responsible for the width of lobes rather than middle-form ones. Figure 3a illustrates the spherical microphone data for the presence of two simultaneous sound sources, processed using Equations (1) and (2), while Figure 3b illustrates the predicted map of the two simultaneous sound sources using the model in Equation (4). The angular range is evaluated over $\Theta = \{0 \leq \theta \leq 180°; 0 \leq \phi \leq 360°\}$.

Figure 2. Beamforming superposition of two sound sources using a spherical order $N = 4$.

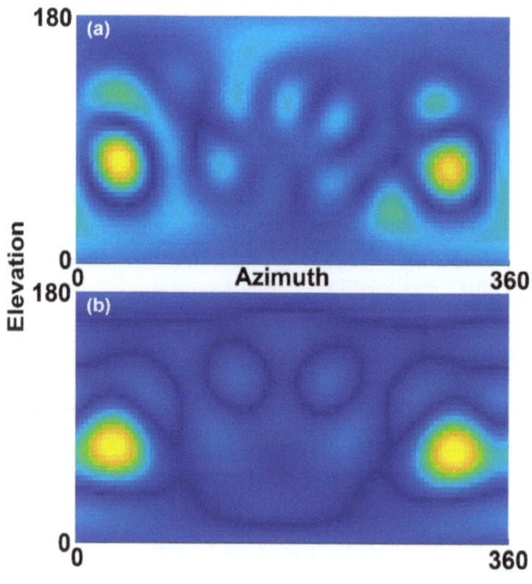

Figure 3. Comparison between the experimental data (**a**) processed according to Equation (1) with the prediction model (**b**) in Equation (4) of two sound sources using a spherical order $N = 4$.

When processing the microphone array data, there is no prior knowledge about the incoming sound field either with the presence or absence of sound sources. It does not make sense to pursue direction of arrival analysis if no sound sources are present in the incoming microphone signals. For the model-based Bayesian detection, we need to establish a background model. Special attention has to be given to the spherical harmonics processing in this case. Specifically, M_0 represents the no-source model for $S = 0$. In this case, the kernel function $g(\Theta_0, \Theta)$ in Equation (3) is only calculated for $N = 0$, namely the zero-order of the spherical harmonics.

$$M_0(\boldsymbol{\Theta}) = \frac{A_0\, g^2(\Theta)}{\max[g^2(\Theta)]}, \tag{5}$$

where the direction of 'no-source' $\boldsymbol{\Theta}_0$ is irrelevant over the angular range $\boldsymbol{\Theta}$. For notation purpose, we collectively denote $\mathbf{M}_S = \{M_0, M_1, \ldots, M_S\}$ as being the prediction models for the directional of arrival analysis, while we denote $\mathbf{M}_D = \{M_0, M_1\}$ for the sound source detection, a small subset of $\mathbf{M}_S$.

3. Bayesian Calculations

Given the data $\mathbf{D}$ as formulated in Section 2.1 and the prediction models $M_S(\boldsymbol{\Theta}_S)$ in Section 2.2, this work relies on Bayes theorem:

$$\begin{array}{ccccccc} p(\boldsymbol{\Theta}_S|\mathbf{D}, M_S) & \times & \mathcal{Z} & = & \mathcal{L}(\boldsymbol{\Theta}_S) & \times & \Pi(\boldsymbol{\Theta}_S), \\ \text{posterior} & \times & \text{evidence} & = & \text{likelihood} & \times & \text{prior}, \end{array} \tag{6}$$

with $\Pi(\boldsymbol{\Theta}_S) = p(\boldsymbol{\Theta}_S|M_S)$ being the *prior* probability and $\mathcal{L}(\boldsymbol{\Theta}_S) = p(\mathbf{D}|\boldsymbol{\Theta}_S, M_S)$ being the *likelihood* function. The prior and the likelihood are both prior probabilistic in nature and need to be assigned a priori. This work applies the principle of maximum entropy (MaxEnt), which leads to a uniform prior and a Student-T distribution for the likelihood (see Ref. [6] for details). The *evidence* $\mathcal{Z}$ in Equation (6) plays a central role for the source detection and source enumeration problems and is determined by

$$\mathcal{Z} = \int_{\boldsymbol{\Theta}_S} \mathcal{L}(\boldsymbol{\Theta}_S)\, \Pi(\boldsymbol{\Theta}_S)\, \mathrm{d}\boldsymbol{\Theta}_S = \int_0^1 \mathcal{L}(\mu)\, \mathrm{d}\mu, \tag{7}$$

where

$$\mu(\mathcal{L}_\epsilon) = \int_{\mathcal{L}(\boldsymbol{\Theta}_S) > \mathcal{L}_\epsilon} \Pi(\boldsymbol{\Theta}_S)\, \mathrm{d}\boldsymbol{\Theta}_S \tag{8}$$

is the *prior mass* with $\mathcal{L}(\mu(\mathcal{L}_\epsilon)) = \mathcal{L}_\epsilon$, and $0 \leq \mathcal{L}_\epsilon \leq \mathcal{L}_{\max}$ as derived by Skilling [7]. The NS algorithm generates a monotonically increasing partition of the likelihood range $[0, \mathcal{L}_{\max}]$

$$0 < \mathcal{L}_0 < \mathcal{L}_1 < \cdots < \mathcal{L}_{T-1} < \mathcal{L}_T < \mathcal{L}_{\max}, \tag{9}$$

via constrained sampling such that $\mathcal{L}_t$ with $t \in \{0, 1, \ldots, T\}$ is sampled from the domain $(\boldsymbol{\Theta}_S : \mathcal{L}(\boldsymbol{\Theta}_S) > \mathcal{L}_{t-1})$. Observe that as $\mathcal{L}_\epsilon$ increases from 0 to $\mathcal{L}_{\max}$, μ_ϵ decreases from 1 to 0, where $\mu_\epsilon = \mu(\mathcal{L}_\epsilon)$ and the partition of Equation (9) generates the monotonically decreasing sequence

$$1 > \mu_0 > \mu_1 > \cdots > \mu_{T-1} > \mu_T > 0. \tag{10}$$

Using the sequences in Equations (9) and (10), the one-dimensional integration on the far-right-hand side of Equation (7) is well approximated by

$$Z \leftarrow \sum_{t=0}^{T} \mathcal{L}_t\, \Delta\mu_t, \tag{11}$$

with

$$\Delta\mu_t = \mu_t - \mu_{t+1}, \quad \text{or} \quad \Delta\mu_t = \mu_{t-1} - \mu_t. \tag{12}$$

Skilling [7] pointed out that the constrained prior mass is a statistical quantity and follows a shrinkage of

$$\mu_t \approx \mathrm{e}^{-t/P}, \tag{13}$$

after t iterations, with P being an integer number for initializing random samples. A detailed proof of this result was given using order statistics in Appendix B of Jasa and Xiang [3].

NS was shown by Jasa and Xiang [3] to be a numerical implementation of *Lebesgue* integration, where Equation (11) represents the sum of weighted integrands of simple functions that are generated by partitioning the range rather than the domain of the function. An early account of this connection of the NS algorithm to *Lebesgue* integration can also be found in Jasa and Xiang [2], published in the Proceedings of the 25th International Workshop on Bayesian Inference and Maximum Entropy Methods in Science and Engineering (MaxEnt 2005).

4. Sound Source Detection, Enumeration, and Localization

The above described calculation is implemented for $P = 500$ [in Equation (13)], the initial population with uniformly distributed prior ranges of all pending parameters, including sound source strength A_s, the elevation θ_S, and azimuth angles ϕ_S. For potentially simultaneous sound sources up to four ($S \leq 4$), the parameter space is of dimensions up to $3 \times S$. The NS is applied to estimating Bayes factors for model order from 1 to 4 via evidence. It is used to rank a potential model accounting for an unknown number of simultaneous sound sources, in a so-called sound source enumeration process. This process has also been described previously in Xiang and Landschoot [6], followed by a DoA analysis based on the selected model M_S that is carried out by Bayesian parameter estimation. When examining this effort critically, the authors recognize that the source enumeration, even if representing a higher level of inference via Bayesian model selection, would still be incomplete if the machine sensory modality, such as in this application of a spherical microphone array, is not notified that the absence of sound sources often represents predominant portions of the sound environment in practical scenarios. It will only make sense to pursue sound source enumeration and DoA analysis if any sound source is ever detected.

The sound source detection is carried out in the scope of this current work using Bayesian model comparison. The prediction models $\mathbf{M}_\mathrm{D} = \{M_0, M_1\}$ solely involve two models, M_0 in Equation (5) and M_1 in Equation (4) for $S = 1$. Note that Equation (5) is separately described because the $g^2(\Theta_0, \Theta)$ needs special attention, in which the spherical order is set to $N = 0$, while ,for M_S for $S \geq 1$, the spherical order $N = 4$ due to the 32-channel spherical microphone array used for this work.

For the source detection, Bayesian evidence is estimated using the NS based on the 'no-source' model M_0 against the 'one-source' model M_1. Specifically, for M_0-based sampling, there is still one pending parameter A_0 to sample. Figure 4 (left) shows an experimental investigation when the microphone array data contain no sources but noisy background signals. The evidence estimation using the NS demonstrates insignificant differences to that of M_1, indicating that the source detection is negative. Figure 4 (right) shows that if the microphone array data contain sound sources, yet an unknown number, the evidence estimation clearly shows significant differences in comparison with those of 'no-sources'. The source detection is positive.

Upon a positive detection of sound sources, a further process involves Bayesian model selection. A set of sound source models from M_0 to M_4 is involved for estimating Bayes factors:

$$B_{i,i-1} = 10\,\mathrm{lg}\left(\frac{\mathcal{Z}_i}{\mathcal{Z}_{i-1}}\right), \quad \text{[decibans]} \tag{14}$$

for $i = 1, 2, \ldots 4$. Figure 5 illustrates one set of Bayes factor estimations. In this work, the evidence and Bayes factors are calculated in units [decibans] denoting 10 times logarithm base 10 [$10\,\mathrm{lg}$] in honor of Thomas Bayes [8]. In this case, the source enumeration using

Bayesian model selection suggests that two sources are contained in the data. At this stage, the interest in specific DoA parameters is pushed into background. Upon the selection of a two-source model using the NS, the exploration samples during the iterative NS process also provide posterior samples for the model M_2 as a byproduct; they are readily available once the evidence $\mathcal{Z}_2$ for two sound sources is sufficiently explored. The posterior samples provide parameter estimates of two sound sources in terms of source strength A_2, and angular parameters $\mathbf{\Theta}_2$. The data processed using Equation (1) and the model prediction according to the posterior estimates are compared in Figure 3.

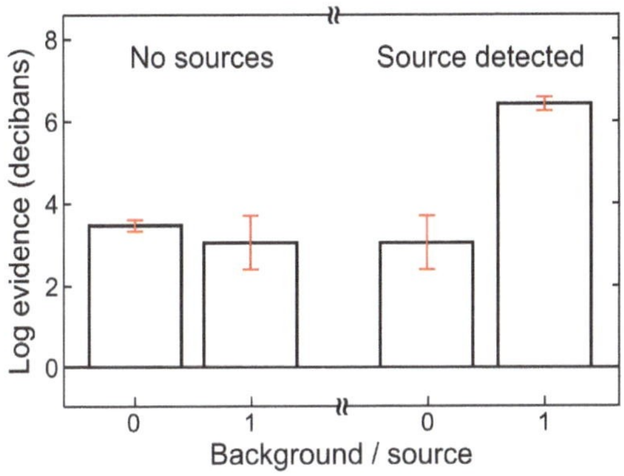

Figure 4. Sound source detection based on Bayesian model comparison. Bayesian evidence is estimated using both 'no-source' model M_0 and one-source model M_1. The evidence is expressed in unit [decibans] in honor of Thomas Bayes [8].

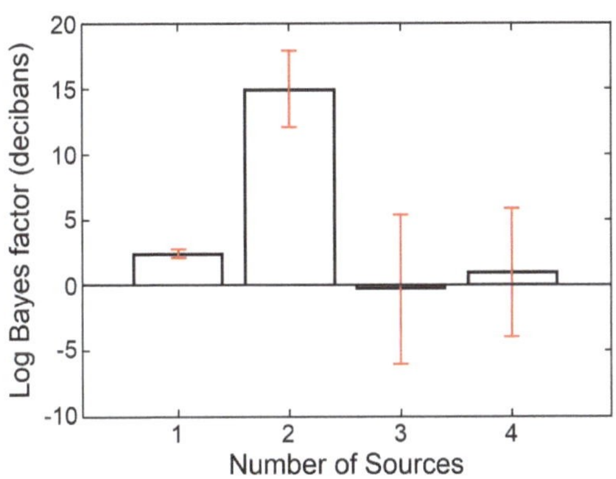

Figure 5. The sound source enumeration based on Bayes factor estimation. The Bayes factors are expressed in unit [decibans] in honor of Thomas Bayes [8]. A two-source model is preferred by the Bayesian model selection. The evidence estimated using nested sampling also provides the posterior as a byproduct.

5. Concluding Remarks

From its introduction into Bayesian calculations, Skilling's nested sampling [1] had an immediate impact on room-acoustic research [2], where an early account of the Lebesgue integral view on the nested sampling was first exposed in the MaxEnt Community in 2005. A thorough handling of its mathematical foundation of the Lebesgue integral was given at a later point [3]; 'Interpreting nested sampling as a statistical approximation of a Lebesgue integral opens the possibility of a large body of existing research to be applied in the analysis and possible extension of the algorithm'. Over the past 15 years, a stream of applications using NS in acoustics science and engineering has emerged. Among others, this paper reports on an acoustic application of nested sampling using a spherical microphone array within the Bayesian framework.

Author Contributions: Conceptualization, N.X.; methodology, N.X., T.J.; software, N.X.; validation, N.X.; formal analysis, N.X., T.J.; investigation, N.X.; resources, N.X.; data curation, N.X.; writing—original draft preparation, N.X.; writing—review and editing, T.J.; visualization, N.X.; supervision, N.X.; project administration, N.X. All authors have read and agreed to the published version of the manuscript.

Funding: This research received no external funding.

Institutional Review Board Statement: Not applicable.

Informed Consent Statement: Not applicable.

Data Availability Statement: The data that support the findings of this study are available from the corresponding author upon reasonable request.

Acknowledgments: The authors are grateful to John Skilling, Paul Goggans, and Kevin Knuth for their stimulating discussions. Stephen Weikel, Christopher Landschoot, and Thomas Metzger have contributed to parts of this work in scope of their MS degree thesis projects.

Conflicts of Interest: The authors declare no conflicts of interest.

Abbreviations

The following abbreviations are used in this manuscript:

DoA	Direction of arrival
NS	Nested sampling
MaxEnt	Principle of maximum entropy

References

1. Skilling, J. Nested sampling. In Proceedings of the Bayesian Inference and Maximum Entropy Methods in Science an Engineering, AIP Conference Proceedings, Garching, Germany, 25–30 July 2004; Volume 735, pp. 395–405.
2. Jasa, T.; Xiang, N. Using nested sampling in the analysis of multi-rate sound energy decay in acoustically coupled rooms. *AI Conf. Proc.* **2005**, *803*, 189–196.
3. Jasa, T.; Xiang, N. Nested sampling applied in Bayesian room-acoustics decay analysis. *J. Acoust. Soc. Am.* **2012**, *132*, 3251–326 [CrossRef] [PubMed]
4. Botts, J.M.; Escolano, J.; Xiang, N. Design of IIR Filters With Bayesian Model Selection and Parameter Estimation. *IEEE Tran ASLP* **2013**, *21*, 669–674. [CrossRef]
5. Fackler, C.J.; Xiang, N.; Horoshenkov, K.V. Bayesian acoustic analysis of multilayer porous media. *J. Acoust. Soc. Am.* **201** *144*, 3582–3592. [CrossRef] [PubMed]
6. Xiang, N.; Landschoot, C. Bayesian Inference for Acoustic Direction of Arrival Analysis Using Spherical Harmonics. *J. Entrop* **2019**, *21*, 579. [CrossRef] [PubMed]
7. Skilling, J. Nested sampling for general Bayesian computation. *Bayesian Anal.* **2006**, *1*, 833–859. [CrossRef]
8. Jeffreys, H. *Theory of Probability;* Oxford University Press: Oxford, NY, USA, 1961; Reprinted by Clarendon Press, Oxford, N USA, 2003.

physical sciences forum

MDPI

Proceeding Paper

Construction of Dimensionless Groups by Entropic Similarity †

Robert K. Niven

School of Engineering and Technology, The University of New South Wales, Canberra, ACT 2600, Australia; r.niven@adfa.edu.au

† Presented at the 42nd International Workshop on Bayesian Inference and Maximum Entropy Methods in Science and Engineering, Garching, Germany, 3–7 July 2023.

Abstract: Since the early 20th century, dimensional analysis and similarity arguments have provided a critical tool for the analysis of scientific, engineering, and thermodynamic systems. Traditionally, the resulting dimensionless groups are categorized into those defined by (i) *geometric similarity*, involving ratios of length scales; (ii) *kinematic similarity*, involving ratios of velocities or accelerations, and (iii) *dynamic similarity*, involving ratios of forces. This study considers an additional category based on *entropic similarity*, with three variants defined by the following: (i) ratios of global or local entropy production terms $\Pi_{\text{entrop}} = \dot{\sigma}_1/\dot{\sigma}_2$ or $\hat{\Pi}_{\text{entrop}} = \hat{\dot{\sigma}}_1/\hat{\dot{\sigma}}_2$; (ii) ratios of entropy flow rates $\Pi_{\text{entrop}} = \mathcal{F}_{S,1}/\mathcal{F}_{S,2}$ or magnitudes of entropy fluxes $\hat{\Pi}_{\text{entrop}} = ||\boldsymbol{j}_{S_1}||/||\boldsymbol{j}_{S_2}||$; and (iii) the ratio of a fluid velocity to that of a carrier of information $\Pi_{\text{info}} = U/c$. Given that all phenomena involving work against friction, dissipation, spreading, chemical reaction, mixing, separation, or the transmission of information are governed by the second law of thermodynamics, these are more appropriately analyzed directly in terms of competing entropic phenomena and the dominant entropic regime, rather than indirectly using ratios of forces. This work presents the entropic dimensionless groups derived for a wide range of diffusion, chemical reaction, dispersion, and wave phenomena, revealing an entropic interpretation for many known dimensionless groups and many new dimensionless groups.

Keywords: dimensionless group; dimensional analysis; similarity; entropy production; entropy transport; information theory

1. Introduction

A *dimensionless group* is a dimensionless (unitless) parameter which represents a property of a physical system, independent of the units used for its measurement. These have a number of applications. First, by the consideration of *similarity* or *similitude*, dimensionless groups can be matched between a system (prototype) and its model [1,2]. This enables smaller-scale experimental models, which can dramatically simplify the experimental requirements. Second, the method of *dimensional analysis* can be used to extract the governing dimensionless groups from the list of parameters for a system [3]. Under the Buckingham Pi theorem, this also reduces the number of variables by the number of dimensions. Third, by the method of *non-dimensionalization*, the governing equation(s) for a scientific or engineering system can be converted into dimensionless form, revealing its dominant dimensionless groups [4–9]. Finally, the one-parameter Lie group of point transformations can be used to non-dimensionalize an equation system [10]. This method reveals the intrinsic dimensions of the system, while the resulting Lie symmetries encompass other known invariances such as Galilean invariance [10].

Since the early 20th century, dimensional methods have been recognized as critical tools for the analysis of natural and engineered systems of all kinds [11–16]. The derived groups are generally categorized into those defined by *geometric similarity*, defined by ratios of length scales; *kinematic similarity*, defined by ratios of velocities or accelerations; and *dynamic similarity*, defined by ratios of forces [5–9,17,18].

The aim of this study is to examine a new category of *entropic similarity* proposed for the construction of dimensionless groups, based on ratios of entropic terms [19,20]. Given that all phenomena involving work against friction, dissipation, spreading, chemical reaction, mixing, separation, or the transmission of information are governed by the second law of thermodynamics, these are more appropriately analyzed directly in terms of competing entropic phenomena and the dominant entropic regime, rather than indirectly using ratios of forces. The entropic perspective is shown to provide a new entropic interpretation for many known dimensionless groups, as well as the foundations for a number of new groups. These extend the utility of dimensional arguments for the analysis of new and existing problems throughout science and engineering. This work presents a summary of two detailed studies on this topic [19,20].

2. Background

We first introduce some theoretical background. Applying the Reynolds transport theorem [5,9] to the flow of a body of fluid containing thermodynamic entropy S [J K^{-1}] through a region of space (control volume, CV), we obtain expressions for the global and local entropy production, derived, respectively, for an integral control volume and for an infinitesimal fluid element [21–23]:

$$\dot{\sigma} = \iiint\limits_{CV} \hat{\dot{\sigma}} dV = \iiint\limits_{CV} \left[\frac{\partial}{\partial t} \rho s + \nabla \cdot (\boldsymbol{j}_S + \rho s \boldsymbol{u}) \right] dV \geq 0 \quad (1)$$

$$\hat{\dot{\sigma}} = \frac{\partial}{\partial t} \rho s + \nabla \cdot (\boldsymbol{j}_S + \rho s \boldsymbol{u}) \geq 0 \quad (2)$$

where all symbols are defined in Table 1. The inequalities in (1) and (2) give the global and local forms of the second law of thermodynamics, defined for an open system.

Table 1. Nomenclature used in this study.

Symbol	Description	SI Units
Mathematical Operators		
∇	Cartesian gradient operator	[m^{-1}]
$^\top$	vector or matrix transpose	
tr	trace of a matrix	
$\boldsymbol{a} \cdot \boldsymbol{b}$	vector scalar product $= \boldsymbol{a}^\top \boldsymbol{b} = \sum_i a_i b_i$	
$\boldsymbol{A} : \boldsymbol{B}$	Frobenius tensor scalar product $= \text{tr}(\boldsymbol{A}^\top \boldsymbol{B}) = \sum_i \sum_j A_{ij} B_{ij}$	
$\vert\vert\boldsymbol{a}\vert\vert$	Euclidean norm $= \sqrt{\boldsymbol{a}^\top \boldsymbol{a}}$ for vector $\boldsymbol{a}$	[units of $\boldsymbol{a}$]
Roman symbols		
a	magnitude of acceleration	[m s^{-2}]
a_s	acoustic velocity	[m s^{-1}]
A_s	cross-sectional area $= \pi d^2/4$ for a sphere	[m^2]
c	index of chemical species; celerity of a wave; speed of light	[-]; [m s^{-1}]; [m s^{-1}]
c_P	specific heat capacity of the fluid at constant pressure	[J K^{-1} kg^{-1}]
$C_c = \rho m_c$	molar concentration of chemical species c per unit volume	[(mol species) m^{-3}]
C_D	Drag coefficient	[-]
C_k	molar concentration of charged species k per unit volume	[(mol species) m^{-3}]
d	index of chemical reactions; pipe diameter or flow length scale	[-]; [m]
$\mathcal{D}_c$	diffusion coefficient for the cth chemical species	[m^2 s^{-1}]
$\mathcal{D}_c'$	thermodynamic species diffusion parameter	[(mol species)^2 K J^{-1} m^{-1} s^{-1}]
D_k	diffusion coefficient	[m^2 s^{-1}]
$D_k{}'$	thermodynamic charge diffusion parameter	[C^2 K J^{-1} m^{-1} s^{-1}]
Da	Damköhler number	[-]
$\boldsymbol{e}$	strain rate tensor $= \frac{1}{2}(\nabla \boldsymbol{u} + (\nabla \boldsymbol{u})^\top)$	[s^{-1}]
f	Darcy friction factor	[-]
f_b	thermodynamic force (gradient or free energy) for process b	
F	magnitude of force; Faraday constant	[N]; [C (mol charge)$^{-1}$]

Table 1. *Cont.*

Symbol	Description	SI Units
F_D	drag force	[N]
$\mathcal{F}_S$	global entropy flow rate	[J K^{-1} s^{-1}]
Fr	Froude number	[-]
g	acceleration due to gravity	[m s^{-2}]
$\Delta\widetilde{G}_d$	change in molar Gibbs free energy for the dth reaction	[J (mol reaction)$^{-1}$]
Gr	Grashof number	[-]
H_L	head loss	[m]
i	index of entropy-producing or entropy-transporting process	[-]
i_k	charge flux or current density of the kth charged species (positive for positive ion flow)	[A m^{-2}]
j_a	thermodynamic flux or reaction rate for process a	
j_c	molar flux of the cth chemical species	[(mol species) m^{-2} s^{-1}]
j_Q	heat flux	[J m^{-2} s^{-1}]
j_S	non-fluid entropy flux	[J K^{-1} m^{-2} s^{-1}]
k	index of charged species; thermal conductivity	[-]; [J K^{-1} m^{-1} s^{-1}]
k_{nd}	rate constant for nth mechanism of dth chemical reaction	[variable]
ℓ	length scale	[m]
L	pipe length	[m]
$L_{a,b}$	phenomenological coefficient between the (a, b) flux and force	[m s^{-2}]
Le	Lewis number	[-]
m_c	molality of chemical species c	[(mol species) kg^{-1}]
M	Mach number	[-]
n	index of chemical reaction mechanisms	[-]
$\boldsymbol{n}$	outwardly directed unit normal	[-]
Pe	Peclet number	[-]
Pr	Prandtl number	[-]
Q	volumetric flow rate	[m^3 s^{-1}]
R	ideal gas constant	[J K^{-1} (mol species)$^{-1}$]
Ra	Rayleigh number	[-]
Re	Reynolds number	[-]
s	specific entropy (per unit mass of fluid)	[J K^{-1} kg^{-1}]
Sc	Schmidt number	[-]
t	time	[s]
T	absolute temperature	[K]
$\boldsymbol{u}$	(mass-average) fluid velocity	[m s^{-1}]
$\boldsymbol{u}_c$	(mass-average) velocity of chemical species c	[m s^{-1}]
U	representative velocity	[m s^{-1}]
V	volume	[m^3]
$\boldsymbol{w}$	spin tensor $= \frac{1}{2}(\nabla \boldsymbol{u} - (\nabla \boldsymbol{u})^\top)$	[s^{-1}]
$\boldsymbol{x}$	Cartesian position vector	[m]
y	depth of water	[m]
z_k	charge number (valency)	[(mol charge) (mol species)$^{-1}$]
Greek symbols		
α	thermal diffusion coefficient	[m^2 s^{-1}]
α'	thermodynamic thermal diffusion parameter	[J K m^{-1} s^{-1}]
β	thermal expansion coefficient	[K^{-1}]
β_{ncd}	power exponent of cth species in nth mechanism of dth chemical reaction	[–]
$\boldsymbol{\delta}$	Kronecker delta tensor	[-]
θ	residence time of flow process	[s]
κ_k	electrical conductivity or specific conductance for the kth charged species	[A V^{-1} m^{-1}]
λ	second viscosity or first Lamé coefficient; wavelength	[Pa s]; [m]
μ	dynamic viscosity	[Pa s]
μ_c	chemical potential of species c	[J (mol species)$^{-1}$]
ν	kinematic viscosity or momentum diffusion coefficient	[m^2 s^{-1}]
$\hat{\xi}_d$	rate of the dth reaction per unit volume	[(mol reaction) m^{-3} s^{-1}]

Table 1. *Cont.*

Symbol	Description	SI Units
π	ratio between the circumference and diameter of a circle	[-]
Π	global or summary dimensionless group	[-]
$\hat{\Pi}$	local dimensionless group	[-]
ρ	fluid density	[kg m^{-3}]
$\dot{\sigma}$	global entropy production	[J K^{-1} s^{-1}]
$\hat{\dot{\sigma}}$	local entropy production	[J K^{-1} m^{-3} s^{-1}]
τ	momentum flux or viscous stress tensor, defined positive in compression	[Pa]
Φ	electrical potential	[V = J C^{-1}]

In most flow systems, distinct entropy production terms $\dot{\sigma}$ or $\hat{\dot{\sigma}}$ and entropy flux terms j_S can be identified with particular entropy-producing or entropy-transporting processes, enabling the ranking of their relative contributions.

3. Similarity

Dimensionless groups Π are usually categorized as follows [5–9,17,18]:

(i) Those which represent *geometric similarity*, defined by ratios of length scales, areas or volumes:

$$\Pi_{\text{geom}} = \frac{\ell_1}{\ell_2} \quad \text{or} \quad \Pi_{\text{geom}} = \frac{\ell_1^2}{\ell_2^2} \quad \text{or} \quad \Pi_{\text{geom}} = \frac{\ell_1^3}{\ell_2^3} \tag{3}$$

(ii) Those which represent *kinematic similarity*, defined by ratios of magnitudes of velocities or accelerations:

$$\Pi_{\text{kinem}} = \frac{U_1}{U_2} \quad \text{or} \quad \Pi_{\text{kinem}} = \frac{a_1}{a_2} \tag{4}$$

(iii) Those which represent *dynamic similarity*, defined by ratios of magnitudes of forces:

$$\Pi_{\text{dynam}} = \frac{F_1}{F_2} \tag{5}$$

We here consider an additional category to represent *entropic similarity* [19,20], for which there are three definitions:

(i) Dimensionless groups defined by ratios of global or local entropy production terms:

$$\Pi_{\text{entrop}} = \frac{\dot{\sigma}_1}{\dot{\sigma}_2} \quad \text{or} \quad \hat{\Pi}_{\text{entrop}} = \frac{\hat{\dot{\sigma}}_1}{\hat{\dot{\sigma}}_2} \tag{6}$$

(ii) Dimensionless groups defined by ratios of global flow rates of thermodynamic entropy, or by magnitudes of their local fluxes:

$$\Pi_{\text{entrop}} = \frac{\mathcal{F}_{S,1}}{\mathcal{F}_{S,2}} \quad \text{or} \quad \hat{\Pi}_{\text{entrop}} = \frac{||j_{S_1}||}{||j_{S_2}||} \tag{7}$$

(iii) Dimensionless groups defined by an information-theoretic criterion, such as the ratio of the magnitude of the fluid velocity U to that of a carrier of information c:

$$\Pi_{\text{info}} = \frac{U}{c} \tag{8}$$

Typically, c corresponds to the celerity of a wave within the fluid, providing a signal for changes in the downstream boundary conditions.

4. Analyses by Entropic Similarity

This study now examines a variety of natural phenomena, including diffusion, chemical reaction, dispersion, and wave propagation. These all involve competition between entropic phenomena, for the production or transport of entropy. In each section, the concept of entropic similarity is applied to construct a number of entropic dimensionless groups. These are compared to the known dimensionless groups used in each field, to examine their similarities and differences and to provide several new insights.

4.1. Diffusion Phenomena

Generally, *diffusion* is defined as the flux of a physical quantity in response to a gradient in a corresponding intensive variable. Usually, this term is applied to phenomena acting at molecular scales, while *dispersion* used to describe mixing processes acting at microscopic to macroscopic scales. Diffusion and dispersion are irreversible processes, governed by the resulting increase in thermodynamic entropy.

We first consider isolated diffusion processes, based on migration of a physical quantity in its conjugate field. Historically, these were identified with named diffusion laws. The practical or empirical relations for these laws, for several simplifying assumptions, are set out in Table 2 [8,23–27]. However, diffusion is a thermodynamic process, in which the flux is driven by its conjugate thermodynamic gradient. The corresponding thermodynamic diffusion relations, based on modified diffusion parameters (labeled $'$), are also listed in Table 2. By comparison and some analysis, the connections between the thermodynamic and practical diffusion parameters can be established [19] and are also listed.

Table 2. Practical and thermodynamic diffusion laws.

Relation	Conserved Quantity	Practical Equation	Thermodynamic Equation
Fourier's	Heat	$j_Q = -k\nabla T = -\alpha\nabla(\rho c_P T)$	$= \alpha' \nabla \frac{1}{T}$ with $\alpha' = kT^2 \simeq \alpha \rho c_p T^2$
Newton's	Momentum	$\tau = -\mu(\nabla u + (\nabla u)^\top)$ $= -\nu(\nabla(\rho u) + [\nabla(\rho u)]^\top)$	$= -\mu(\nabla u + (\nabla u)^\top) - \lambda\, \delta\, \nabla \cdot u$ $= -2\mu e - \lambda\, \delta\, \mathrm{tr}(e)$
Fick's	Chemical species c	$j_c = -\mathcal{D}_c \nabla C_c = -\mathcal{D}_c \nabla(\rho m_c)$	$= -\mathcal{D}'_c \nabla \frac{\mu_c}{T}$ with $\mathcal{D}'_c \simeq \frac{\rho m_c \mathcal{D}_c}{R}$
Ohm's	Charged species k (in solution)	$i_k = -\kappa_k \nabla\Phi = -D_k \nabla\left(\frac{z_k^2 F^2 C_k \Phi}{RT}\right)$	$= -D_k' \frac{\nabla\Phi}{T}$ with $D_k' \simeq \kappa_k T \simeq \frac{z_k^2 F^2 C_k D_k}{R}$

The local entropy production equation for each isolated diffusion process is then listed in Table 3 [21–23]. Substituting the thermodynamic equations from Table 2, expressed either in terms of a fixed gradient or a fixed flux, gives the forms listed in the third and fourth columns of Table 3.

Table 3. Entropy production of isolated diffusion processes.

Conserved Quantity	Definition	Fixed Gradients	Fixed Fluxes
Heat	$\hat{\sigma}_\alpha = j_Q \cdot \nabla \frac{1}{T}$	$= \alpha \rho c_p T^2 \left\|\nabla \frac{1}{T}\right\|^2$	$= \frac{1}{\alpha \rho c_p T^2} \|\|j_Q\|\|^2$
Momentum	$\hat{\sigma}_\nu = -\tau : \frac{\nabla u}{T}$	$= \frac{1}{T}[2\mu e + \lambda\, \delta\, \mathrm{tr}(e)] : \nabla u$	$= \frac{1}{2\rho\nu T}[\|\|\tau\|\|^2 + \lambda\, \mathrm{tr}(\tau)\, \mathrm{tr}(e)] - \frac{\tau : w}{T}$
Chemical species c	$\hat{\sigma}_{\mathcal{D}_c} = -j_c \cdot \nabla \frac{\mu_c}{T}$	$= \frac{\rho m_c \mathcal{D}_c}{R} \left\|\nabla \frac{\mu_c}{T}\right\|^2$	$= \frac{R}{\rho m_c \mathcal{D}_c} \|\|j_c\|\|^2$
Charged species k	$\hat{\sigma}_{D_k} = -i_k \cdot \frac{\nabla\Phi}{T}$	$= \frac{z_k^2 F^2 C_k D_k}{RT^2} \|\|\nabla\Phi\|\|^2$	$= \frac{R}{z_k^2 F^2 C_k D_k} \|\|i_k\|\|^2$

Invoking entropic similarity, entropic dimensionless groups can be defined to rank pairwise diffusion processes:

$$\begin{aligned}
\hat{\Pi}_{\nu/\alpha} &= \frac{\hat{\hat{\sigma}}_\nu}{\hat{\hat{\sigma}}_\alpha} \to Pr = \frac{\nu}{\alpha}, &\quad \hat{\Pi}_{\nu/\mathcal{D}_c} &= \frac{\hat{\hat{\sigma}}_\nu}{\hat{\hat{\sigma}}_{\mathcal{D}_c}} \to Sc_c = \frac{\nu}{\mathcal{D}_c} \\
\hat{\Pi}_{\nu/D_k} &= \frac{\hat{\hat{\sigma}}_\nu}{\hat{\hat{\sigma}}_{D_k}} \to Sc_k = \frac{\nu}{D_k}, &\quad \hat{\Pi}_{\alpha/\mathcal{D}_c} &= \frac{\hat{\hat{\sigma}}_\alpha}{\hat{\hat{\sigma}}_{\mathcal{D}_c}} \to Le_c = \frac{\alpha}{\mathcal{D}_c} \\
\hat{\Pi}_{\alpha/D_k} &= \frac{\hat{\hat{\sigma}}_\alpha}{\hat{\hat{\sigma}}_{D_k}} \to Le_k = \frac{\alpha}{D_k}, &\quad \hat{\Pi}_{\mathcal{D}_c/D_k} &= \frac{\hat{\hat{\sigma}}_{\mathcal{D}_c}}{\hat{\hat{\sigma}}_{D_k}} \to \frac{\mathcal{D}_c}{D_k}
\end{aligned} \tag{9}$$

Substituting the relations in Table 3 for fixed gradients and assuming constant other properties, these reduce to ratios of diffusion coefficients, including the well-known Prandtl, Schmidt and Lewis numbers, the latter two for chemical species c or charged species k. Alternatively, for fixed fluxes and other constant properties, the above definitions yield the reciprocals of these groups. Similar dimensionless groups can be derived from ratios of entropy fluxes [19].

Commonly, the diffusion groups in (9) are derived by dimensional analysis [7,24], by non-dimensionalization of the governing equations [4,8,28], or by construction [4,8,22,24,27,29]. The application of entropic similarity therefore provides an entropic basis for these groups, more representative of their origin in the maximization of entropy.

The concept of entropic similarity can also be used to construct hybrid groups with a fixed flux and a fixed gradient, for example:

$$\hat{\Pi}_{\alpha/\mathcal{D}_c} = \frac{\hat{\hat{\sigma}}_\alpha}{\hat{\hat{\sigma}}_{\mathcal{D}_c}} = \frac{\alpha T^2 \rho^2 c_P m_c \mathcal{D}_c}{R} \frac{||\nabla T^{-1}||^2}{||j_c||^2} \tag{10}$$

Such groups are not readily obtained by kinematic or dynamic similarity. For thermodynamic cross-phenomena governed by the Onsager [30,31] relations $j_a = \sum_b L_{a,b} f_b$, entropic similarity gives dimensionless groups of the form:

$$\hat{\Pi}_{L_{c,d}/L_{a,b}} = \frac{\hat{\hat{\sigma}}_{L_{c,d}}}{\hat{\hat{\sigma}}_{L_{a,b}}} \to \frac{L_{c,d}}{L_{a,b}} \tag{11}$$

These enable the ranking of different cross-phenomenological processes.

4.2. *Chemical Reaction Phenomena*

Chemical reactions are governed by their entropy production:

$$\hat{\sigma}_d = -\hat{\hat{\xi}}_d \frac{\Delta \widetilde{G}_d}{T} = -\left(\sum_n k_{nd} \prod_c C_c^{\beta_{ncd}}\right) \frac{\Delta \widetilde{G}_d}{T} \tag{12}$$

which is calculated for the dth reaction over all reaction mechanisms n and chemical species c. By entropic similarity, dimensionless groups can be constructed for the competition between chemical reactions and diffusion phenomena:

$$\begin{aligned}
\hat{\Pi}_{d/\alpha} &= \frac{\hat{\dot{\sigma}}_d}{\hat{\dot{\sigma}}_\alpha} = \frac{\hat{\dot{\xi}}_d \Delta\widetilde{G}_d}{\hat{\dot{\sigma}}_\alpha} = \frac{-(\sum_n k_{nd} \prod_c C_c{}^{\beta_{ncd}}) \Delta\widetilde{G}_d}{\alpha \rho c_p T^3 \,||\nabla T^{-1}||^2} \\
\hat{\Pi}_{d/v} &= \frac{\hat{\dot{\sigma}}_d}{\hat{\dot{\sigma}}_v} = \frac{\hat{\dot{\xi}}_d \Delta\widetilde{G}_d}{\hat{\dot{\sigma}}_v} = \frac{-(\sum_n k_{nd} \prod_c C_c{}^{\beta_{ncd}}) \Delta\widetilde{G}_d}{\rho v[||\nabla \boldsymbol{u}||^2 + \mathrm{tr}((\nabla \boldsymbol{u})^2)] + \lambda(\nabla \cdot \boldsymbol{u})^2} \\
\hat{\Pi}_{d/\mathcal{D}_c} &= \frac{\hat{\dot{\sigma}}_d}{\hat{\dot{\sigma}}_{\mathcal{D}_c}} = \frac{\hat{\dot{\xi}}_d \Delta\widetilde{G}_d}{\hat{\dot{\sigma}}_{\mathcal{D}_c}} = \frac{-R(\sum_n k_{nd} \prod_c C_c{}^{\beta_{ncd}}) \Delta\widetilde{G}_d}{C_c \mathcal{D}_c T \left|\left|\nabla \frac{\mu_c}{T}\right|\right|^2} \\
\hat{\Pi}_{d/D_k} &= \frac{\hat{\dot{\sigma}}_d}{\hat{\dot{\sigma}}_{D_k}} = \frac{\hat{\dot{\xi}}_d \Delta\widetilde{G}_d}{\hat{\dot{\sigma}}_{D_k}} = \frac{-RT(\sum_n k_{nd} \prod_c C_c{}^{\beta_{ncd}}) \Delta\widetilde{G}_d}{z_k^2 F^2 C_k D_k ||\nabla \Phi||^2}
\end{aligned} \tag{13}$$

These groups bear some similarities to the Damköhler number, used in chemical engineering to rank the rate of a chemical reaction with the residence time [22,29]:

$$\hat{\Pi}_{d/\theta} = \frac{\hat{\dot{\xi}}_d \theta}{\prod_c C_c} = \Big(\sum_n k_{nd} \prod_c C_c{}^{\beta_{ncd}-1}\Big)\theta \quad \rightsquigarrow Da_d = k_d\, C_c^{\beta_{cd}-1}\theta \tag{14}$$

where the squiggly arrow denotes the limit of a single reaction mechanism with single-species kinetics.

4.3. Dispersion Phenomena

Fluid flow systems exhibit *inertial dispersion*, the spreading of momentum, and other fluid-borne properties (heat, chemical species or charge) due to inertial flow. For the *internal flow* of a fluid within a cylindrical pipe, the inertial head loss [6–9,17,18] and entropy production [10,32,33] are given by

$$H_{L,I} = \frac{fL}{2gd} U^2, \qquad \dot{\sigma}_{\mathrm{int},I} = \frac{\rho g H_L Q}{T} = \frac{\pi \rho d f L}{8T} U^3 \tag{15}$$

For purely viscous flow, the head loss and entropy production are given by [4,6,17]

$$H_{L,v} = \frac{128 \mu L}{\pi \rho g d^4} Q = \frac{32 v L}{g d^2} U, \qquad \dot{\sigma}_{\mathrm{int},v} = \frac{16 \pi \rho v L}{T} U^2 \tag{16}$$

The competition between inertial dispersion and viscous diffusion for an internal flow can be represented by the entropic group:

$$\Pi_{\mathrm{int},I/v} = \frac{\dot{\sigma}_{\mathrm{int},I}}{\dot{\sigma}_{\mathrm{int},v}} = \frac{f\,Ud}{64 v} \sim f Re, \tag{17}$$

We recover the Reynolds number $Re = Ud/v$, commonly derived by dynamic similarity as the ratio of the inertial and viscous forces. In the entropic formulation (17), this is multiplied by the friction factor, which represents the resistance of the pipe to the imposed flow.

For the *external flow* of a fluid around a sphere, the drag coefficient and entropy production due to inertial dispersion are given by

$$C_D = \frac{F_D}{\frac{1}{2}\rho A_s U^2}, \qquad \dot{\sigma}_{\mathrm{ext},I} = \frac{F_D U}{T} = \frac{\pi \rho d^2 C_D U^3}{8T} \tag{18}$$

For purely viscous flow, the drag force and entropy production are given by [4,6,8,17]

$$F_D = 3\pi\rho\nu dU, \qquad \dot{\sigma}_{\text{ext},\nu} = \frac{3\pi\rho\nu dU^2}{T} \tag{19}$$

The competition between inertial dispersion and viscous diffusion for an external flow can be represented by the entropic group [34]:

$$\Pi_{\text{ext},I/\nu} = \frac{\dot{\sigma}_{\text{ext},I}}{\dot{\sigma}_{\text{ext},\nu}} = \frac{C_D\, Ud}{24\nu} \sim C_D Re, \tag{20}$$

We again recover the Reynolds number. For external flow, this is modified by the drag coefficient, expressing the resistance of the object to the flow.

The Reynolds number $Re = \frac{Ud}{\nu}$ can therefore be interpreted by entropic similarity as the ratio of the inertial dispersion coefficient Ud to the viscous diffusion coefficient ν. When Re exceeds a critical threshold, the inertial dispersion will dominate, leading to flows of increasing turbulent character. This interpretation extends to other groups commonly used to rank inertial dispersion and diffusion processes, including the Peclet numbers for heat, species, and charge [4,22,24,29]:

$$\begin{aligned}
\Pi_{\text{int},I/\alpha} &= \frac{\dot{\sigma}_{\text{int},I}}{\dot{\sigma}_{\text{int},\alpha}} \sim f Pe_\alpha, &\quad \text{with } Pe_\alpha &= \frac{Ud}{\alpha} = RePr, \\
\Pi_{\text{int},I/c} &= \frac{\dot{\sigma}_{\text{int},I}}{\dot{\sigma}_{\text{int},\mathcal{D}_c}} \sim f Pe_c, &\quad \text{with } Pe_c &= \frac{Ud}{\mathcal{D}_c} = ReSc_c, \\
\Pi_{\text{int},I/k} &= \frac{\dot{\sigma}_{\text{int},I}}{\dot{\sigma}_{\text{int},D_k}} \sim f Pe_k, &\quad \text{with } Pe_k &= \frac{Ud}{D_k} = ReSc_k
\end{aligned} \tag{21}$$

For inertial dispersion by free convection, involving heat- or density-driven flow with a convective velocity U_{conv} [8], similar formulations give the Grashof and Rayleigh numbers:

$$\begin{aligned}
\Pi'_{\text{int},I/\nu} &= \left(\frac{\dot{\sigma}_{\text{int},I}}{\dot{\sigma}_{\text{int},\nu}}\right)^2 = f^2 Gr, &\quad \text{with } Gr &= \left(\frac{U_{\text{conv}} d}{\nu}\right)^2 = \frac{g d^3 |\Delta\rho|}{\rho\nu^2} = \frac{g\beta d^3 |\Delta T|}{\nu^2} \\
\Pi'_{\text{int},I/\alpha,\nu} &= \frac{\dot{\sigma}^2_{\text{int},I}}{\dot{\sigma}_{\text{int},\nu}\,\dot{\sigma}_{\text{int},\alpha}} \sim f^2 Ra, &\quad \text{with } Ra &= \frac{(U_{\text{conv}} d)^2}{\nu\alpha} = \frac{g d^3 |\Delta\rho|}{\rho\nu\alpha} = \frac{g\beta d^3 |\Delta T|}{\nu\alpha}
\end{aligned} \tag{22}$$

For heat, mass, or charge transfer processes, either with forced convection by fluid flow or in competition, entropic similarity arguments can be used to construct many dimensionless groups including the Nusselt, Sherwood, Biot, Fourier, Stefan, Eckert, Brinkman, and Stanton numbers [19]. Related arguments can be used to construct entropic dimensionless groups—some of which are known—for turbulent, hydrodynamic, shear-flow, and multiphase dispersion processes [19].

4.4. Wave Phenomena

Finally, the information-theoretic definition of similarity (8), based on the ratio of a representative velocity to the celerity of a wave, can be used to derive entropic groups for a variety of flow systems. Some examples are listed in Table 4, which gives the traditional interpretation by dynamic similarity and the entropic formulation. In each case, the dimensionless group provides a discriminator between two flow regimes, characterized by flows which can ($\Pi < 1$) or cannot ($\Pi > 1$) be influenced by downstream disturbances, such as a change in boundary conditions. In many instances, a disturbance to the higher regime will be manifested as a sharp transition (shock wave or hydraulic jump) in the flow field.

Table 4 reports the well-known Mach number for compressible flow, and two forms of the Froude number for shallow and deepwater flows. The Mach number discriminates between the occurrence of subsonic ($M < 1$) or supersonic ($M > 1$) flow, while the Froude numbers discriminate between the occurrence of subcritical ($Fr < 1$) or supercritical

($Fr > 1$) flow. Additional Froude numbers for flows with surface transitional gravity waves, surface capillary waves, internal gravity waves, or inertial waves can also be derived [20]. The unnamed group for the flow of subatomic particles in the last line of Table 4—for which no dynamic similarity derivation was identified—provides a discriminator between flows at less than the speed of light ($\Pi_c < 1$) and flows at greater than the speed of light ($\Pi_c > 1$). The latter effect is known, revealed by Vavilov–Cherenkov radiation, caused by the emission of charged particles at greater than the speed of light in a dielectric medium [35,36].

Table 4. Dimensionless groups defined by wave motion in different flows.

Flow	Dynamic Similarity	Waves	Entropic Similarity	Group
Compressible	$M = \sqrt{\frac{\text{inertial force}}{\text{elastic force}}} = \frac{U}{a_s}$	Acoustic	$\Pi_{a_s} = M = \frac{U}{a_s}$	Mach number
Shallow water	$Fr_y = \sqrt{\frac{\text{inertial force}}{\text{gravity force}}} = \frac{U}{\sqrt{gy}}$	Shallow water	$\Pi_y = Fr_y = \frac{U}{\sqrt{gy}}$	Froude number (shallow)
Deep water	$Fr_\lambda = \sqrt{\frac{\text{inertial force}}{\text{gravity force}}} = U\sqrt{\frac{2\pi}{\lambda g}}$	Deep water	$\Pi_\lambda = Fr_\lambda = U/\sqrt{\frac{\lambda g}{2\pi}}$	Froude number (deep)
Subatomic particles	?	Electromagnetic	$\Pi_c = \frac{U}{c}$	

5. Conclusions

This study examines the concept of *entropic similarity* for the construction of dimensionless groups from ratios of entropic terms. These include (i) ratios of entropy production terms, (ii) ratios of entropy flow rates or fluxes, and (iii) the ratio of a flow velocity to that of a signal within the flow. Through the analysis of different entropic phenomena, including diffusion, chemical reaction, dispersion, and wave propagation, it is shown that many known dimensionless groups have an entropic interpretation. These include the diffusion ratios (Prandtl, Schmidt, and Lewis numbers), the competition between inertial dispersion and diffusion (Reynolds, Peclet, Grashof, and Rayleigh numbers), and the competition between flow and wave propagation in different flow systems (Mach and Froude numbers). New entropic groups are also derived for the competition between hybrid diffusion processes (one fixed gradient and one fixed flux), cross-phenomenological diffusion processes (Onsager coefficients), the competition between chemical reaction and diffusion (to give groups related to Damköhler numbers), and the flow of subatomic particles relative to the speed of light. Further analyses also yield new entropic interpretations or new groups for other dispersion phenomena [19]. The analyses provide a new entropic foundation for dimensionless groups, applicable to a wide variety of natural and engineered systems.

Funding: This research received no external funding.

Institutional Review Board Statement: Not applicable.

Informed Consent Statement: Not applicable.

Data Availability Statement: No data were generated in this study.

Acknowledgments: We thank Harald Kleine, Matthias Kramer, and Richard Manasseh for comments on parts of this work.

Conflicts of Interest: The author declares no conflict of interest. The funders had no role in the design of the study; in the collection, analyses, or interpretation of data; in the writing of the manuscript, or in the decision to publish the results.

References

1. Carvallo, E. Sur une similitude dans les fonctions des machines. *J. Phys. Theor. Appl.* **1892**, *1*, 209–212. [CrossRef]
2. Vaschy, A. Sur les lois de similitude en physique. *Ann. Télégraphiques* **1892**, *19*, 25–28.
3. Buckingham, E. On physically similar systems; illustrations of the use of dimensional equations. *Phys. Rev.* **1914**, *4*, 345–376. [CrossRef]

4. Schlichting, H. *Boundary-Layer Theory*, 6th ed.; McGraw-Hill: New York, NY, USA, 1968.
5. White, F.M. *Fluid Mechanics*, 2nd ed.; McGraw-Hill Higher Education: New York, NY, USA, 1986.
6. Street, R.L.; Watters, G.Z.; Vennard, J.K. *Elementary Fluid Mechanics*, 7th ed.; John Wiley: New York, NY, USA, 1996.
7. Streeter, V.L.; Wylie, E.B.; Bedford, K.W. *Fluid Mechanics*, 9th ed.; McGraw-Hill: Boston, MA, USA, 1998.
8. White, F.M. *Viscous Fluid Flow*, 3rd ed.; McGraw-Hill: New York, NY, USA, 2006.
9. Munson, B.R.; Young, D.F.; Okiishi, T.H.; Huebsch, W.W. *Fundamentals of Fluid Mechanics*, 6th ed.; John Wiley: New York, N USA, 2010.
10. Niven, R.K. Invariance properties of the entropy production, and the entropic pairing of inertial frames of reference by shear-flo systems. *Entropy* **2021**, *23*, 1515. [CrossRef] [PubMed]
11. Langhaar, H.L. *Dimensional Analysis and Theory of Models*; John Wiley: New York, NY, USA, 1951.
12. Sedov, L.I. *Similarity and Dimensional Methods in Mechanics*; Infosearch Ltd.: London, UK, 1959.
13. Birkhoff, G. *Hydrodynamics, a Study in Logic, Fact and Similitude*, 2nd ed.; Princeton University Press: Princeton, NJ, USA, 1960.
14. Gratton, J. Similarity and self similarity in fluid dynamics. *Fundam. Cosm. Phys.* **1991**, *15*, 1–106.
15. Pope, S.B. *Turbulent Flows*; Cambridge University Press: Cambridge, UK, 2000.
16. Barenblatt, G.I. *Scaling*; Cambridge University Press: Cambridge UK, 2003.
17. Pao, H.F. *Fluid Mechanics*; John Wiley: New York, NY, USA, 1961.
18. Douglas, J.F.; Gasiorek, J.M.; Swaffield, J.A.; Jack, L.B. *Fluid Mechanics*, 6th ed.; Prentice Hall: Harlow, UK, 2011.
19. Niven, R.K. Dimensionless groups by entropic similarity: I—Diffusion, chemical reaction and dispersion processes. *Entropy* **202** *25*, 617. [CrossRef] [PubMed]
20. Niven, R.K. Dimensionless groups by entropic similarity: II—Wave phenomena and information-theoretic flow regimes. *Entrop* **2023**, *25*, 1538. [CrossRef] [PubMed]
21. de Groot, S.R.; Mazur, P. *Non-Equilibrium Thermodynamics*; Dover Publ.: Mineola, NY, USA, 1962.
22. Bird, R.B.; Stewart, W.E.; Lightfoot, E.N. *Transport Phenomena*, 2nd ed.; John Wiley & Sons: New York, NY, USA, 2006.
23. Kondepudi, D.; Prigogine, I. *Modern Thermodynamics: From Heat Engines to Dissipative Structures*, 2nd ed.; John Wiley: Chicheste UK, 2015.
24. Bosworth, R.C.L. *Transport Processes in Applied Chemistry*; Horwitz Publ.: Sydney, Australia, 1956.
25. Guggenheim, E.A. *Thermodynamics: An Advanced Treatment for Chemists and Physicists*; North-Holland Publ. Co.: Amsterdam The Netherlands, 1967.
26. Levine, I.N. *Physical Chemistry*; McGraw-Hill: New York, NY, USA, 1978.
27. Newman, J.S. *Electrochemical Systems*, 2nd ed.; Prentice Hall: Englewood Cliffs, NJ, USA, 1991.
28. Eckert, E.R.G. *Introduction to Heat and Mass Transfer*; McGraw-Hill: New York, NY, USA, 1963.
29. Fogler, H.S. *Elements of Chemical Reaction Engineering*, 2nd ed.; Prentice Hall: Englewood Cliffs, NJ, USA, 1992.
30. Onsager, L. Reciprocal relations in irreversible processes, I. *Phys. Rev.* **1931**, *37*, 405–426. [CrossRef]
31. Onsager, L. Reciprocal relations in irreversible processes, II. *Phys. Rev.* **1931**, *38*, 2265–2279. [CrossRef]
32. Bejan, A. *Entropy Generation Through Heat and Fluid Flow*; John Wiley: New York, NY, USA, 1982.
33. Niven, R.K. Simultaneous extrema in the entropy production for steady-state fluid flow in parallel pipes. *J. Non-Equil. Thermodyn* **2010**, *35*, 347–378. [CrossRef]
34. Duan, Z.; He, B.; Duan, Y. Sphere drag and heat transfer. *Sci. Rep.* **2015**, *5*, 12304. [CrossRef] [PubMed]
35. Cherenkov, R.A. Visible radiation produced by electrons moving in a medium with velocities exceeding that of light. *Phys. Rev* **1937**, *52*, 378–379. [CrossRef]
36. Frank, I.; Tamm, I. Coherent visible radiation of fast electrons passing through matter. *Dokl. Akad. Nauk. SSSR* **1937**, *14*, 109–114

MDPI AG
Grosspeteranlage 5
4052 Basel
Switzerland
Tel.: +41 61 683 77 34

Physical Sciences Forum Editorial Office
E-mail: psf@mdpi.com
www.mdpi.com/journal/psf

www.ingramcontent.com/pod-product-compliance
Lightning Source LLC
Chambersburg PA
CBHW041412210926
51850CB00013B/3248
* 9 7 8 3 7 2 5 8 4 1 7 9 0 *